ROCK IDENTIFICATION TABLE

Dr. Eugene Mitacek
New York Institute of Technology
Old Westbury, New York

CONTACT METAMORPHISM

by extrusive magma	by intrusive magma

Crystalloblastic

Fabric random or retained from original sediment

Original Rock	Degree of metamorphism	Metamorphic Rock	Mineral Particles — Essential	Mineral Particles — Accessory	Texture	Fabric
SHALE, GRAYWACKE SHALE	outer part of contact aureole	Contact SLATE	Matrix of microcrystalline quartz, muscovite, chlorite; sometimes biotite, albite	Knots of andalusite or cordierite	dense to porphyroblastic	directed (schistose)
	middle part of contact aureole	SPOTTED SLATE	Matrix of microcrystalline quartz, muscovite, chlorite; graphite	Spots of mica, chlorite	dense	
	inner part of contact aureole	KNOTTED SLATE	biotite, cordierite, graphite	andalusite, muscovite, graphite	dense to fine-grained	
		CONTACT HORNFELS	quartz; plagioclase, orthoclase; diopside, hypersthene	epidote, calcite, garnet, wollastonite, etc.		
MARLY SEDIMENTS	outer to inner part of contact aureole	SKARN	quartz; oligoclase, anorthite, ±K feldspar; diopside, pyroxene	quartz, micas, graphite, brucite, pyrrhotite, etc.	fine to coarse-grained	
CALCITE OR DOLOMITE	inner part of contact aureole	MARBLE	calcite	calcite, quartz, etc.		random, sometimes with features of the original stratification
	in direct contact	TACTITE	wollastonite, diopside, garnet, vesuvianite			
SHALE OR MARLY SEDIMENTS	in direct contact and vicinity	PORCELLANITE	some modification of SiO_2	microscopic silicates: mullite, cordierite; etc.; glass	dense	

PHYSICAL GEOLOGY
Principles, Processes, and Problems

Charles J. Cazeau
DEPARTMENT OF GEOLOGICAL SCIENCES
STATE UNIVERSITY OF NEW YORK AT BUFFALO

Robert D. Hatcher, Jr.
DEPARTMENT OF GEOLOGY
CLEMSON UNIVERSITY

Francis T. Siemankowski
DEPARTMENT OF SCIENCE EDUCATION
STATE UNIVERSITY COLLEGE AT BUFFALO

HARPER & ROW, PUBLISHERS
New York, Evanston, San Francisco, London

Dr. Eugene Mitacek, author of the *Rock Identification Table,* has taught subjects in chemical and earth sciences since 1958 at universities throughout the world. In 1967 he won an international competition to head the earth science department of the International School of the United Nations in New York immediately prior to his present position as Associate Professor in the Division of Sciences at the New York Institute of Technology. He has been the recipient of honors for gifted teaching and numerous awards in the form of fellowships from the universities in Poland, Bulgaria, Czechoslovakia, Yugoslavia, England—University of London, United States—University of California, and The University of West Indies. Dr. Mitacek authored four books, several study guides, published 24 articles, and made significant contributions to his area of science. He is also a member of leading scientific organizations in the United States.

The *Rock Identification Table* has been tested, published, and recognized by the international scientific community as an innovation in the field. Dr. Mitacek's original work has been highly evaluated and honored by universities in several countries. The project has also stimulated additional theoretical and experimental activities at universities abroad. Many institutions have used the Table in conjunction with field exercises and have found it to be a valuable reference.

Students always feel more of the relevance of their studies when handling samples of natural objects such as rocks, and must be provided with means to identify and classify them from the very beginning. The *Rock Identification Table* complements this text and will therefore form an excellent nucleus in which to structure geology or earth science courses.

Sponsoring Editors: Ronald K. Taylor, Dale Tharp
Project Editor: Pamela Landau
Designer: Rita Naughton
Production Supervisor: Stefania J. Taflinska
Compositor: Ruttle, Shaw & Wetherill, Inc.
Printer: Kingsport Press
Binder: Kingsport Press
Art Studio: J & R Technical Services, Inc.

PHYSICAL GEOLOGY
Principles, Processes, and Problems

Library of Congress Cataloging in Publication Data
Cazeau, Charles J.
    Physical geology.

    Includes index.
    1. Physical geology.   I. Hatcher, Robert D.,
1940–   joint author.   II. Siemankowski, Francis T.,
joint author.   III. Title.
QE28.2.C38   551   75–25962
ISBN 0-06-041209-7

To the late Michael Siemankowski

Contents

Preface

Although this book was written with the nongeology major in mind, we hope it will serve potential geologists equally well in providing sufficient introductory background to the subject. This book attempts to present the processes and principles of physical geology in a readable format and to employ a writing style that is open and easy to understand but still maintains technical quality and accuracy.

We hope to show that geologic science is a dynamic subject, constantly changing as knowledge increases. This is exemplified by the recent development of plate tectonics theory. Although a chapter is devoted to this subject in which both sides of the issue are considered, plate tectonics theory is integrated throughout the text but only in such a way that it is *not* used as a solution to all geologic problems.

The contributions of geology to environmental and resources problems are potentially monumental. The increased realization of the importance of geology and geologists to humankind has placed geology in a new role in our society. Environmental problems stemming from increases in the world population and our own increased consumption of fuels and other resources are discussed.

Each chapter begins with a set of goals for the student and ends with a summary. This emphasis on important points at the beginning and reemphasis at the end of the chapter should assist in the learning process. Many of the questions that follow each chapter cannot be answered simply by flipping through the text and copying the answer. The answers in some cases will call for some imagination and initiative on the part of the student.

The suggested readings were selected with the hope that many students will become interested enough in some topics to want to pursue them further. The readings are, for the most part, not difficult, but most of them contain references to the technical literature of geologic journals and more advanced texts.

C.J.C.
R.D.H.
F.T.S.

ACKNOWLEDGMENTS

The writers would like to recognize the assistance, suggestions and constructive criticisms of the following individuals: Fred Amos, Peter Avery, Paul Birkhead, P. E. Calkin, C. Craddock, Anita Epstein, S. D. Heron, Jr., George Hazelton, K. S. LaFleur, E. McFarlan, Jr., Jan Patterson, Bill R. Smith, D. S. Snipes, H. G. Spencer, E. Randolph Stone, Don and Mary Ann Thompson, Bill Valente, Charles Welby, and the durable secretaries of the departments of geology at State University of New York at Buffalo and Clemson University.

Those who assisted in the preparation of

art work were: Francis Lestingi, Michael Siemankowski, and Wojciech Skrzydlewski. Clerical assistance was given by Florence Blasik, Judith Petrusek, and Helene Arnone.

Special thanks are extended to Ron Taylor and his associates at Harper & Row as well as to Reginald H. Pegrum (Professor Emeritus) and John S. King of the State University of New York; Meta Smith, an editorial consultant of Chicago; and the late Virgil I. Mann of the University of North Carolina.

The authors are grateful for the photos provided by: William Dize of the U.S. Geological Survey; A. Vitaliano of American Airlines; Robert W. Mitchell of Texas Tech; Albert Moldvay of Monte Blanco, Mexico; Utah Travel Council; United Press International; Barbara A. Shattuck of National Geographic; and J. D. Martinez, Louisiana State University.

The Publisher wishes to thank the following individuals for their careful reviews and helpful suggestions: J. S. Pittman, San Antonio College; Warren D. Huff, University of Cincinnati; John Nicholas, University of Bridgeport; Charles C. Plummer, California State University at Sacramento; James B. Johnson, Mesa College.

Part
I

Earth, space, and cycles

1 Geology and geologists

STUDENT OBJECTIVES

At the conclusion of this chapter the student should be aware of:

1. the nature of the science of geology

2. some of the ways geology influences our lives

3. how the scientific method works

4. some basic concepts in geology as they have developed since the time of the Greeks

*Science is simply common sense at its best—
that is, rigidly accurate in observation, and
merciless to fallacy in logic.*

T. H. Huxley

ON GEOLOGY

A familiar scene. It is morning and you are having breakfast in the kitchen. You test your cup of coffee to see how hot it is and leisurely unfold the newspaper. As usual, you think, there is bad news in the world. An earthquake in Chile has left many dead and thousands homeless. They say the intensity of the earthquake was 8.1 on the Richter scale, whatever that means. Hawaii is bracing for the arrival of large sea waves triggered by the quake. You wonder, If they can predict the arrival of these big waves in Hawaii, then why can't they predict earthquakes? Another item catches your eye. Flooding in the Midwest. Several towns inundated, and men paddling boats through the streets. At the bottom of the page the newspaper notes that scientists are still observing a new volcano that appeared a few days ago in the north Atlantic. The volcano is spewing out ash and cinders amid thunderous explosions, they say.

You turn the page and take another sip of coffee. Ah, here are, perhaps, more relevant matters on the local news front. In addition to the furor over the severe pollution of the lake, a manufacturer is accused of polluting the groundwater supply by pumping acids and other wastes into a disposal well. You make a mental note to avoid drinking any more water. The new medical-arts building is sinking at one end, because it was built on a layer of soft clay. The owners are trying to sue the engineering firm. Another item: A small landslide has blocked Highway 22 eight miles south of the city. And, finally, before turning to the comics page, you read that a local official has been appointed to a statewide committee on the energy crisis. The committee plans to look into the shortage of natural gas and fuel oil.

In this hypothetical situation you started your day, perhaps without realizing it, by reading a series of geological reports. All the items noted share one thing in common: They involve *geology*. Geology is the study of the earth. The overall objective of the geologist is to try to answer questions concerning the earth's physical nature, both past and present. Of equal concern to the geologist is the application of this knowledge to certain problems that beset human beings as they wrest from the earth the things that they need (e.g., oil, gas, water, metals, construction materials) while still trying to maintain harmony with the environment.

In attempting to understand the nature of the earth, geologists study the rocks that make up its outer crust. These rocks might be thought of as documents that have survived through millions of years yet carry within them the clues to past events (Figure 1–1). Geologists seek to unlock these secrets by careful study of rock records, not only in their natural outdoor settings, but by subjecting samples of these rocks to further scrutiny in the laboratory. In addition, the geologist must pay close attention to forces operative at the earth's surface (e.g., wind, wave, and

Figure 1–1. Fossil of a eurypterid preserved in the Bertie Limestone (Silurian Period). Remains such as these help geologists to interpret conditions on the earth millions of years ago. (Photo by Jan Patterson.)

stream action) and forces operative within the earth (e.g., those that produce volcanic eruptions and earthquakes). There is constant interplay between these forces and the rocks they affect. These *materials* (rocks) and the external–internal forces (*processes*) that act on them constitute the central theme of *physical geology*.

THE SCIENTIFIC METHOD

Although the investigation of earth materials and the processes affecting them may seem complicated, a great deal of common sense is involved in their study. Like other scientists, geologists use the *scientific method* in their approach to the solution of earth problems. Webster's dictionary defines scientific method as, "a method of research in which a problem is identified, relevant data are gathered, a hypothesis is formulated from these data, and the hypothesis is empirically tested." Detectives use this method in their work. When the detective comes upon the scene of a crime, he, like the geologist, is confronted by a series of happenings in the past. By intelligent observation, the search for clues, and the use of laboratory findings, the detective may skillfully reconstruct the chronology of the crime. The geologist follows a similar pattern of gathering clues, using laboratory investigation, and making inferences.

We all use scientific method, consciously or unconsciously, in our daily lives. For example, suppose that you share an apartment with George. George snores loudly and likes to sleep late. You typically rise early. One morning you awaken and are greeted by silence from George's room. No snoring. Even before getting out of bed, you hypothesize that George got up early and went out. While you are dressing, you look out of the window and notice that George's car is gone from the driveway. This strengthens your original hypothesis that George departed early. You tentatively predict that George's

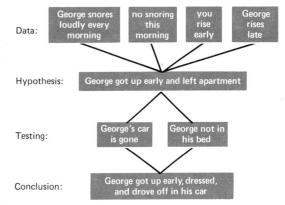

Figure 1–2. This diagram shows how you were using the scientific method in your thinking to describe the early-morning behavior of George.

bedroom will be empty. You go directly there and verify that George's room is indeed empty. Figure 1–2 shows schematically how you were thinking scientifically.

The situation, however, may be more complicated than this. Assume that you cannot verify that George is gone, because he always locks his bedroom door. Furthermore, George's cousin has a key to the car and often borrows it in the morning. Other bits of information include: (1) lately George has grown increasingly despondent over his latest love affair, and (2) he is poorly prepared to take an important quiz later this afternoon. Certainly your first hypothesis might still stand up, but you are not fully able to test it. You then consider alternative possibilities or hypotheses (Figure 1–3).

Not only are you now entertaining more than one hypothesis. You are using other hypotheses or assumptions in your thinking, that is, whether or not George or his cousin took the car. You have now been drawn into the method of multiple working hypotheses based on the data at hand. You have a complex problem driving you to test each hypothesis, eliminating those that do not explain all the facts. At this stage you would formulate a series of tests toward this objective, such as: (1) calling aloud to George, (2)

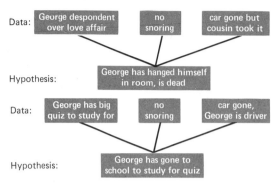

Figure 1-3. Alternative interpretations of what happened to George.

breaking down the bedroom door, (3) climbing up a ladder to look in his outside window, (4) calling his cousin, (5) looking for suicide notes, and (6) going to places where George usually studies. The point is, in this absurd example, that you are behaving no differently than the average scientist. From your own experience you can undoubtedly think of many parallels to this example.

In summary, geologists, like all scientists, use scientific method in their approach to earth problems, but the mechanics of thinking and reasoning are not much different than those we use ourselves from day to day.

HISTORICAL DEVELOPMENT
OF GEOLOGY

We can gain further perspective of what geology is all about by tracing the historical development of some of the fundamental ideas on which geology is based. Who was the first geologist? Nobody knows. We have archaeological evidence that primitive peoples who lived more than 12,000 years ago used certain rocks and minerals to make tools and weapons. More attractive materials were used for adornment. At first the acquisition of these materials was probably casual or accidental. Later the search became methodical. Quarries and mines thousands of years old have been discovered. Flint mines that were worked 4,000 years ago have been found in Great Britain. Copper mines in Cyprus are even older.

Among the first to set down their observations of geological happenings were the Greeks. Thales (640–546 B.C.) noted that wave action along sea coasts broke up and moved rock fragments, and that the Nile River dumped its waterborne sediment load at its mouth. He thus implied that the pattern of lands and seas may not be permanent, and that moving water has the power to make shorelines retreat by *erosion* (literally, "to eat away") or advance by building them seaward (Figure 1–4).

Xenophanes (570–480 B.C.), along with other Greek thinkers, observed fossil seashells entombed in rock and correctly deduced that much dry land had once been under the sea. Aristotle (384–322 B.C.) attached a time significance to such happenings. He believed that changes in land and sea patterns took place little by little over such long periods of time that they were undetectable within the span of a human lifetime. Despite this shrewd observation Aristotle was wrong about the origin of earthquakes. He taught that the interior of the earth was hollow and that earthquakes

Figure 1-4. This photo of the Nile delta, the Red Sea in the upper right, was taken from orbit during the *Gemini* program. The Suez Canal can also be seen. (NASA photograph.)

were caused by the rushing of winds through subterranean caverns.

Two Romans, Strabo (64–24 B.C.) and Pliny the Elder (A.D. 23–79), followed the example of the Greeks as observers of nature. Strabo, one of the first geographers, called Mount Vesuvius an extinct volcano. He was right about its volcanic origin, but it was not really extinct. A century or so later it erupted, burying Pompeii and Herculaneum under a thick layer of volcanic ash. During the same eruption Pliny the Elder lost his life while attempting to observe the volcano at close range. Pliny had been an avid and tireless collector of miscellaneous information about nature. He gathered all of this in several volumes, which he called *Historia naturalis*. Although this work contained many errors, it was one of the few inclusive works on natural phenomena that scholars could consult during the several hundred years following Pliny's death.

During the Middle Ages, scientific inquiry declined markedly. Dogma stifled new ideas, and much of the findings of the Greeks was lost. Only Aristotle's teachings were permitted. A rigid interpretation of biblical accounts created dogmatic single-mindedness, even in those individuals inclined to inquire independently. A stumbling block to critical inquiry was the widespread assumption that the earth was only 6,000 years old. (This time span was calculated from biblical data.) Using this premise, it was inconceivable that millions of years had been required to form mountains and valleys. Thus from about the time of Christ to the end of the fifteenth century, the development of geology as a systematic inquiry was seriously retarded. The other sciences suffered as well.

As the Middle Ages drew to a close, Leonardo da Vinci (1452–1519) was active as a renowned painter, sculptor, and engineer in Florence, Italy. While he was a military engineer for Cesare Borgia, Da Vinci had been able to study the Italian countryside in some detail. Some of his important observations revived ideas of the Greeks 20 centuries before him. Da Vinci claimed that rivers, fed by rain and meltwaters of snow, could change the landscape and carve mountains out of highlands. Fossil shells in northern Italy were, he said, the remains of marine animals that lived and died when that part of Italy was covered by salt water. As Da Vinci expressed it, "Sufficient for us is the testimony of things produced in the salt water and now found again in the high mountains, sometimes far from the sea." Da Vinci was also aware that water sank into the ground and could travel long distances in the subsurface through porous, inclined rock layers. Many of Da Vinci's writings were in a code that he devised himself. Some of his papers, still extant, are not decipherable. It would be interesting to discover what ideas, inventions, and observations the Italian genius had hidden in these papers.

Da Vinci was 42 years old when Georg Bauer (1494–1555) was born. Bauer is better known as Agricola, a German scholar who earned his living as a physician and diplomat but was more genuinely interested in natural processes. He wrote that " . . . water mightily alters the appearance of the earth's surface . . . "and such processes " . . . have been in operation since the most remote antiquity." In 1553 his efforts in research and experimental work on minerals resulted in a 12-volume work entitled *De re metallica*. For that time this classic was the most up-to-date compendium of mining, metallurgy, and geology.

Despite the contributions of Agricola, unorthodox views about the earth were not popular. Giordano Bruno (1548–1600) entered the Dominican Order but left it after continual conflict with his superiors. His attitude was to rely less on authority and to substitute direct observation of nature. He was arrested and imprisoned by the Venetian Inquisition and was eventually burned at the stake on February 17, 1600. Among his indiscretions were the stated beliefs that the flood of Noah never occurred and that the positions of land and seas could change.

Nearly 70 years after the death of Bruno, Nicholas Steno, a Dane, published an important work on the nature of layered, or stratified, rock. Like Bruno, Steno was a churchman. Unlike Bruno, he did not leave the Church and was not burned at the stake. In fact, Steno became the vicar-apostolic in the north of Europe. His contribution to geology can be considered a breakthrough. His publication, the *Prodromus* (1669), which he wrote while living in Florence, served as a model for future geologists seeking to condense their data into the form of written reports. More importantly Steno stated that stratified sedimentary rock was originally formed as horizontal layers one atop the other such that the bottom layers were older than those above. This principle, known as

Figure 1–5. Exposure of layered sedimentary rocks. These rocks were formed originally on a sea bottom. The lower layers were formed first and are older. Those above were formed later and are younger. (Photo by S. D. Heron, Jr.)

the *law of superposition* wherein younger layers lie atop older layers, forms the basis for modern historical geology (Figure 1–5).

Although this concept seems self-evident, someone had to be the first to note it formally. Steno's law of superposition of strata permitted other geologists to begin to frame a chronology of events based on succession of strata from oldest to youngest. It was soon realized that assemblages of fossils within these strata would show the progression of life on the earth from primitive to modern types. Much later this assisted Charles Darwin and others in formulating the theory of evolution.

The eighteenth century was a significant period in the maturing of geologic thought. More and more ideas were generated as a direct result of work in the field. In addition, geologists in several countries began the task of subdividing local sequences of rock strata into a meaningful chronological order, utilizing the law of superposition.

Giovanni Arduino in Italy, Johann Lehmann and A. G. Werner in Germany, and John Strachey in England were all involved in attempts to subdivide the geologic record. Arduino, for example, believed that a threefold division of rock units could be made. He called these simply *Primitive, Secondary,* and *Tertiary* from oldest to youngest. Lehmann's classification was similar. He believed that his intermediate grouping of rocks was formed during the flood of Noah. These rocks contained numerous fossils which he thought represented remains of creatures destroyed during the flood. Although Lehmann's intermediate rock group (the *Flotz-Schichten*) represented nothing of the sort (these rocks are millions of years old), it might be noted that he did relate his rock groupings to a chain of events in the past. Lehmann should not be construed as either stupid or naive simply because he considered his intermediate rock group to have been formed during Noah's flood. With the data available to Lehmann at that time, his assumption was a perfectly valid hypothesis, in keeping with the scientific method.

Werner (1749–1817) was an interesting character. At Freiberg, Germany, he developed what amounted to a cult of personality. His popular and dynamic lectures in geology and mineralogy attracted students from all over Europe. Although Werner unquestionably excelled as a teacher (many of his students later became competent geologists), he violated a basic rule of science: He became too enamored of his own ideas to the point of ignoring verifiable facts and the work and observations of colleagues whose data conflicted with his pet theories.

When he was 38, Werner tried to explain the origin of the earth and all of its rocks. He conceived of a universal ocean enveloping the entire globe early in the earth's history. Over a short period of time, he said, four distinct episodes of precipitation of rock materials from this ocean took place, thus accounting for all rocks in existence. Werner casually dismissed the molten lava erupting from volcanoes (obviously not a precipitate out of seawater) as heated rock material caused by coal beds burning in the subsurface.

Werner mistakenly avoided any field observation beyond the restricted area where he lived and worked. Other geologists of the time were not so short-sighted. A Frenchman named Jean Guettard (1715–1786) and a contemporary, Nicholas Desmarest (1725–1815), were field-oriented. They worked independently in the province of Auvergne, France. In this area dotted with extinct volcanoes and multiple lava flows, the Frenchmen studied these features at first hand and then published maps and reports showing that this region of France was once one of live volcanoes—a novel idea at the time, and not entirely believed by many. The philosophy these geologists represented was summed up by Desmarest, who said, "Go and see."

Meanwhile Scotsman James Hutton (1726–1797) was working on his *Theory of the Earth,* to be published in 1795, a time when Werner's popularity was widespread. Hutton was trained in law and medicine; he liked neither and turned to farming. His financial success in this pursuit permitted an early retirement at age 42. He spent most of the rest of his life studying geology. Hutton was perceptive, sensing a relationship between the solidified sandstones he had observed and the recent sands accumulating on beaches and along streams. It finally occurred to him that all of the same processes operating in his world had also operated inexorably throughout the past. Laws of nature invoked millions of years ago were still in force now. In other words, water ran downhill millions of years ago just as it does today; waves beat upon beaches then as now; and the forces that created and destroyed rock materials are still operable (Figure 1–6).

Figure 1–6. Waves crashing on a modern beach. Such waves have performed geologic work for millions of years. (U.S. Navy photograph.)

The implication of Hutton's great insight is that, if we take a close look at how rocks are forming today, we can better interpret the meaning of rocks formed under parallel conditions in the past. This important geologic concept is called the *law of uniformitarianism*, or, stated another way, "The present is the key to the past." Hutton's book, putting forth these ideas, aroused little immediate attention: He was a poor writer. A friend, John Playfair, recognized the value of Hutton's contribution. A more lucid writer, Playfair amplified Hutton's ideas in a book of his own but gave due credit to Hutton. Thus the idea of uniformitarianism began to take root.

At about the same time, an English surveyor named William "Strata" Smith (1769–1839) was on the verge of making contributions to geology equally as important as those of Hutton. Smith was a young man when he began to supervise the digging of canals. He noticed that layered rock strata cut through by his workmen contained very interesting and beautiful fossils. At first Smith collected these fossils as a hobby. Soon after, he realized that a specific rock layer yielded a unique assemblage of fossils different from those in layers above and below. It occurred to him that a certain rock layer could be recognized miles away by its unique aggregation of fossils even though the ravages of erosion had disrupted the continuity of that rock layer.

Thus, before Smith was 30 years old, he had developed two basic principles of geology: (1) the *principle of correlation*, meaning that sedimentary rock layers could be compared and matched up using their fossil content, even though separate exposures of the same rock might be miles apart; and (2) the *principle of faunal succession*, meaning that, at a certain time in the past, a group of organisms peculiar to that time had lived, died, and become entombed in the rocks forming then, to be followed later by different groups of organisms that would suffer the same fate but again leave the record of their passing imprinted into the rock record. Using these principles, Smith criss-crossed the island of Great Britain by horseback, covering more than 10,000 miles, to make the first geologic map of England and Wales, which he completed in 1815. This map showed the pattern of distribution of rock types across these two countries. Smith later became known as the father of stratigraphy (an important branch of geology dealing with the description and interpretation of rock strata), but his efforts were not recognized officially until he was about 62 years old, 8 years before his death.

Charles Lyell (1797–1875) was born in the year of Hutton's death and was 18 years old when William Smith published his geologic map of England and Wales. Lyell summarized and reinforced the geologic accomplishments of his predecessors in the form of a three-volume work, *Principles of Geology* (1833). In this book he hammered home the ideas of: (1) the immensity of geologic time, (2) correlation and faunal succession, and (3) uniformitarianism. The book was amazingly popular and was published in numerous editions. Lyell was still writing prefaces for later editions as late as 1868, 35 years after it first appeared.

By the time of Lyell, most of the basic ideas and concepts of modern geology had been established. Yet this did not mean that stagnation would set in. On the contrary, geology continues to be a very dynamic and exciting science up to today. The steady flow of new data about the earth is constantly modifying our understanding of the earth. Sometimes surprises happen. For example, in 1938 a strange-looking fish was caught off the coast of South Africa. It was a *coelacanth*, a fish that supposedly had been extinct for 70 million years.

In 1915 a German scientist, Alfred Wegener, suggested that the continents were once joined as one large supercontinent in the south polar region. Subsequently it broke up into the present, smaller continents. According to Wegener, the continents have drifted to their present positions from the South Pole. This theory of *continental drift* was rejected by other scientists and was even laughed at. New evidence to support this idea emerged during the 1940s, and today continental drift is an integral part of a stimulating theory on the nature of the earth called *plate tectonics*. (This will be discussed in detail later.) Meanwhile new vistas for geology have opened up as humankind explores the planets. Also, concern is growing regarding limited natural resources in a world with a burgeoning population threatened with the demise of environmental quality. The challenge of geology today is perhaps even greater than it was during the times of Steno, Smith, and Hutton.

SUMMARY

Geology is the study of the earth. The data of geology satisfy intellectual curiosity on such matters as the causes of earthquakes and volcanic eruptions, but geology also has practical applications for finding sufficient natural resources and solving problems of water pollution. Geologists, like other scientists, use the scientific method in their approach to problems of the earth. Scientific method involves formulating a hypothesis and then gathering data to test that hypothesis. Each of us uses the scientific method routinely in our daily lives.

Fundamental ideas about the earth were first developed by the Greeks, although primitive peoples thousands of years before the Greeks had sought out and had mined natural materials that they needed for weapons and tools. After the stagnation of the Middle Ages, basic principles of geology were announced by Steno (law of superposition), Hutton (law of uniformitarianism), and Smith (correlation and faunal succession). Today geology is still a dynamic science as humankind reaches out to explore other planets while trying to solve problems of pollution and energy resources on this planet.

Questions for thought and review

1. What is physical geology?
2. List several types of earth materials (rocks and minerals) that are important in our civilization.
3. In a daily newspaper, try to find as many articles as you can that seem to relate to geology.
4. What geologic activities of the earth do you think would have had the greatest impact on primitive peoples insofar as their welfare was concerned?

Selected readings

Agar, W. M., Flint, R. F., and Longwell, C. R., 1929, *Geology from Original Sources*, Holt, Rinehart and Winston, New York.

Fenton, C. L., and Fenton, M. A., 1945, *Giants of Geology*, Doubleday, Garden City, N.Y.

Flesch, Rudolph, 1951, *The Art of Clear Thinking*, Collier Books, New York.

2 General survey of the earth

STUDENT OBJECTIVES

At the end of this chapter the student should know something about:

1. the size and shape of the earth

2. the nature, composition, and origin of the atmosphere and the oceans

3. the gross internal structure of the earth

4. the nature and importance of the earth's gravitational and magnetic fields

The world is a beautiful book, but of little
use to him who cannot read it.

Goldini

In the previous chapter we saw how early geologists such as Smith, Hutton, and Lehmann groped for an understanding of the earth. They developed basic laws, such as uniformitarianism, that govern the way the earth changes and evolves. What would an eighteenth-century geologist such as Hutton have given to see photographs of the earth taken from orbit by astronauts, or to examine rocks brought back from the moon? No doubt Hutton would have given a great deal to be able to grasp some grand overview of the earth such as its precise size and shape, the nature of the atmosphere and the oceans, and the internal structure and composition of the earth, as well as knowledge of the earth's gravity and magnetic field. Hutton probably would have agreed that an understanding of the earth is best achieved first by knowing something about these gross aspects of this planet before focusing attention on more specific phenomena. In this chapter, given the knowledge acquired during the nearly 200 years since Hutton, and because of scientists like him, we are able to consider some of these larger features of the earth.

THE EARTH'S SIZE AND SHAPE

The earth is a nearly perfect sphere 8,000 mi (12,900 km) in diameter. The now-familiar view of the earth as seen and photographed from space by moonbound astronauts dramatizes this major feature of the earth (Figure 2-1). The earth deviates from perfect sphericity in the flattening in the polar regions by about 27 mi (43 km), with a corresponding bulge at the Equator. The earth thus could be more accurately described as an oblate spheroid. These deviations, however, are quite minor when compared with the overall size of the earth.

We should also consider the surficial irregularities of the earth. The vertical distance from the top of Mount Everest, the highest point on earth at 29,002 ft (8,840 m),

to the deeper parts of the ocean such as the Marianas Trench in the Pacific at 36,204 ft (11,034 m) is only 13 mi (20 km). Compared with the 4,000-mi (6,451-km) distance to the center of the earth, such irregularities, although considerable by human standards, are quite trivial. We can make another comparison to underline this point. If we could shrink the earth to the size of a billiard ball, keeping all surface irregularities proportional, then the earth so reduced would be smoother than an average billiard ball. Furthermore, as impressive as the atmosphere and hydrosphere seem to us, they comprise the merest films of air and moisture on the surface of the earth.

THE ATMOSPHERE

The atmosphere constitutes the accumulation of gases surrounding the earth. It is most dense near the surface. We are aware that meteors entering the atmosphere from outer space begin to heat up and glow because of friction at heights of 75–100 mi (120–160 km) above the earth; this may be considered the effective outer limit of the atmosphere. Yet it is known that atmospheric gases, although extremely rarefied, extend several hundred miles (kilometers) beyond the earth.

The atmosphere is made up of approximately three-fourths nitrogen and one-fourth oxygen by volume, as well as other components in much smaller quantities. Most important of these are carbon dioxide and water vapor. Table 2-1 shows the details of atmospheric composition; Figure 2-2 illustrates the subdivisions of the atmosphere.

In the *troposphere*, extending to a height of 7–10 mi (11–16 km), air density is greatest; most of our weather occurs here. Few clouds form above this layer. With increasing altitude within the troposphere, temperature decreases to a frigid −55°C. Above the troposphere lies the *stratosphere*. A distinctive feature of the stratosphere is the pre-

Figure 2–1. Our planet as seen from space.
(NASA photograph.)

Table 2–1. Composition of the earth's atmosphere as dry air at ground level. (Constituents listed in order of abundance.)

Constituent	Volume (%)
Nitrogen	78.08
Oxygen	20.95
Argon	0.93
Carbon dioxide	0.03
Neon	trace
Helium	trace
Methane	trace
Krypton	trace
Nitrous oxide	trace
Carbon monoxide	trace
Xenon	trace

SOURCE: Modified from D. R. Bates, 1961, *The Earth and Its Atmosphere.* Science Editions, New York.

sence of the *ozone layer*, encountered at an approximate height of 15 mi (24 km). At this point temperature increases take place as a result of interaction with solar ultraviolet radiation. The ozone layer absorbs these harmful radiations. Above the ozone layer the stratosphere merges with the *ionosphere* or upper atmosphere, a series of zones of ionized particles notable for their ability to reflect radio waves. As shown in Figure 2–2, the lower limit of the ionosphere rises at night and descends during the day. (This accounts for better radio reception at night.) In the ionosphere, displays of the aurora borealis, or northern lights (and aurora australis, or southern lights, in the Southern Hemisphere), take place. Temperature declines sharply to −70°C at a height of 50 mi (80 km), but beyond that, the temperature again rises to about 2000°C at 300 mi (484 km) above the earth.

Because of unequal heating of the air and other but subordinate factors, the atmo-

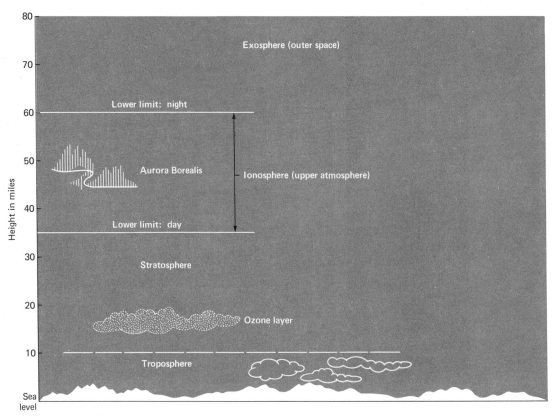

Figure 2–2. Subdivisions of the atmosphere.

sphere is in constant motion. The general planetary circulation of the air is well known: *Trade winds* generally move toward the Equator, while *westerlies* blow toward the polar regions. It is worth noting that the atmosphere makes it possible for the earth to utilize much of the energy received from the sun. The heat energy from the sun is transformed by the atmosphere into winds that affect the hydrosphere, ultimately causing infall of moisture to the continents. In other words, the sun's energy makes possible the intimate interaction of the atmosphere and hydrosphere by controlling climate as well as by providing for utilization of wind and hydraulic energy by humankind.

The origin of the atmosphere is somewhat a matter of assumptions and conjecture, but in all likelihood it did not always have its present composition of nitrogen and oxygen. It has probably evolved over millions of years from an original composition of either methane and ammonia or nitrogen and carbon dioxide. Plausible lines of reasoning in either case can lead to the present condition of the atmosphere. Consider the latter situation, for example, that of an original atmosphere of nitrogen and carbon dioxide. Rubey, a geologist who supports this view, relies on a process known as *degassing*. This refers to the transfer of gases within the earth to the surface by means of volcanic eruption. Gases from volcanoes contain considerable amounts of nitrogen and carbon dioxide. It would follow that, with continued volcanic activity for hundreds of millions of years, the atmosphere would reflect ultimately this dominance. Few scientists dispute that most of the oxygen was produced (and is still being produced) by the breakdown of carbon dioxide by plants in photosynthesis.

THE HYDROSPHERE

The hydrosphere includes all surface waters of the earth as well as that contained in the subsurface as groundwater. More water actually is contained in the subsurface than in all lakes and streams combined. But the bulk of the hydrosphere is, of course, contained in the oceans, which cover 71% of the earth's surface to an average depth of 12,000 ft (3640 m). Oceanic waters are in constant motion. Winds generate waves and superficial movement. Tides produce currents in the shallower waters adjoining coastal areas; and systematic main currents in the oceans include the Gulf Stream, the Labrador Current, and others.

The importance of the oceans to the human race is obvious. It is worthwhile here to sketch their probable origins. Volcanoes spew out large amounts of water during a typical eruption. Evidence of ancient lava flows shows that volcanic activity has been taking place for more than 3500 million years. This suggests that the ultimate source of oceanic water was the earth's interior, and that the hydrosphere formed by the same mechanism by which we accounted for the earth's atmosphere—that is, degassing. The picture is one of a primordial earth with a newly solidified but weak crust through which break repeated exhalations of molten rock and gases containing water. The water is released into the atmosphere where it is cooled, condensed, and precipitated into preexisting basins. In this way the hydrosphere would form gradually. We can visualize this process as continuing to the present but at lessening rates of volcanic activity as the crust becomes thicker and more resistant to these breakthroughs from the interior.

The water thus accumulated must have been fresh or nearly so. Its present saline condition is the result of delivery by streams and rivers of dissolved substances derived from rocks on the continents throughout many epochs of earth history. Most of the salinity of the oceans is attributable to salt—sodium chloride ($NaCl$). Other dissolved ingredients occur in smaller amounts (Table 2-2). It is believed that salinity would progressively increase with time.

Table 2–2. Major constituents of seawater. (Trace = less than 1%.)

Constituent	Weight (%)
Average dissolved solids	2.5
Dissolved gases	
(oxygen and carbon dioxide)	variable
Salts	
Sodium chloride (NaCl)	68.1
Magnesium chloride ($MgCl_2$)	14.5
Sodium sulfate (Na_2SO_4)	11.5
Calcium chloride ($CaCl_2$)	3.2
Potassium chloride (KCl)	1.8
Sodium bicarbonate ($NaHCO_3$)	trace
Potassium bromide (KBr)	trace
Boric acid (H_3BO_3)	trace
Strontium chloride ($SrCl_2$)	trace

SOURCE: Modified from H. U. Sverdrup et al., 1942, *The Oceans*. Prentice-Hall, Englewood Cliffs, N.J.

The hydrosphere is of critical importance to geology. Many geological phenomena are produced by the action of streams, lakes, and glaciers and along shorelines where oceans meet the continents. Weathering and alteration of rock materials could not take place without the involvement of water. Also, in oceanic waters much of the physical and biological history of the earth has been preserved in ancient sediments.

STRUCTURE AND COMPOSITION

It is well known from the study of earthquake records that the gross internal structure of the earth resembles that of an onion, that is, a series of nested, concentric shells or layers (Figure 2-3). The outermost layer — the crust — is thin, analogous to the skin of an apple. It ranges in thickness from about 4–40 mi (6–64 km), with an average of about 20 mi (32 km). It is usually thickest under continental masses and thinnest under oceanic basins. The crust consists of two zones: an underlying dense material of basaltic composition (*basalt* is a dark, fine-grained rock often ejected from volcanoes as lava) and a lighter material of granitic composition making up the bulk of the continents.

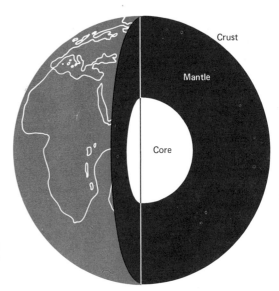

Figure 2–3. Cutaway of the earth showing the general internal structure.

In the detailed composition of the earth's crust, its elements have combined to form *minerals* (Table 2–3). These naturally occurring inorganic chemical compounds usually are crystalline solids. Familiar minerals include *diamond, quartz,* and *mica. Rocks,* in turn, are made up of aggregates of minerals. All rocks fall into one of three major groups: *igneous, sedimentary,* and *metamorphic,* ac-

Table 2–3. The 10 most abundant elements in the earth's crust.

Symbol	Name	Weight (%)
O	Oxygen	46.60
Si	Silicon	27.72
Al	Aluminum	8.13
Fe	Iron	5.00
Ca	Calcium	3.63
Na	Sodium	2.83
K	Potassium	2.59
Mg	Magnesium	2.09
Ti	Titanium	0.44
H	Hydrogen	0.14
Total		99.17

SOURCE: After Brian Mason, 1970, *Principles of Geochemistry*. 3rd ed., Wiley, New York.

cording to how they were formed. Wherever you might pick up a rock from the earth's surface, that rock usually can be assigned to one of these three categories. Igneous rocks have formed by solidification from a molten state. Sedimentary rocks form by the accumulation (usually under water) of rock debris derived from any preexisting rock. They may also result from the precipitation of dissolved substances in water. In either case, sedimentary rocks form layers or *strata* that originally are flat-lying. Metamorphic rocks result when preexisting rocks are subjected to pronounced changes in temperatures and pressures, producing new minerals and texture.

The earth's mantle lies beneath the crust and extends to a depth of about 1800 mi (2900 km). We can only surmise its probable composition, but most scientists believe that the mantle consists of rocks rich in compounds of iron and magnesium. The core of the earth has a radius of 2150 mi (3470 km) and is at least in a partially molten state. The core is thought to be composed of iron and nickel.

We said before that the general picture of the earth's structure was deduced from study of earthquake records. Earthquakes produce waves of vibratory energy known as *seismic waves*. The more rigid the rock, the faster these seismic waves can travel through them. Seismic waves usually speed up as they travel deeper into the earth; abrupt changes in velocity can be measured at the junctions of the crust and mantle and mantle and core. Hence we presume a change in the nature of the material at these points. The overall density of the earth is 5.5 grams per cubic centimeter (gm/cm³) or, in other words, 5.5 times denser than water. The density of the crust is about 2.7 gm/cm³. The mantle ranges from 3.4 gm/cm³ to 7.0 gm/cm³ with increasing depth. The density of the core may be as high as 13.7 gm/cm³, and the pressures at such depths must exceed 20,000 tons/in².

The earth becomes hotter with increasing depth. The rate of increase in temperature is called the *geothermal gradient*. Although the geothermal gradient varies from place to place on the earth, an average figure would be 30°C per mile of depth. This rate, however, cannot be maintained to any great depth, or the rocks of the mantle 200 mi (320 km) down would be melted. We know that the mantle is solid, because some seismic waves cannot travel through liquid; yet they travel through the mantle. We cannot be sure of these deeper temperatures, but some estimates of the temperature of the earth's core range from 2,000°C to 10,000°C.

GRAVITY

Gravitation is the physical force of mutual attraction governing all bodies in the universe. Sir Isaac Newton formulated exactly this gravitational law. His formula is:

$$F = \frac{G \times m_1 m_2}{d^2}$$

where

$m_1 m_2$ = the masses of the two bodies involved
d^2 = the square of their distance
G = the gravitational constant

This means that the greater the mass of either body (that is, the greater the amount of matter it contains), the greater the force of attraction will be. Furthermore, the farther the two bodies are apart, the less the force of attraction will be. Gravity is therefore variable. The earth-trained muscles of astronauts allowed them to bound about like gazelles in the weak gravitational field of the moon (which has a smaller mass than the earth), but they probably would have been crushed by their own weight had they landed on a large planet such as Jupiter.

Even on the surface of the earth itself, the force of gravity is not uniform. You will weigh slightly less, for example, at the Equator than you will at the North Pole because of the outward propelling centrifugal force caused by the earth's rotation. Variation in mass of material within the earth's crust will also cause small but detectable changes in gravity from place to place. The instrument used to measure these small changes is called a *gravimeter*. It is essentially a highly sensitive spring balance with an attached weight. Changes in gravity are detectable by the amount of stretch in the spring from place to place. This instrument and the changes it records in gravity have been put to practical use in the exploration for oil.

The general significance of gravity is, of course, considerably more than searching for oil. Gravity is the all-pervading force that holds the moon in its orbit around the earth, the earth in its orbit around the sun, and the sun in its course through space. Gravity holds each of us to the earth and is the force without which such varied geologic activity as landslides, oceanic tides, and stream flow with resulting erosion would be impossible.

THE MAGNETIC FIELD

The earth behaves as though there were a giant bar magnet embedded within it, aligned approximately along the axis of rotation. We can visualize lines of magnetic force enveloping the earth in much the same way as iron filings will align themselves around an ordinary bar magnet (Figure 2–4). In fact, for many years it was believed that some large metallic object was indeed responsible. Once it was realized, however, that such a metallic mass would be too hot to retain its magnetism, the idea was discarded.

Today one of the best explanations for the origin of the earth's magnetic field is the *dynamo theory* of Elsasser. A dynamo is an electric generator that changes mechanical energy into electrical energy. This electric energy, in turn, can generate magnetic fields as in the case of passing an electric current through a wire wrapped around an iron bar.

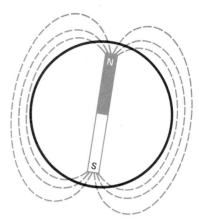

Figure 2–4. The earth's magnetic field is similar to that around a bar magnet.

In the dynamo theory it is thought that the motion of fluids within the earth's metallic core generate and maintain an electric current that sustains the earth's magnetic field.

Besides guiding mariners with their compasses across the oceans, the magnetic field, like the atmosphere, acts as a shielding mechanism against harmful radiation from the sun. Shorter wavelengths, or particles emitted from the sun, are either deflected by the magnetic field or are retained within the field as radiation belts. Although it is difficult to determine what might happen if the magnetic field were to suddenly disappear, the results might be harmful to many forms of life.

SUMMARY

The earth is a nearly perfect sphere 8,000 mi (12,900 km) in diameter. Surface irregularities are minor. The earth can be thought of as a series of onionlike concentric shells. The outermost shells make up the atmosphere, composed mainly of nitrogen and oxygen, and the hydrosphere, which consists of all the surficial waters of the planet. Both the atmosphere and the hydrosphere probably originated as a result of degassing from the earth's interior by volcanic eruptions throughout millions of years of earth history.

The crust or lithosphere is a thin ring enveloping the earth and is composed of rocks and minerals. Its structure is one of lighter continental masses set within a denser substratum. The interior beneath the crust includes a mantle of relatively heavy material and a partially molten metallic core. Knowledge of the earth's interior comes from study of seismic waves produced by earthquakes.

The earth is subject to the law of gravity, as are all bodies within the universe. But this attractive force is not everywhere uniform on the earth. Gravity is necessary for the operation of a variety of geologic processes. The magnetic field of the earth resembles that of an ordinary bar magnet and is perhaps the result of movement of the fluid metallic core within the confines of a weak electrical current inside the earth. The magnetic field serves as a shield against harmful solar radiation.

Questions for thought and review

1. Why might the earth be flattened in the polar regions?

2. How did the atmosphere possibly evolve?

3. What is the probable source of the water that makes up the oceans?

4. What is a mineral? Of what practical value are minerals?

5. How are sedimentary rocks formed?

6. Why does the force of gravity vary from place to place on the earth?

7. If the earth's magnetic field were to disappear, what would you guess had happened in the earth's interior to cause this?

Selected readings

Bates, D. R. (ed.), 1961, *The Earth and Its Atmosphere*, Science Editions, New York.

Bullen, K. E., 1955, The Interior of the Earth, *Scientific American*, vol. 193, no. 3, pp. 56–61.

Elsasser, W. M., 1958, The Earth as a Dynamo, *Scientific American*, vol. 198, no. 5, pp. 44–48.

Heiskanen, W. A., 1955, The Earth's Gravity, *Scientific American*, vol. 193, no. 3, pp. 164–174.

King-Hele, Desmond, 1967, The Shape of the Earth, *Scientific American*, vol. 217, no. 4, pp. 67–76.

3 The earth's spatial setting

STUDENT OBJECTIVES

The objective of this chapter is to enhance the student's perspective of the earth as it relates to the universe. The following topics are considered:

1. the solar system

2. size and number of stars, their distribution and motions

3. interstellar distances

4. how stars and planetary systems may originate

The eternal stars shine out as soon as it is dark enough.

Carlyle

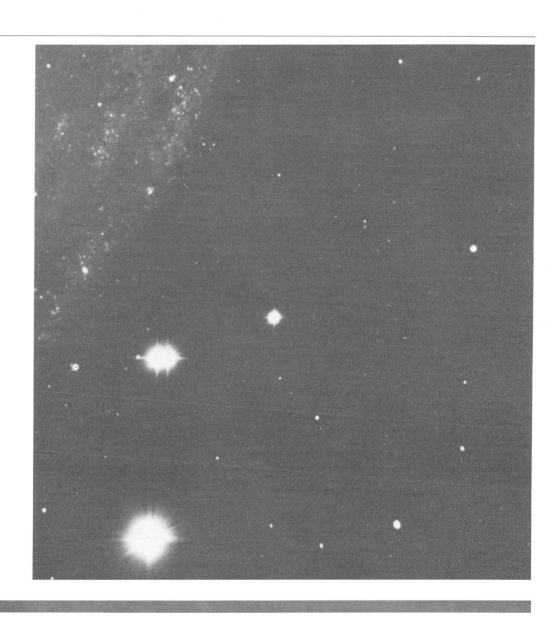

During the past two decades unmanned satellites have been launched by the dozen; astronauts have orbited the earth and walked on the moon; space probes have reached Mars, Venus, Mercury, and Jupiter; and scientists have lived and worked in space for protracted periods of time (*Skylab*). These activities have increased our awareness of the universe around us. In this chapter we will focus attention on the nature of the universe beyond the solar system in order to appreciate further the earth's setting in space. The rapid accumulation of data on the moon and the planets as a result of manned and unmanned space efforts permits the inclusion of Chapter 22 on planetology to detail some of the exciting findings of lunar and planetary exploration.

THE SOLAR SYSTEM

The solar system includes the sun and its retinue of 9 planets, 32 natural satellites of these planets, and numerous asteroids, meteors, and comets, all of which lie within the gravitational jurisdiction of the sun (Figure 3-1). The planets are grouped into two categories: *inner* (also called *terrestrial*) planets, including Mercury, Venus, Earth, and Mars in order outward from the sun, and *outer* (also called *major*) planets, including Jupiter, Saturn, Uranus, and Neptune in order beyond Mars. All of the planets move in the same direction and in the same plane around the sun. Their axial tilt, densities, and velocities in orbit vary (Table 3–1).

The inner planets are grouped together because of their small size, high densities, and proximity to the sun. The outer planets are characterized by their great size, low densities, and greater distance from the sun. In addition, the outer planets retain most of the satellites in the solar system, a few of which are larger than Mercury; for example, Saturn's Titan is 3220 mi (5193 km) in diameter. The outermost planet, Pluto, does not fit in with the major planets. It more closely resembles a minor planet and is thought by many astronomers to be a runaway satellite that escaped Neptune's gravitational influence. During its journey around the sun, Pluto crosses Neptune's orbit.

The belt of asteroids lies between the orbits of Mars and Jupiter. Asteroids are relatively small bodies of irregular shape ranging in size from Ceres (488 mi or 787 km in di-

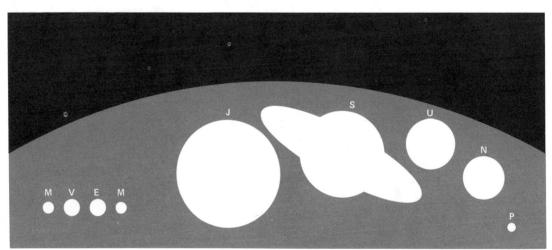

Figure 3–1. Comparison of sizes of planets against the sun's disc. Planets are shown in their proper order (*left to right*) outward from the sun.

Table 3–1. Some characteristics of the nine planets. Values are approximate.

Planet	Diameter (mi)	Density	Velocity (mi/sec)	Satellites	Distance from sun (millions of mi)	Axial rotation
Mercury	3,100	3.8	30	none	36	?
Venus	7,700	4.9	22	none	67	32°
Earth	8,000	5.5	18	1	93	23°
Mars	4,200	4.0	15	2	142	24°
Jupiter	89,000	1.3	8	12	484	3°
Saturn	74,500	0.7	6	10	887	27°
Uranus	32,400	1.2	4	5	1,785	98°
Neptune	31,000	1.6	3	2	2,797	29°
Pluto	3,600	4.0	3	none	3,670	?

ameter) to bodies less than 1 mi (1.67 km) in diameter. More than 1600 asteroids have been noted by astronomers (Table 3–2). In addition countless thousands of smaller pieces of space debris revolve around the sun within the belt of asteroids. Most astronomers regard the belt of asteroids as the remains of either a former planet that unaccountably disintegrated or a planet that never coagulated.

The sun is the central and principal member of the solar system. It is an average yellow star making up 99% of the total matter contained within the solar system. It is a luminous whirling sphere of hot gas 864,000 mi (1,393,548 km) in diameter and consists chiefly of hydrogen (82%) and helium (18%). It also possesses small traces of all the other elements normally found on earth (Figure 3-2).

The heat and light given off by the sun

result from nuclear *fusion*—hydrogen atoms colliding and uniting in complex reactions to form helium. When this takes place, energy is released. In short, the sun acts like a slow-motion hydrogen bomb. The sun pours out almost all of its huge energy into the depths of space. The tiny bit of this energy intercepted by the earth is still so enormous

Figure 3–2. The surface of the sun is turbulent. Plumes of gas may rise many thousands of miles (kilometers) above the surface. (NASA photograph.)

Table 3–2. Some larger asteroids.

Name	Diameter (miles)	Year discovered
Ceres	488	1801
Pallas	304	1802
Vesta	248	1807
Juno	118	1804
Hidalgo	30	1924
Eros	20	1898
Amor	1.5	1932
Apollo	1	1932

that the consumption of all our natural energy reserves (oil, coal, natural gas, wood) will scarcely equal a three-day solar output.

THE STARS

Size

Stars are large, self-illuminating gaseous bodies. Our sun is a relatively large star when compared with the earth. When placed against the disc of the sun, the earth shows up as a mere speck. In fact, 108 earths could be strung out along the sun's equator (Figure 3-3). If the sun were a hollow sphere, more than 1 million earths would be needed to fill it up. Despite this, astronomers refer to the sun as an average-sized star. They have measured much larger stars such as *Antares*. Antares, a red giant in the constellation Scorpio, is often used by navigators to find longitude. This star is estimated to have a 300-million-mi (483-million-km) diameter, which means that nearly 350 of our suns could be strung out along *its* equator. If we could replace the sun with Antares as the center of the solar system, then all of the inner planets, including Earth would be engulfed into its interior (Figure 3-4).

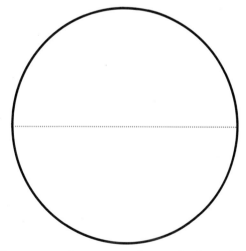

Figure 3–3. About 108 earths can fit along the sun's equator.

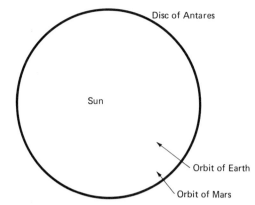

Figure 3–4. Antares would engulf half the solar system if it were our sun.

It should not be assumed that there are no small stars. Smaller stars such as the sun are actually more plentiful than the giants. Some dwarf stars near the end of their careers are no larger than planets the size of Jupiter and perhaps are smaller. The matter in these dwarfs is so closely packed that the densities may be on the order of 40,000 times that of water, and a cubic inch of such material would weigh several tons.

Distances

The moon lies 240,000 mi (387,000 km) from the earth. Compared with the 93-million-mi (150-million-km) distance from the earth to the sun, lunar trips of astronauts are like short hops. If we imagine New York City as the earth, then the journey to the moon is like a 10-mi (16-km) trip beyond the city limits, whereas to travel to the sun is comparable with a trip from New York to Los Angeles.

When we speak of distances and travel times among the planets in our solar system, impressive as the figures might seem, we are still talking about rather local journeys. Once we pass beyond the orbit of Pluto, interstellar distances expressed in miles or kilometers lose all meaning. We must resort to another yardstick of measurement: the speed of light. Light travels at the rate of 186,000 mi (300,000 km) per second. Most interstellar distances are measured in *light-years* —the distance that light travels in one year.

At the speed of light, the moon is less than 2 seconds distant from the earth and the sun 8 minutes away. Approximately 40 stars lie within a distance of 17 light-years. The immense gulf of space takes on significance if we imagine a trip by astronauts to the nearest star outside the solar system, Proxima Centauri, only 4.3 light-years away. If a spaceship could travel at 100,000 mi (161,000 km) per hour, it would fly by the moon in less than 2.5 hours and reach the edge of the solar system in about four years. But then the astronauts would face another 476 years of uninterrupted travel before reaching Proxima Centauri. Another way to put it is to imagine that such a trip could have been launched at the time Columbus sailed for America in 1492. The astronauts

would just now be getting there, and they would still have to face the return journey. It is, of course, improbable that such a trip would be undertaken unless travel close to the speed of light were possible or groups of men and women were willing to go, with their descendents completing the trip.

Numbers and distribution

On a hilltop away from urban centers on a dark, clear night, the average observer might be able to see as many as 2500 stars with the naked eye. Probably half of these will be double stars whose duality cannot be distinguished without a telescope. If one views the heavens with a good pair of binoculars, the number of visible stars becomes impossible to count. Photographs taken through large telescopes reveal so many stars that they appear as large, luminous clouds (Figure 3-5). The number of stars runs into the trillions.

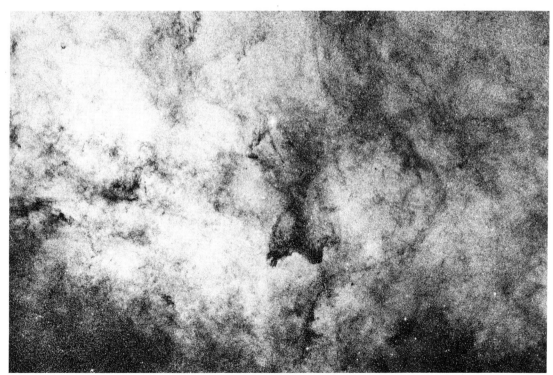

Figure 3–5. Star cloud in the region of Sagittarius. (Courtesy of Mount Wilson and Palomar Observatories.)

Individual stars are not evenly distributed throughout space but, rather, are concentrated into relatively dense *galaxies* (or island universes, as they are sometimes called). The sun and its planets are located in the outer part of the *Milky Way* galaxy, a disc-shaped spiral type similar to M81 in Ursa Major (Figure 3-6).

The dimensions of the Milky Way have been estimated at between 80,000 and 100,000 light-years in diameter and about 10,000 light-years thick. The number of individual stars in such galactic concentrations can only be approximated. The Milky Way probably contains somewhere between 100

and 200 billion stars. But it is by no means the largest galaxy. Galaxies may take on forms other than the spiral shape shown in Figure 3-6. The estimated number of galaxies of all types lying within range of our instruments may be as high as 6 billion. Despite these staggering numbers of stars, collisions are extremely rare, indicating the incredible dimensions of the universe.

Motions

All bodies within the universe undergo rapid and complex motions. To the hilltop observer on a dark night, the stars appear to be unmoving. This appears so because of the great distances between the observer and the star. The stars do indeed shift position, but this becomes apparent only after centuries of observation. It is comparable with watching a jet airplane high in the sky and at a great distance. The jet appears to be standing almost still, especially if it is moving away from the observer.

The complexities of movement are exemplified by the earth. It spins on its axis at 1000 mi (16,130 km) per hour as it moves along its orbital path around the sun at about 18 mi (29 km) per second. Small but measurable wobbles in the axial rotation are also present. The sun itself is moving through space at 11 mi (17 km) per second, carrying the earth and the other planets along with it. Thus the pathway of the earth through space is not a simple elliptical orbit around the sun in a confined region of space, but, rather, a series of long, swinging loops within the galaxy. In addition, the Milky Way itself is rotating and moving through space, and its vast population of stars is participating in these motions.

Astronomers tell us that the universe is expanding; that is, all galaxies move outward from a common center and at the same time recede from each other in much the same way as dots on an inflating balloon. It seems that, the more distant is a galactic system, the faster it moves. Remote galaxies

Figure 3–6. The spiral galaxy M81. (Courtesy of Hale Observatories.)

move at velocities greater than 60,000 mi (97,000 km) per second. Many astronomers assume on this basis that galaxies lying beyond detectable range (as yet) may approach the speed of light. If we project present galactic routes back to whatever their point of origin, they tend to converge at a time range of 6–10 billion years ago. This may approximate the age of the universe. Why all this matter in the universe began to expand outward that long ago is conjectural. Some interesting theories have been suggested, but discussion of these is beyond the scope of this chapter.

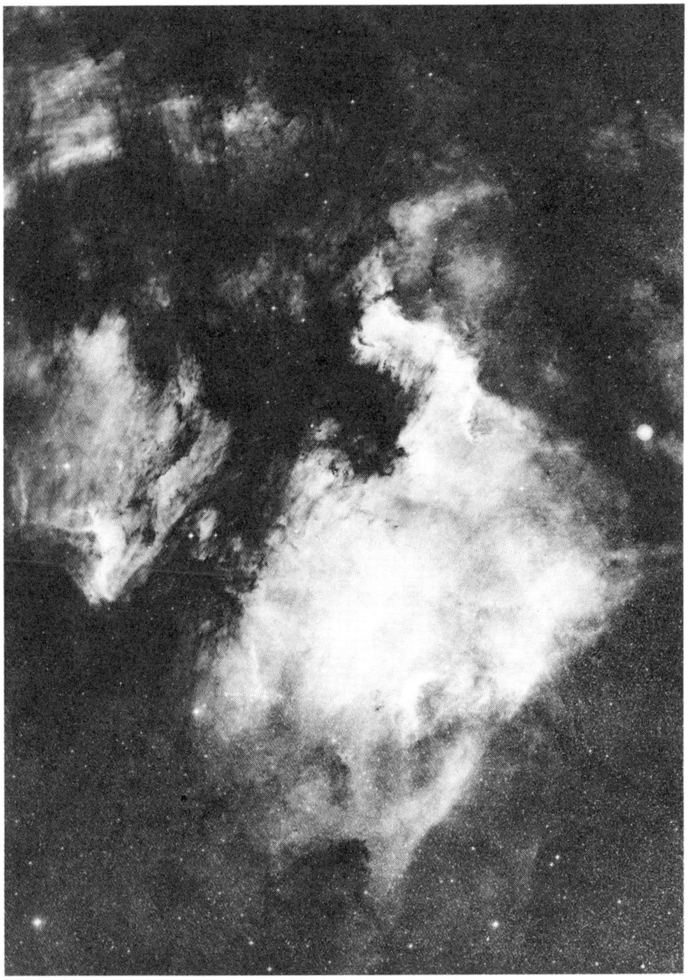

Origin

It would seem to most of us that outer space is virtually empty, especially when we are informed that a cubic mile (cubic kilometer) of space contains only a few milligrams or so of matter (the weight of a few grains of sand). Yet, the universe is so large that this small amount of matter adds up to such an enormous quantity that it obscures telescopic observation of distant stars just like fog. Most of this "fog" consists of hydrogen atoms. Some astronomers maintain that this interstellar matter is equal to that contained in all the stars (Figure 3-7).

There is general agreement that interstellar matter is the raw material that goes into the creation of new galaxies and stars. This rarefied gas, under the influence of gravity, condenses into clouds that evolve into galaxies. Local eddies within the galactic cloud become individual stars. If our own sun was originally one of these local eddies, then it is probable that the sun, earth, and other planets had a common origin.

This situation was considered as long ago as 1796 when Pierre LaPlace, a French astronomer, proposed his *nebular hypothesis*. He visualized the early solar system as a giant rotating cloud of hot gas, progressively condensing and increasing in its speed of rotation. When centrifugal force finally exceeded the gravitational attraction holding the cloud together, a ring of gas was detached and became stabilized at the equator of the cloud. Similar rings were expelled as the cloud continued to contract. Each ring gathered together to form a planet, while the central mass became the sun (Figure 3-8).

A current theory concerning the origin of the sun and planets has been modified from LaPlace's nebular concept. In essence, the planets are now thought to have originated as local eddies in the rotating solar cloud rather than as rings. At this stage the sun

Figure 3–7. Gas clouds in Cygnus. (Courtesy of Hale Observatories.)

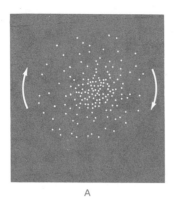 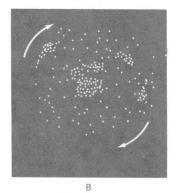 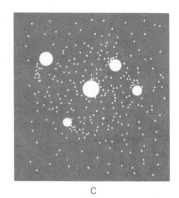

A   B   C

Figure 3–8. The nebular hypothesis of LaPlace.

would still be dark. With continued contraction, however, pressure and temperature would increase so much at the center of the sun that thermonuclear reactions would begin. The sun would then become luminous. If stars form this way, then planetary systems must be commonplace rather than accidental rarities. The astronomer Hoyle estimated the number of planetary systems in each galaxy to be approximately 1 million. Hoyle concluded, "I find myself wondering whether somewhere among them there is a cricket team that could beat the Australians."

SUMMARY

The sun, its nine planets, and their satellites make up the solar system together with asteroids, meteors, and comets. The sun generates its heat and light by nuclear fusion — the conversion of hydrogen to helium, the two chief constituents of the sun. The sun is so large that the earth becomes a small speck when silhouetted against its disc. Yet compared with a red giant star such as Antares, the sun itself is an even smaller speck. The solar system is a relatively crowded region of space compared with enormous interstellar distances. These distances are best measured by the speed of light, 186,000 mi (300,000 km) per second. The nearest star is more than 4 light-years away.

Stars are not evenly distributed throughout space but are concentrated into large rotating galaxies. The solar system belongs to the Milky Way galaxy, which contains 100–200 billion stars. As many as 6 billion galaxies may exist. All of these astral bodies are moving at high velocities outward and away from each other as though the universe were exploding.

Space between stars is not empty but contains extremely rarefied hydrogen gas. It is believed that condensation of this gas into huge clouds eventually creates galaxies. Local eddies within these clouds condense to form individual stars and planetary systems. Solar systems such as ours may be common.

Questions for thought and review

1. Can you explain why the planets should move in the same direction and in the same plane around the sun?
2. In what ways do the sun's energy emissions manifest themselves on earth?
3. If intelligent inhabitants inhabited a planet circling Proxima Centauri, what means would you suggest for trying to communicate with them?
4. When you look at a distant star, you are told that you are not seeing the star as it

is right now but as it appeared a given number of years ago. Can you explain why this is so?

5. Besides the current nebular concept for the origin of the solar system, there are older theories that involve the near collision of our sun with another star. Why do you think these theories have been discarded?

Selected readings

Adler, Irving, 1956, *The Stars: Steppingstones into Space,* New American Library, New York.

Baker, R. H., and Frederick, L. W., 1974, *An Introduction to Astronomy,* 7th ed., Van Nostrand Reinhold, New York.

Bondi, Hermann, 1960, *The Universe at Large,* Doubleday, Garden City, N.Y.

Brown, Harrison, 1957, The Age of the Solar System, *Scientific American,* vol. 196, no. 4, pp. 80–94.

Hoyle, Fred, 1950, *The Nature of the Universe,* New American Library, New York.

Reynolds, J. H., 1960, The Age of the Solar System, *Scientific American,* vol. 203, no. 5, pp. 171–182.

Shapley, Harlow, 1960, *Of Stars and Men,* Washington Square Press, New York.

Struve, Otto, and Zebergs, Velta, 1962, *Astronomy of the 20th Century,* Macmillan, New York.

Urey, H. C., 1952, The Origin of the Earth, *Scientific American,* vol. 187, no. 4, pp. 53–60.

4 Cycles and equilibrium

STUDENT OBJECTIVES

This chapter introduces the student to some very basic concepts in geology. After reading this chapter the student should:

1. understand the concept of equilibrium in nature and its underlying cause

2. be able to describe the workings of the hydrologic cycle

3. know what the rock cycle is and how rocks form within this cycle

4. know what tectonism is and be able to explain its significance

5. be able to relate geologic cycles to the concept of equilibrium states

6. have an appreciation of the immensity of geologic time and how it is measured

*Everything in nature is a cause from
which flows some effect.*

Spinoza

THE DYNAMIC EARTH

In the preceding two chapters we have depicted the larger features of the earth's architecture and the setting of the earth in space. In describing such tangibles as the earth's size and shape, there is always the danger of conveying the impression that we are talking about some kind of immobile, unchanging piece of inert matter. Just the opposite is true. The earth is a dynamic system that is constantly undergoing change.

At the surface of the earth we see waves crashing against the seashore, sand and pebbles bouncing and rolling along a stream bed, and rainfall washing soil from an improperly constructed highway fill. Below the surface, the earth seems to vibrate all the time. Sensitive instruments may record as many as 1 million earthquakes per year, most of them, of course, too small to be felt (Figure 4–1). Smashed and folded rocks of great age attest to the fact that the earth has been very active since its earliest beginnings. We well might ask, What is the basic cause of

Figure 4–1. Seismograph records show that the earth's interior is in a continuous state of dynamic activity. (U.S. Geological Survey.)

these and other natural processes? Why have they operated continually throughout the history of the earth? In answering these questions we come to grips with fundamental concepts about the nature of the earth. One of these fundamental concepts is *equilibrium*.

THE STRUGGLE FOR EQUILIBRIUM

It is a cold winter's night. The temperature is below freezing. You decide that it is time to bring in the dog, so you walk outside to call him. The door accidentally closes and locks. You have no coat, and you begin to shiver. Finally someone lets you (and the dog) back inside, and you stand next to the fireplace to get warm. After a while you get too warm and begin to perspire. In both cases, of course, your body was trying to maintain a balance of heat. Stated another way, attempts were made, both outdoors and inside, to restore a state of equilibrium that had been upset. During your shivering and perspiring, your body was in a dynamic condition, first attempting to warm and then to cool.

This episode is somewhat comparable with the condition of the earth itself. The earth is in a dynamic state, both externally and internally, because of an unequal distribution of energy, principally in the form of *heat*, on and within the earth. The energy that drives surface processes is derived from the sun, whereas the energy that motivates internal processes is provided by decay of radioactive elements and perhaps primeval heat remaining from the time the earth originated. Unequal heating of the surface and an uneven distribution of heat in the interior provide the earth with its nonuniform energy distribution and dynamic character.

All processes occur because of energy differences from place to place. These processes act to bring about a more equal energy distribution on and within the earth. This

Figure 4–2. The waves breaking along this beach were generated by winds, which in turn were created by unequal heating of the air by the sun. (U.S. Navy photograph.)

striving to establish or restore equal energy distribution is an attempt by nature to change the earth into an equilibrium system.

Despite the persistent effort to achieve this goal of equilibrium, success is only temporary, because nonrandom factors are constantly introduced into given equilibrium systems. Consider a grain of sand in a stream: It attempts to achieve equilibrium by traveling downstream to the sea, thus exhausting all of its energy. However, when it reaches a beach next to the sea, a strong onshore wind blows it a mile (kilometer) inland and lodges it on the side of a high hill. The sand particle must begin its struggle for equilibrium again. At the same time, the wind responsible for displacing the sand grain was also seeking a state of equilibrium because of unequal heating of the air by the sun (Figure 4-2). Thus it can be seen that the earth is a complex of equilibrium systems, each interacting with and influencing the other.

CYCLES

The expression, "There is nothing new under the sun," may have been inspired by the fact that events have a habit of repeating themselves. This is as true of the daily reappearance of the sun as it is of wars and clothing fashions. So universal is this periodic governance of things that it is difficult to find anything—living or nonliving—that is not involved in or influenced by cyclical events. The most obvious examples can be drawn from astronomy. The earth's rotation and circuits of the sun produce the days and years. Lunar phases and solar eclipses are predictable recurrences. Halley's comet returns to the solar system every 76 years. Here on earth, tides and hurricanes, bird migrations, and the spawning of salmon all follow some repeated pattern.

Cycles are not always simple or predictable in detail. They may be thousands or millions of years in duration so that their repetitive nature is not readily detectable. Cycles may also be interrupted or accelerated. Cycles within cycles, such as the months within the year, may exist. Some cycles are so enmeshed within other cycles that, without the one, the other cannot operate.

GEOLOGIC CYCLES

The geologist is well aware that most of the geologic phenomena that are studied are cyclic. It is also evident that certain fundamental cycles in geology form a set of processes that are interrelated by changes in physical conditions that dictate arrival at different equilibrium states. In other words, we see cyclical geologic events obeying the decree of nature to seek equilibrium conditions.

Why cyclic?—this is not always an easy question to answer. However, differences in energy in the system generally are responsible for the progress from one stage in a cycle to another. Heat energy, as in heating and cooling, causes changes in many cycles. Mechanical energy differences also may motivate portions of cycles. In general, if the same forces or circumstances that activated a particular cycle are extant throughout the duration of the cycle, the cycle may be renewed. The simplest example is the *hydrologic cycle*, activated by heat energy from the sun. As long as the sun exists and radiates heat, the hydrologic cycle repeats itself. If the sun were to disappear, of course, the hydrologic cycle would end.

The hydrologic cycle is one of three geologic cycles fundamental to an understanding of the earth. The other two are the *rock cycle* and the *tectonic cycle*. An understanding of these three cycles and their interrela-

tionships will contribute greatly to a knowledge of geology.

The hydrologic cycle

The hydrologic cycle is the most apparent of the three cycles in its operation. Wherever we see water, we see part of this cycle. Large rivers like the Mississippi and the Amazon constantly pour billions of gallons (liters) of water into the sea. Yet the continents do not dry up, nor do the oceans fill up. Evaporation from the ocean's surface eventually results in replenishment of water by rain or snow. Thus the hydrologic cycle expresses the way in which water leaves the great reservoir of the oceans, covering more than 70% of the earth's surface, and eventually returns to the oceans (Figure 4-3). This cycle reflects a series of changes in physical conditions and corresponding readjustments by the material involved (water in this case) to restore an equilibrium state.

Initially water evaporates from the ocean into the atmosphere. When dry, undersaturated air moves over the water surface, water evaporates in order to establish a saturated state at the prevailing atmospheric temperature. When the air becomes saturated, a state of equilibrium exists. This saturated air mass may be in motion; as it moves, its temperature may change because of passage from daylight to darkness or formation of clouds by cooling in the upper atmosphere. The air mass, through cooling, may become oversaturated, and water will precipitate out. If the precipitation (as rain or snow) occurs over the ocean, the route of the completed cycle will be shorter. However, if the moist air has moved over the land and precipitation occurs there, the moisture will be disposed of in one of several ways.

Precipitation may become *surface runoff*, as filmy layers and streamlets of water flow over the surface. These join with other streamlets to form small streams to form larger streams to flow into lakes and/or eventually into the ocean, thereby completing the

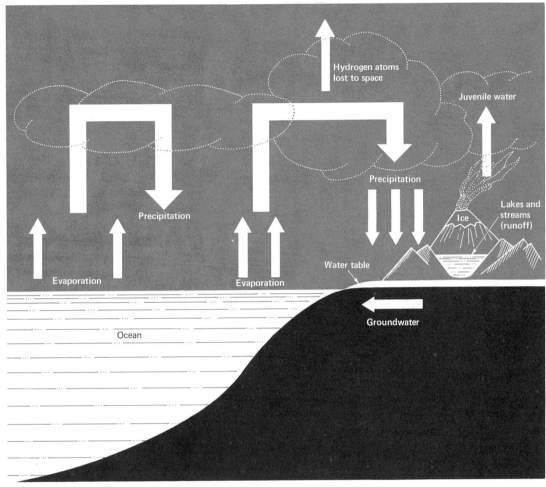

Figure 4–3. The hydrologic cycle.

cycle (Figure 4-4). A large portion of runoff evaporates into undersaturated air as it flows over the ground surface. This readmits the moisture back to the atmosphere, where it will precipitate later.

Another portion of precipitation may soak into the ground and become infiltrated moisture. This groundwater supply is the source of most drinking water. However, some of this water is taken into the roots of plants; that which is not used in the life processes of the plants is evaporated (*transpired*) from their leaves, returning it to the atmosphere. Groundwater moves slowly in the subsurface and may reemerge as springs or join with surface water in streams and lakes. The hydrologic cycle has many subcycles, because so much water is evaporated, precipitated, and evaporated again and again before it finally reaches the oceans.

Within the confines of the hydrologic cycle, streams, groundwater, and moving ice (glaciers) mold the continents, while sediment transported and deposited by moving water creates new rock material out of old. Valuable resources such as oil and gas are formed within rocks as a result of aqueous deposition. Many geologists would find

A

B

Figure 4–4. *A.* Filmy layers of water flow over the ground surface during a rainstorm as sheetwash. *B.* Runoff in the form of a rapidly moving stream.

themselves unemployed without the benefits of the hydrologic cycle.

For humankind in general, the hydrologic cycle furnishes our freshwater supplies. Unfortunately this cycle is highly susceptible to human influence (Figure 4-5); that is, we come face-to-face with water pollution.

Water pollution involves the introduction of foreign substances that deteriorate the quality of natural waters so that these waters are unfit for either aquatic or human life. It is ourselves—mostly unwittingly—who have brought about this situation on a large scale in the twentieth century.

Pollutants spread from a source when they are introduced and can infect entire stream networks. The hydrologic cycle is a self-cleansing system. As pollutants circulate, most of them should be dispersed or otherwise rendered harmless. A major point, however, is that the intensification of industrial activity, coupled with the population explosion, has outstripped nature's ability to cleanse itself. Thus pollution is spreading through the entire hydrologic system. In addition to streams and lakes, oceans are beginning to change, as indicated by measurable increases in the amount of lead (from automobile exhaust and other pollutants) in the North Atlantic. In some localities even rain and snow no longer fall as pure substances but pick up foreign matter during their descent through a polluted atmosphere.

Increasingly and ominously, pollution goes underground as deep wells are drilled to dispose of noxious substances. Underground waters face a more lingering contamination, because subsurface waters circulate so much more slowly than do surface waters.

Pollutants introduced into the hydrologic cycle must eventually spread (and have in some instances) into plants and animals and so infest links in the food chain. DDT and mercury are examples that have received a great deal of publicity and attention. It has not been shown conclusively that DDT is lethal to humans. Indeed, some scientists point to the beneficial effects of DDT in saving lives otherwise lost to diseases carried by insects such as the mosquito. But, although this is true, it should be borne in mind that years may pass before a particular substance is judged dangerous. In such cases our abil-

Figure 4–5. Many streams have become polluted by solid wastes and chemicals. (Courtesy of H. S. Johnson, Jr.)

ity to reverse immediately the effects would be minimal.

We might ask, What if all the waste products of our society—municipal and industrial—were to cease at once? In such a case the self-cleansing activities of the hydrologic cycle would remain operative, and some stream systems would be restored to normal in a comparatively short time. It would probably take many decades, however, to restore large bodies of water such as Lake Erie, the shallowest of the Great Lakes. Here the dumping of wastes into water systems will not cease immediately. Even with abatement procedures by cities and factories, the continued and rapid growth of society seems to assure us of a pollution problem for decades to come. We will explore these and other questions connected with the quality of our environment in later chapters.

The rock cycle

The rock cycle is essential to an understanding of physical geology. It epitomizes those geologic processes that bring about the formation of all rocks found in the earth's crust. It also illustrates, as you might expect, a number of equilibrium states. The rock cycle is shown schematically in Figure 4-6. Note here that *magma* is the hot, molten material from which all rocks are eventually formed. Lava from volcanoes is merely magma that appears at the surface of the earth. Cooled and hardened magma becomes *igneous* rock. At higher temperatures magma is stable (in equilibrium) as a melt. As heat is lost from the molten mass, the temperature is lowered, causing minerals to begin to crystallize in order to restore a state of equilibrium.

An important subcycle, also shown in Figure 4-6, involves gases and water vapor released from magma. Gases may make their way to the surface and become part of the atmosphere. Water vapor may do likewise and become part of the hydrosphere. The atmosphere and hydrosphere interact to produce the *biosphere*, which includes all of the life forms occurring on the earth. The biosphere contributes sediment that becomes part of the main cycle.

When exposed at the surface of the earth, igneous rocks begin to *weather*. Certain minerals break down chemically, while others separate mechanically from the rock. This occurs because the surface of the earth is a low-temperature, low-pressure, mainly oxidizing environment quite different from that in which these rocks formed. Therefore igneous rocks are no longer in equilibrium at the surface and thus begin to interact with the surface environment to reestablish equilibrium. In this sense we can define weathering as the response of earth materials in achieving equilibrium under surface conditions. Note that in this definition we include other rock types as well as igneous rocks as being susceptible to the weathering process (see Chapter 11).

Materials broken down by weathering may undergo downslope readjustment by

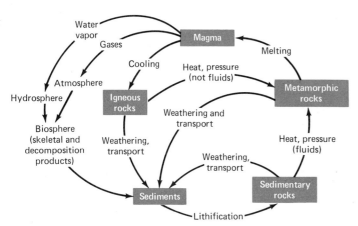

Figure 4–6. The rock cycle showing the inter-relationship of the materials of the earth's crust (in boxes) and the processes (arrows) affecting them.

gravity or may be carried into streams and even travel as far as the oceans. Whether transported a short or a long distance from its point of origin, a mass of unconsolidated material will be deposited and come to rest as *sediment*. Sediment may consist of not only fragmental material but chemical precipitates and organic matter as well. In any case the sediment tends to arrange itself in layers. As time passes, the weight of overlying sediment causes compaction of the more deeply buried layers. Grains will move about to fill spaces more efficiently and relieve some of the pressure applied from above (a form of *equilibration*). Through tighter compaction, partial dissolution of grains (at points of stress) and/or chemical precipitation of dissolved material from water trapped in pore spaces between grains, the sediment may become *lithified* (literally, turned to stone) as *sedimentary rock*.

If sedimentary rocks are more deeply buried, and temperatures become substantially higher than those under which they were formed, the rock mass will again undergo readjustment. Grain reorientation may occur first, and new minerals possessing greater stability under the elevated temperature and pressure conditions may appear.

This entire process is called *metamorphism*, and the new rocks formed are called *metamorphic rocks*. If the metamorphic process is carried to an extreme of high temperature, the melting point is reached and magma regenerated. Thus a close connection between igneous and metamorphic processes is demonstrated. And all steps in the rock cycle result from attempts to achieve equilibrium under changing physical and chemical conditions.

The cross-arrows in Figure 4–6 show that the rock cycle can be (and often is) interrupted. Igneous rocks can be converted directly into metamorphic rocks, whereas either sedimentary or metamorphic rocks can be recycled back into sediments if they are exposed, weathered, and transported at the earth's surface. In the rock cycle any type of rock can be theoretically converted to any other type of rock with the exception of sedimentary rock, which cannot become igneous without first being metamorphosed.

The tectonic cycle

The term *tectonism* refers to the instability of the earth's crust. Forces within the earth act to uplift, downwarp, or otherwise move and deform segments of the crust. Evidence of deformation in the rocks themselves is commonplace (Figure 4-7). Usually, these *tectonic forces* work slowly over long periods of time, changing the outline of seas and continents by uplift and subsidence. These same forces, however, may announce their presence vividly when earthquakes occur.

Instability of the earth's crust is recurrent. Many areas of the land surface contain layers of sedimentary rock that show successive remains of marine animals alternating with remains of land animals and plants or other evidence of emergence. In such areas the sea has invaded and retreated several times over millions of years in response to different kinds of tectonic movements. A tectonic cycle might involve deposition of sediment

40

Figure 4–7. These rock layers have been folded by tectonic forces while deep within the earth. Later uplift and erosion have exposed them at the surface.

and volcanic material on the edge of a continent and be followed by uplift by lateral compression, forming a mountain chain. Then the mountain chain will be eroded. A second cycle of deposition of the eroded products of the older mountain chain on the continental edges begins a new cycle. One cycle will take several hundred million years.

Without tectonic activity, we would have no rock cycle. The uplift of large portions of the crust allows erosional forces to convert rock masses into sediments and to carry them away. Downwarping of the crust creates areas in which sedimentary rocks may form. Tectonic forces are also instrumental in changing preexisting rocks into metamorphic rocks. If tectonism were

nonexistent, then, quite early in the earth's history, original elevations above sea level would have been reduced to flat plains awash by the sea or somewhat below sea level. In that case development of the human race as it is would have been most unlikely.

Within recent years the concept of *plate tectonics* has gained widespread favor. Its special appeal lies in its ability to relate large-scale tectonic phenomena around the world to a few unifying ideas about the earth. The interest in these intriguing ideas has been so intense that they have been the subject of popular books and special television programs. In later chapters the evidence for continental drift and plate tectonics will be presented in some detail. So also will be the arguments against. Here we will merely sketch the essential points of plate tectonics theory.

Advocates of plate tectonics theory visualize the crust and upper mantle as being composed of a number of large plates or slabs capable of independent movement. The continents are situated and are passengers on these plates. The continents are, in a sense, carried along a "conveyer belt" outward from oceanic ridges, the source of the new crustal material.

Where two plates come into contact, one plate slides beneath the other, and the leading edge of the descending plate is deflected downward (called *subduction*) and absorbed into the mantle. Thus new ocean floor is created along the flank, while the leading edges of many plates are destroyed. The cyclic nature of oceanic crust and mountain building processes, that is, the tectonic cycle, have been recognized to be related within plate tectonics theory. Those who espouse global tectonics have an array of impressive evidence to support their conclusions. On the other hand, some scientists find weaknesses in the theory, and some reject it outright. This is a healthy condition in science, because it keeps both sides actively seeking the truth rather than vegetating in self-satisfaction. (Additional discussion of the tectonic cycle may be found in Chapter 21.)

TIME AND CHANGE

From previous discussion we can see that our earth is an active earth. Geologic cycles are consistently in operation: the swift rush of a mountain stream, the crumbling of ancient rocks by weathering, the piling up of sediments on the sea bottom, the recurrence of earthquakes, the melting and ejection of rocks from volcanoes, and many other examples. Yet within one human lifetime how much change can you see? As you approach Denver, Colorado, from the east, the silhouette of the Rocky Mountains can be seen against the skyline (Figure 4–8). Does it look any different than when it was viewed by the first explorers pushing west? In another example, we can compare photographs taken a century apart of the hills at Gettysburg, Pennsylvania, the scene of intense fighting during the Civil War. After a century the hills retain the same essential contours. When we look on mountains, valleys, plateaus, or other natural features, it would

seem they are eternal. They are not. Changes too small to be detectable within a human lifetime have taken place, whether in the Rockies, at Gettysburg, or elsewhere on earth.

If given enough time, many small changes add up to big changes. Geologic cycles can best be understood if we relate them to the framework of time. The earth is about 4.5 billion years old. Such an enormous span of time can most easily be grasped if we relate it to some other scale. For example, suppose we let one second equal one year. Then the average human lifetime would be 72 or 73 seconds; the American Revolution would have occurred more than three minutes ago; and Columbus would have discovered America about eight minutes ago. Going back further, the time of Christ would be only 33 minutes ago, and the entire recorded history of human civilizations encompasses less than two hours on our scale. Living things became abundant 20 years ago but first appeared in the sea 100 years ago. On this scale the earth was formed about 150 years ago. We humans are latecomers on the scene, occupying but a brief instant of time in the span of earth history.

Figure 4–8. The topography of the Rocky Mountains appears the same today as it did to early settlers. (T. S. Lovering, U.S. Geological Survey.)

Profound change can be wrought through the operation of geologic cycles over a 4.5-billion year period. For example, if tectonic action could thrust up the Florida peninsula at the rate of 1 in. (2.54 cm) per century, we would have a mountain range (assuming no erosion) to rival the Himalayas after 35 million years. If we assume that such a mountain range could be flattened by erosion at the same rate of 1 in. (2.54 cm) per century, then 65 generations of Himalaya-sized mountains could have grown up and been destroyed since the earth was formed.

In another hypothetical situation we can speed up the rate of change. If the state of Louisiana were to subside at the rate of 12 in. (30.48 cm) per century, it would take only 50,000 years for the advancing sea to put the entire state under water. New Orleans would be a submerged city more than 200 mi (322 km) offshore. If the sea retreated at the same rate, and the cycle kept repeating itself, it would still take 45,000 such advances and retreats of the sea across Louisiana to equal the age of the earth.

While we speak knowledgeably about the age of the earth today, this was not always so. As noted in Chapter 1, people 300 years ago thought that the earth was about 6000 years old, based on interpretation of biblical data. Since that time, and until well into this century, scientists have painstakingly searched for an age for the earth and for some way to measure the duration of past events. Chronology can be determined in two ways: by *relative* methods and by *absolute* methods.

Relative-age dating has been used by the field geologist for many years. It is based in part on the law of Superposition (see Chapter 1). Relative methods permit us to set up a series of events as recorded in the rocks in their proper chronological order, but we are unable to assign an actual number of years to each event.

The quest for so-called absolute methods of age determination led first to crude attempts to learn the age of the earth in years. By the turn of the century two approaches had been devised. One calculated the rate at which sodium is added to the world's oceans. By dividing the annual sodium increment into its present content, the age of the oceans was determined to be 90–100 million years. It was assumed that the amount of sodium added annually remained the same, and also that the oceans were initially composed of fresh water. A second approach was based on rates of sedimentation. Assuming that sediments pile up at a fixed rate, we can take the maximum thickness of sedimentary rocks laid down on the earth and arrive at a figure for the approximate age of the earth. One such figure, derived by Sollas in 1905, was 30 million years. These methods fail, because the rates of sodium and sediment accumulation vary and could not be determined accurately. Another method, determined by Kelvin, was based on the rate of heat loss by the earth. Results in this case also fell short of the actual age of the earth, but was nearer the mark than other methods.

It was clear that reliable age determinations could not be realized unless a geologic "clock" could be found in which the rate of elapsed time was constant or nearly so. Such a clock was discovered in 1905 by Boltwood at Yale University. He suggested that uranium changes (or decays) into lead by radioactive emission of particles. The rate at which this process takes place is constant and is unaffected by pressure and temperature changes. Thus, if the ratio of lead to uranium could be measured accurately, a fairly precise age could be determined. The more lead (the higher the ratio) present, the older the rock. This set the stage for later exploitation of this and other similar *radiometric methods*. The use of radiometric age dating of rocks containing radioactive minerals gathered momentum rather slowly for two reasons: (1) some of Boltwood's dates—for example, 2 billion years for a rock from Ceylon—were greeted with complete disbelief by most geologists; and (2) early techniques resulted in conflicting ages, suggesting that the method was not reliable. Nevertheless, the uranium–lead method gained acceptance as techniques improved and as other radiometric methods were developed.

Table 4-1 summarizes radiometric dating methods, several of the elements involved, and their decay products. The minerals indicated in the table are common ones used, but there are others. *Half-life* refers to the length of time needed for a radioactive isotope to convert one-half of its weight into the decay product. *Isotopes* are various forms of an element, each of which has a different atomic mass. Some of these isotopes are radioactive. A long half-life such as that of rubidium-87 or uranium-238 allows dating of very old, but not younger, rocks. A short half-life such as that of carbon-14 permits dating of organic material no older than 50,000–60,000 years. Carbon-14 dating of organic material such as bone, ashes, wood, and cloth has been of much value to both archeologists and geologists.

Long before radiometric age dating came into use, geologists had subdivided geologic time into eras, periods, and epochs, based on painstaking study of the geologic record and especially on changes in the nature of ancient organisms preserved as fossils in rocks. Modification and refinement of this geologic time scale to its present form have taken two centuries. Prior to employment of radiometric methods to define an absolute time scale, the duration in years of these subdivisions was a matter of conjecture. In Table 4-2, we indicate a few radiometric dates for these periods.

SUMMARY

The earth is not immobile and unchanging but rather is in a dynamic condition attributable chiefly to an unequal heat distribution on and within the earth. This situation evokes a response by nature to achieve a state of equilibrium and accounts for many of the geologic phenomena (processes) operative on and within the earth. These are expressed in three interrelated geologic cycles that constitute the broad foundation of geology: (1) the hydrologic

Table 4–1. Radiometric dating methods.

Method	Decay process	Half-life (millions of years)	Minerals used	Rock types dated	Limits
Uranium–lead[1]}	$U^{238} \rightarrow Pb^{206}$ $U^{235} \rightarrow Pb^{207}$ $Th^{232} \rightarrow Pb^{208}$	4,500 713 1,300	Zircon Uraninite	Igneous Metamorphic	> 100 m.y.
Potassium–argon	$K^{40} \rightarrow Ar^{40}$	1,200	Biotite Hornblende Feldspars Glauconite Whole rock[2]	Igneous Metamorphic Sedimentary	> 10 m.y.
Rubidium–strontium	$Rb^{87} \rightarrow Sr^{87}$	50,000	Whole rock[2] Biotite Feldspars	Igneous Metamorphic	> 100 m.y.
Radiocarbon or carbon-14	$C^{14} \rightarrow N^{14}$	0.006570	Organic Matter	Sedimentary	100–60,000 yrs

[1] All three decay processes may be used to obtain three ages for comparison.

[2] Whole rock ages are obtained by measuring the respective isotopes separated from the entire rock sample, not a specific mineral.

Table 4–2. The geologic time scale.

Era	Period	Years before present (in millions)	Important developments
Cenozoic	Quaternary	0–2	Ice ages, hominoids
	Tertiary	2–65	Abundant mammals
Mesozoic	Cretaceous	65–220	Extinction of dinosaurs
	Jurassic		First birds, many dinosaurs
	Triassic		Mammals, reptiles
Paleozoic	Permian	220–600	Many extinctions
	Pennsylvanian		Coal-forming swamps and first reptiles
	Mississippian		
	Devonian		Abundant fishes
	Silurian		First land plants
	Ordovician		First fishes
	Cambrian		Most phyla established
Precambrian		600–3000	Algae, primitive marine animals
Azoic		> 3000	No life

cycle, (2) the rock cycle, and (3) the tectonic cycle.

The hydrologic cycle involves the circulation of waters from the oceans to the land and back to the oceans. Surface and subsurface waters, in returning to the oceans, perform geologic work such as erosion, transportation, and deposition of sediment. The rock cycle shows the interrelationships of the materials of the earth's crust (igneous, sedimentary, and metamorphic rocks) and the processes that affect them. In the tectonic cycle forces can uplift and downwarp the crust, activating or influencing other geologic cycles.

The hydrologic cycle has been affected by pollutants resulting from expanding population and industrial activity. Although the hydrologic cycle is a self-cleansing system, it has been overtaxed in this capability. The problem of water pollution cannot be ended immediately and will continue for years to come.

Geologic cycles have operated over a span of 4.5 billion years. Given this enormous amount of time, great changes are possible on the earth, even though the process of change operates very slowly. The age of the earth has been estimated by radiometric methods. These methods rely on the fact that each natural radioactive material decays into another substance at a fixed and known rate. Knowing the ratio of parent to daughter materials in a rock provides the approximate number of years since the process of decay began.

Questions for thought and review

1. What are some factors that might contribute to unequal heating of the earth's surface by the sun?

2. Can you think of any natural activities in your own geographic area that illustrate the concept of equilibrium?

3. List several examples of cycles in nature other than those mentioned in this chapter.

4. What differences would you expect in the operation of the hydrologic cycle in a desert region versus a moist, well-vegetated region?

5. What happens to most precipitation after it falls onto the land?

6. In your geographic region is there any evidence to support the fact that it was once under the sea? What evidence?

7. If tectonic forces in your region were downwarping the crust at a rate of 12 in. (30.48 cm) per century, how much time would be needed to submerge the land below sea level?

8. What is meant by absolute methods of age determination? Are they really absolute?

9. Bone, wood, cloth, and ashes are susceptible to dating by the carbon-14 method. Are these the only materials that could be used? Can you think of others?

Selected readings

Davis, K. S., and Day, J. A., 1961, *Water, the Mirror of Science,* Doubleday, Garden City, N.Y.

Eicher, D. L., 1968, *Geologic Time,* Prentice-Hall, Englewood Cliffs, N. J.

Janssen, R. E., 1952, The History of a River, *Scientific American,* vol. 186, no. 6, pp. 74–80.

Sayre, A. N., 1950, Groundwater, *Scientific American,* vol. 183, no. 5, pp. 134–140.

Siever, Raymond, 1974, The Steady State of the Earth's Crust, Atmosphere and Oceans, *Scientific American,* vol. 230, no. 6, pp. 72–79.

Turekian, K. K., 1968, *Oceans,* Prentice-Hall, Englewood Cliffs, N.J.

The face of the earth

5 Building the earth: minerals

STUDENT OBJECTIVES

At the end of this chapter the student should:

1. understand what minerals are and their relationship to chemical compounds

2. understand the makeup of minerals and how they are held together

3. be able to use techniques of mineral identification

4. know something of the origin of minerals

5. know the common mineral groups and something about them

Crystallization is . . . the organizing . . .
principal of inorganic nature, producing
definite forms for each species of inorganic
matter as life does for each living species.

H. P. Whitlock

WHY STUDY MINERALS?

Minerals form the essentials and luxuries of our everyday lives. They may be the raw material from which one or more metals are extracted. The home in which you live, the car you drive on highways made of asphalt or concrete, and many of the fertilizers used on crops all are directly dependent on minerals as the raw materials that form part or all of these products. Minerals may be used directly as gemstones (Figure 5-1). In fact, you may be wearing one or more minerals in a ring or other piece of jewelry.

Civilizations have sought out particular minerals for many centuries for the same reasons that we still seek them today. Ancient Greeks and Romans mined and refined several metals. They valued silver and gold and recognized the beauty of many gems, including diamond, opal, emerald, and sapphire. The presence of marble (made of the mineral calcite) in Greece enabled sculpture and ornate architecture to reach an advanced stage there. These ancient civilizations also believed in the medicinal and magical powers of amulets made of minerals and stones. Some of these ideas were held through the Middle Ages and, indeed, are still believed in some quarters today.

Figure 5–1. Rough gem diamond in rock. (Courtesy of Ward's Natural Science Establishment.)

Throughout history countries have gone to war over minerals. And they have won or lost wars, because they utilized certain minerals or lacked them. The Greeks, for example, in their wars against the Persians, used metal armor, while the Persians used leather armor. This may be the reason that the Persians lost, although the Greeks might insist it was a matter of skill. Most of the Spanish exploration of the New World during the sixteenth century was aimed at obtaining gold and other precious commodities for the mother country.

Mineralogy, a subdivision of geology, is the science that deals with study of minerals. It had its beginnings in ancient Greece where many minerals were examined and catalogued. Much was known by the Greeks about the occurrence of minerals of commercial value.

Modern mineralogy had its origins in the sixteenth century when people like Conrad Gesner (1516–1565) and Agricola (see Chapter 1) abandoned the medieval ideas of the miraculous properties of minerals and placed the subject on a more scientific foundation. Agricola, regarded as the father of mineralogy, made the first attempt to classify minerals based on their physical properties.

Today we are more concerned with environmental problems than we were in past centuries. Increasing concern also is being voiced over questions of how long our essential mineral deposits will last. The United States, a country in which only 2% of the world population uses as much as 40% of the world's raw materials produced each year, is becoming increasingly aware of this fact, because more and more raw materials must be imported. The days of living "high off the hawg" may be a thing of the past. We will discuss this in greater detail in Chapter 23.

WHAT IS A MINERAL?

A *mineral* is a naturally occurring crystalline inorganic material that possesses specific physical and chemical properties which are either constant or vary within certain limits. Minerals are composed of atoms, ions, and occasionally molecules. The atoms, ions, or molecules that make up minerals are held together by different types of chemical bonds. The particular constituents of minerals, along with the type or types of bonds holding them together, and the internal arrangement of their constituents determine the various physical properties characteristic of given minerals. These include color, streak, luster, density and specific gravity, hardness, cleavage and fracture, and magnetic properties. Reaction with acids is a chemical property of some minerals. All of these properties are determined by a mineral's composition, bonding, and structure. Therefore, before discussing individual physical properties of minerals, we must consider their composition, bonding, and structure.

BUILDING ATOMS AND ELEMENTS

A chemical element is the smallest part of a compound, or group of the same elements, that will still behave chemically as other elements of the same species. An atom is the smallest part of an element that maintains the identity of that element. Of 104 discovered elements, 92 of these occur in nature. However, fewer than 12 elements make up the most common minerals (Table 5-1). Likewise, we know of about 2000 different minerals, but fewer than 200 are common.

The atoms in chemical elements are made up principally of protons, neutrons, and electrons. A proton is an elementary particle with a positive charge and an atomic mass number (approximately the weight) of 1. A neutron has no charge but, like a proton, has a mass of 1. An electron has a negative charge and an almost negligible mass (1/1845 of the mass of a proton). Protons and neutrons make up the nucleus of an atom, while electrons occur as a "cloud" encircling the nucleus. Therefore most of the mass of an atom is in the nucleus, but most of its volume is actually space with the electron cloud spread throughout. The cloud is denser in certain portions of this space.

The number of protons in an atom determines its atomic number (Z). The mass number (A) is determined by adding the number of protons and the number of neutrons (N). The 104 elements may be related to each other graphically by a diagram called the Periodic Table of Elements (Figure 5-2). The most common elements and the minerals they form in the earth's crust have low atomic and mass numbers—that is, are rela-

Table 5-1. Abundances of common elements in the earth.

Atomic number (Z)	Symbol	Element	Abundance in the crust		Abundance in the earth[1]
			Weight (%)	Volume (%)	
8	O	Oxygen	46.60	93.77	29.53
14	Si	Silicon	27.72	0.86	15.20
13	Al	Aluminum	8.13	0.47	1.09
26	Fe	Iron	5.00	0.43	34.63
20	Ca	Calcium	3.63	1.03	1.13
11	Na	Sodium	2.83	0.97	0.57
19	K	Potassium	2.59	1.83	0.07
12	Mg	Magnesium	2.09	0.29	12.70
		Totals	98.59	99.65	94.92

[1] Assumes that the earth has an iron–nickel core, a silicate–peridotite mantle and a basaltic–granitic crust. Several other elements are also abundant in the earth: Ni — 2.39%, S — 1.93%, Mn — 0.22%, and Co — 0.13%.

SOURCE: After Brian Mason, 1970, *Principles of Geochemistry*, 3rd ed., Wiley, New York.

1																1	2
H 1.008																**H** 1.008	**He** 4.003
3 **Li** 6.940	4 **Be** 9.013											5 **B** 10.82	6 **C** 12.010	7 **N** 14.008	8 **O** 16.000	9 **F** 19.00	10 **Ne** 20.183
11 **Na** 22.997	12 **Mg** 24.32											13 **Al** 26.97	14 **Si** 28.06	15 **P** 30.98	16 **S** 32.066	17 **Cl** 35.457	18 **Ar** 39.944
19 **K** 39.096	20 **Ca** 40.08	21 **Sc** 45.10	22 **Ti** 47.90	23 **V** 50.95	24 **Cr** 52.01	25 **Mn** 54.93	26 **Fe** 55.85	27 **Co** 58.94	28 **Ni** 58.69	29 **Cu** 63.54	30 **Zn** 65.38	31 **Ga** 69.72	32 **Ge** 72.60	33 **As** 74.91	34 **Se** 78.96	35 **Br** 79.916	36 **Kr** 83.7
37 **Rb** 85.48	38 **Sr** 87.63	39 **Y** 88.92	40 **Zr** 91.22	41 **Nb** 92.91	42 **Mo** 95.95	43 **Tc** (99)	44 **Ru** 101.7	45 **Rh** 102.91	46 **Pd** 106.7	47 **Ag** 107.88	48 **Cd** 112.41	49 **In** 114.76	50 **Sn** 118.70	51 **Sb** 121.76	52 **Te** 127.61	53 **I** 126.92	54 **Xe** 131.3
55 **Cs** 132.91	56 **Ba** 137.36	57 ****La** 138.92	72 **Hf** 178.6	73 **Ta** 180.88	74 **W** 183.92	75 **Re** 186.31	76 **Os** 190.2	77 **Ir** 193.1	78 **Pt** 195.23	79 **Au** 197.2	80 **Hg** 200.61	81 **Tl** 204.39	82 **Pb** 207.21	83 **Bi** 209.00	84 **Po** 210.	85 **At** (210)	86 **Rn** 222.
87 **Fr** (223)	88 **Ra** 226.05	89 †**Ac** 227.0															

*Lanthanum series

58 **Ce** 140.13	59 **Pr** 140.92	60 **Nd** 144.27	61 **Pm** (147)	62 **Sm** 50.43	63 **Eu** 152.0	64 **Gd** 156.9	65 **Tb** 159.2	66 **Dy** 162.46	67 **Ho** 164.94	68 **Er** 167.2	69 **Tm** 169.4	70 **Yb** 173.04	71 **Lu** 174.99
90 **Th** 232.12	91 **Pa** 231.	92 **U** 238.07	93 **Np** (237)	94 **Pu** (244)	95 **Am** (243)	96 **Cm** (245)	97 **Bk** (245)	98 **Cf** (248)	99 **Es** (253)	100 **Fm** (254)	101 **Md**	102 **No**	103 **Lw**

†Actinium series

( ) Nos. in parentheses indicate longest-lived isotope.

Figure 5–2. Periodic table of the elements.

tively light elements — and occur in the upper part of the Periodic Table (Table 5-1).

means are ordinarily located near the left side of the Periodic Table (this includes sodium, potassium, calcium, magnesium, and

ATOMS, IONS, MOLECULES, AND BONDS

Atoms possess no charge and are said to be "electrically neutral." Ions are formed when an atom gains or loses one or more electrons, thereby creating an excess or deficient net charge. An atom may give up an electron (become positively charged), if one or more of the outermost electrons are loosely held by the oppositely charged nucleus. A state of greater stability may be achieved by reducing the number of electrons so that the ion may achieve the electronic structure of one of the noble gases such as helium, neon, or argon (Figure 5–3). The noble-gas structure is stable, because electrons in elements forming ions by this

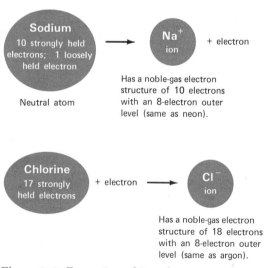

Figure 5–3. Formation of ions from neutral atoms.

so on). Likewise electrons in certain other elements are strongly attracted to their nuclei, and additional electrons may become attached to the atom to form negatively charged ions. This is also a condition of greater stability in many elements; these ions also achieve a noble-gas structure of eight electrons in the outermost level.

Some atoms may form ions with positive charges; others form negatively charged ions. Even though they may have achieved a more stable electronic arrangement, such charged species are quite reactive and readily combine with oppositely charged ions to form ionic bonds. Ions can be considered to be spherical. So when an ionic compound is formed, one ion surrounds itself with ions of opposite charge. Ionic bonding is dominant in the minerals halite ($NaCl$), sylvite (KCl), and fluorite (CaF_2). Ionic bonding occurs in combination with other bond types in many other minerals.

Certain elements do not lose or gain electrons readily. These include elements of the same kind that form compounds (such as H_2, Cl_2, N_2) or elements of different types that may also combine (as water, H_2O; carbon dioxide, CO_2; and methane, CH_4). These are called covalent compounds, and the basic building block of these compounds is the molecule. Covalent bonds are formed by sharing of one or more electrons between two atoms. Both atoms must have empty slots where the shared electrons may reside. Many atoms achieve the eight-electron noble-gas structure by sharing electrons and forming covalent bonds. Covalent bonds are directional, because particular electrons are shared in a definite space between two atoms (Figure 5–4). An individual atom may form several covalent bonds with other atoms, depending on the number of electrons and empty sites available for sharing. Covalent bonds are important in minerals. Diamond (C) and sphalerite (ZnS) are bound by predominantly covalent bonds. Many other minerals, such as many members of the

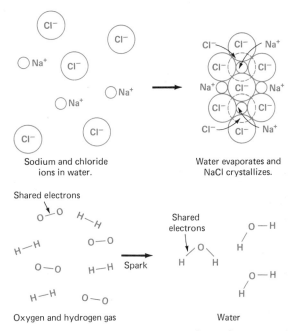

Sodium and chloride ions in water.

Water evaporates and NaCl crystallizes.

Oxygen and hydrogen gas

Water

Figure 5–4. Formation of ionic and covalent bonds.

silicate group, contain both ionic and covalent bonds.

Metallic elements may bond together (metallic bonding) by forming tightly packed structures of spherical atoms in which one or more electrons are free to migrate throughout the structure—hence the high electrical conductivity of metals. In addition to their presence in native metals (gold, copper, platinum, and others), metallic bonds dominate in other minerals such as pyrite (FeS_2) and chalcopyrite ($CuFeS_2$).

Certain very weak bonds occur in some minerals. These provide some minerals with particular unique properties, such as the ease with which the sheets of mica are split apart. These bonds involve not sharing, movement, or gain or loss of electrons but a minor charge imbalance brought about by peculiarities in the crystal structure. A weak bond is formed by neutralizing weak charge differences. These bonds are present in graphite, clay minerals, and talc as well as micas.

CRYSTALS AND THE CRYSTALLINE STATE

Minerals have an orderly arrangement of the ions, atoms, or molecules that compose them; that is, minerals are crystalline materials. Materials that are enclosed by smooth plane faces are called crystals (Figure 5–5). Not all crystalline materials are crystals, because crystal faces did not form during crystallization. But, by definition, all minerals are crystalline solids. Crystal faces may not form because of crowding of the mineral and competition for space during its formation or because of competition with other minerals which have already crystallized or which form crystal faces more easily. Or the mineral may crystallize in such a fine-grained or poorly crystallized form that crystal faces are either not visible or are lacking.

The atoms, ions, or molecules in minerals

A crystal has a crystal structure and is crystalline.

A crystalline material has *no* crystal faces but does have a crystal structure, so it is crystalline.

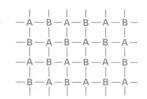

Regular arrangement of atoms or ions in a crystal structure or lattice.

Figure 5–5. Comparison of crystals and crystalline materials and their relationship to a crystal lattice.

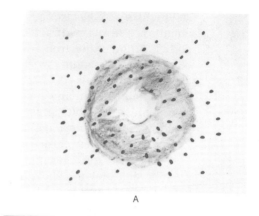

A

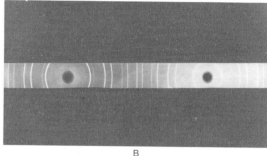

B

Figure 5–6. A. Laue x-ray photograph of sphalerite. (D. P. Miller, Clemson University.) B. X-ray powder photograph of sphalerite.

are spaced a few angstroms—1 angstrom (Å) = 10^{-8} cm—apart. If the head of a pin is 1/10 cm in diameter, this is 10,000,000 Å. This atomic spacing happens to be about the same as the wavelength of x rays (a form of electromagnetic radiation similar to light), so a crystal lattice (the orderly arrangement of atoms or ions in a crystal structure) acts as a diffraction grating to split up x rays into long and short wavelengths, similar to a prism splitting light. The manner in which x rays are diffracted (similar to reflected) from a crystal lattice is a measure of the spacing between atoms or ions in the lattice. The precise structure of crystals may be determined in this way (Figure 5–6). Max von Laue in 1912 and William Henry Bragg and William Laurence Bragg in 1914 developed the first methods for determining certain distances in crystals by using x rays. Since then, other more advanced methods have been devised so that we now know the internal structures of most common minerals.

Just as the internal arrangement of atoms or ions in a crystal structure is regular and repetitious for a given mineral, so is the external set of crystal faces, wherever they appear. The crystal form (set of crystal faces) is determined by the internal structure, and both may be related to six crystal systems (Figure 5–7). These crystal systems are defined by three or four axes in various combinations of different lengths and at different angles to one another. A given mineral always crystallizes within a particular system. Thus by observing the arrangement of crystal faces, provided they are well developed, an important clue to the identity of the mineral is obtained.

ISOMORPHISM AND POLYMORPHISM

Variation in the composition of minerals is the rule rather than the exception. Much of this variation takes the form of restricted substitutions by particular elements. This substitution of ions, atoms, or molecules that occurs without changing the crystal structure is called *isomorphism*. Isomorphism is favored where potential substituents have a similar ionic size and charge (Table 5-2). When sizes and charges of substituting ions

Table 5–2. Ionic radii and charges of some common ions.

Element	Ion	Ionic radius (Å)
Oxygen	O^{2-}	1.40
Silicon	Si^{4+}	0.41
Aluminum	Al^{3+}	0.50
Iron	Fe^{2+}	0.74
Iron	Fe^{3+}	0.64
Calcium	Ca^{2+}	0.99
Sodium	Na^{+}	0.97
Potassium	K^{+}	1.33
Magnesium	Mg^{2+}	0.66

SOURCE: From Brian Mason, 1970, *Principles of Geochemistry*, 3rd ed., Wiley, New York.

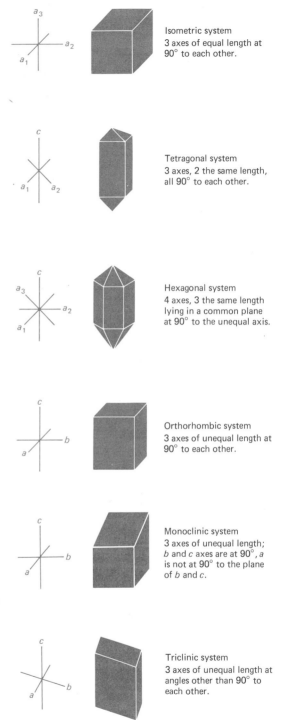

Isometric system
3 axes of equal length at 90° to each other.

Tetragonal system
3 axes, 2 the same length, all 90° to each other.

Hexagonal system
4 axes, 3 the same length lying in a common plane at 90° to the unequal axis.

Orthorhombic system
3 axes of unequal length at 90° to each other.

Monoclinic system
3 axes of unequal length; *b* and *c* axes are at 90°, *a* is not at 90° to the plane of *b* and *c*.

Triclinic system
3 axes of unequal length at angles other than 90° to each other.

Figure 5–7. The coordinate axes of the six crystal systems.

are very close to one another, isomorphism may be unlimited; that is, one ion may substitute for another in any proportion. But if there is a significant difference between sizes or charges (or both) of substituting ions, only partial isomorphic substitution may occur.

In olivine, $(Mg, Fe)_2SiO_4$, magnesium and iron may substitute for each other in any proportion. The reason is that both ions have the same charge $(+2)$, and there is little difference in the sizes of the ions (the ionic radius of $Mg^{2+} = 0.66$ Å, that of $Fe^{2+} = 0.74$ Å). In contrast, magnesium may substitute for calcium in the calcite structure, but this substitution is limited to less than 20%. Above 20% the calcite structure breaks down. Although calcium and magnesium have identical charges, their ionic sizes differ significantly (the ionic radius of $Ca^{2+} = 0.99$ Å).

There are numerous examples of minerals whose compositions are the same but whose crystal structures are different. Such a variation of structure with the same composition is called *polymorphism*. Diamond and graphite (both composed of a single element —carbon) are excellent examples of polymorphs whose different properties are immediately evident.

PHYSICAL PROPERTIES

Physical properties of minerals include their color, streak, luster, density and specific gravity, hardness, cleavage and fracture, and other special properties (including magnetism, taste, reactivity with acids, and double refraction). All of these are derived properties, because they are determined by the chemical composition and crystal structure of a given mineral. We may use physical properties to identify many common minerals.

Color

The *color* of any material is the appearance of light that is incident on the material and then is partially or completely absorbed and reemitted. Black materials absorb all incident light; white materials reemit all incident wavelengths. Various colors arise when certain wavelengths are absorbed while others are not.

The color of a mineral is one of its most obvious properties. Although color is diagnostic of certain minerals, like brassy yellow pyrite and bright green malachite, it may be misleading in attempting to identify others. For example, fluorite may be purple, green, yellow, blue, or colorless. The color of quartz may be even more variable.

Streak

The *streak* of a mineral is the color of its powder, commonly obtained by scratching the mineral on an unglazed porcelain plate. Streak is a more constant and diagnostic property than color and is therefore quite useful in the identification of minerals whose color is variable. In the example mentioned above, the mineral fluorite has quite a variable color. However, the streak of fluorite is always white. Although the mineral hematite may be red, brown, or silvery metallic in color, its streak is always reddish brown. Streak is particularly useful in tracking down the heavier metallic minerals; their streaks are almost always darker than their color.

One precaution should be observed when determining the streak of a mineral: The mineral must be softer than the streak plate. If it is harder than the plate, the color observed will be that of powdered porcelain and not that of the powdered mineral.

Luster

The *luster* of a mineral is the appearance of light that is reflected from a fresh surface. This is only indirectly related to the color of a mineral, because luster is a surface property dependent on the quality of the surface, its regularity, and its smoothness as dictated by the crystal structure. Color is an absorptive property and is not related to the surface of a mineral.

The lusters of minerals fall into two major groups, metallic and nonmetallic, with a smaller group of minerals with submetallic lusters existing between (Table 5–3). A metallic luster is the appearance of light that is reflected from steel or some other metallic object. Many types of nonmetallic luster have been described. Earthy, dull, adamantine (brilliant), vitreous (glassy), silky, pearly, greasy, and waxy are several of the terms we can apply to different types of nonmetallic luster.

Table 5–3. Common lusters and some examples.

Luster		Mineral
Metallic		Pyrite
		Chalcopyrite
		Galena
		Chalcocite
		Graphite
		Hematite
		Magnetite
Nonmetallic	Vitreous	Quartz
		Beryl
	Adamantine	Diamond
	Dull	Kaolinite
	Silky	Gypsum (satin spar)
		Asbestos
	Resinous	Sphalerite
	Pearly	Talc
	Greasy	Nepheline
	Waxy	Serpentine

Density and specific gravity

The *density* of a substance is its mass (weight on earth) per unit volume. The *specific gravity* is the weight of the substance divided by the weight of the same volume of water. This is also a measure of density. Some minerals, such as kaolinite, are very light and thus have a low density; others,

such as galena and barite, feel heavy and actually have a higher density. Quartz, the feldspars, muscovite, and calcite feel neither very light nor very heavy. They have an average density and impart an average density to the rocks that contain them. The average specific gravity of the crust of the earth is about 2.65. This is very close to that of all the "average" minerals mentioned above. Although there are precise methods for determining the density (or specific gravity) of a mineral, the "heft" or how heavy a mineral feels is sufficient for determining this property in the identification of most common minerals.

Hardness

Hardness describes resistance to being scratched. The hardness of a mineral is a function of its crystal structure and bonding and not of its composition. Diamond and graphite have the same composition (carbon), but their hardnesses differ markedly: Each has a different structure.

Hardness is conveniently related to some scale, and the one commonly used in the study of minerals is Mohs' scale of hardness (Table 5–4). Hardness is a property that varies only slightly for a given mineral, provided it has not been weathered or otherwise altered. Determination of the hardness of a mineral may be made by scratching the un-

Table 5–4. Mohs' Scale of Hardness.

Hardness	Mineral
1	Talc
2	Gypsum
3	Calcite
4	Fluorite
5	Apatite
6	Orthoclase (K-feldspar)
7	Quartz
8	Topaz
9	Corundum
10	Diamond

known by members of Mohs' scale of hardness and bracketing the mineral by finding a mineral in the series that it will scratch while not scratching the next one up the scale. Less precise but useful determination of hardness may be made by using one's fingernail (hardness 2–2.5), a copper penny (3), a piece of window glass (5–5.5), a streak plate (5.5–6), a knife blade (5.5–6), an iron nail (5), and a steel file (6.5–7).

Mohs' hardness scale is actually a relative scale. Gypsum is about twice as hard as talc, calcite three times as hard as talc, and so on up to the jump from corundum to diamond, where there is a very large change in actual hardness. On an absolute scale, if corundum has a hardness of 9, diamond would have a hardness of about 40.

Cleavage and fracture

Whenever a mineral is broken, it will produce surfaces that are nearly alike time after time. If the mineral crystallizes possessing one or more aligned weaknesses in its crystal structure, it will break along smooth plane surfaces following these weaknesses (Figure 5–8). One weakness zone will produce a *cleavage* (cleavage direction). All parallel planar surfaces produced from breaking along a particular weakness zone comprise one cleavage direction (Figure 5–9). Other zones of weakness in the crystal lattice may produce other cleavage directions. A given

1 cleavage direction

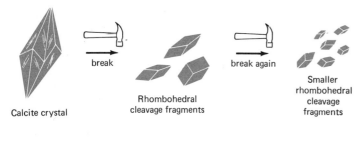

Calcite crystal — break — Rhombohedral cleavage fragments — break again — Smaller rhombohedral cleavage fragments

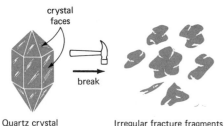

crystal faces

Quartz crystal — break — Irregular fracture fragments

Figure 5–8. Differences between cleavages, crystal faces and fractures.

A. Muscovite

Figure 5–9. Some minerals may have one cleavage direction, others may have several.

2 cleavage directions

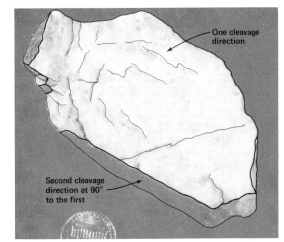

B. Plagioclase

3 cleavage directions

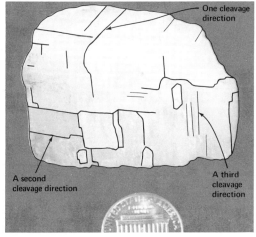

C. Calcite

4 cleavage directions

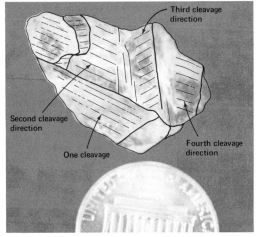

D. Fluorite

59

Figure 5–10. Conchoidal fracture in quartz.

mineral will always possess the same number of cleavage planes; this varies from zero to six directions.

Minerals are three-dimensional objects. Those specimens that are not bound by crystal faces and have zero, one, or two cleavage directions will be terminated by a *fracture* surface or by a common boundary with an adjacent crystal. Fractures may be irregular, needlelike, smooth, conchoidal, (Figure 5–10), and so on. Irregular fractures are the most common. Fracture is dictated by the crystal structure.

Special properties

Magnetism. All materials interact in some way with a magnetic field, and certain materials interact very strongly. Minerals that interact very strongly with a magnetic field may be more easily identified by determining this property with a hand magnet. Magnetite and pyrrhotite are two common minerals that exhibit magnetism.

Taste. Halite (NaCl) and sylvite (KCl) are two common minerals that are soluble in water and may be distinguished by their taste. Halite is distinguished by its salty taste; sylvite has a bitter salty taste.

Feel. Certain minerals may be distinguished by their feel. Talc has a greasy feel; graphite feels slippery.

Double refraction. The property of double refraction arises when light travels through a mineral at different speeds in different directions. Clear calcite best exhibits double refraction (Figure 5–11). One dot on a piece of paper will appear as two when viewed through a cleavage fragment of this mineral.

Reaction with acids. Many minerals will react with various types of acids. The most obvious and most easily determined reaction is effervescence with the evolution of a gas, such as carbon dioxide. Carbonates will react in this manner to yield carbon dioxide. Some carbonates, such as dolomite, $(CaMg(CO_3)_2)$, react slowly or not at all unless some of the mineral is powdered. Effervescence and evolution of carbon dioxide thus indicate that the mineral is a carbonate; the rate of reac-

Figure 5–11. Double refraction in calcite. (Courtesy of Ward's Natural Science Establishment.)

tion (how fast it effervesces) will aid in determining which carbonate is present.

Odor. Some minerals under proper circumstances possess an odor. Sphalerite when crushed or powdered will release a sulfurous odor. Kaolinite when damp will have an earthy smell.

ORIGIN OF MINERALS

Minerals crystallize in a variety of environments and may change into other minerals, if environmental conditions change. Minerals may crystallize from molten rock material; form as chemical precipitates in oceans, streams, or groundwater; crystallize as fresh- or seawater evaporates; form directly from gases as sublimates; form during weathering; or recrystallize from other minerals during metamorphism.

Some minerals form under a very narrow range of physical and chemical conditions and are therefore restricted in occurrence and abundance. Other minerals, such as quartz, form under a variety of conditions and therefore occur as stable minerals in many environs.

Minerals are forming today in different environments. They are forming on or beneath the surface as molten rock material cools. Sulfur forms in the throats of volcanoes; the sulfur deposits in and around Vesuvius have been mined since Roman times. Calcium carbonate as the mineral aragonite is forming today in the ocean off the Bahamas. At times the water there is clear, while at others it is clouded with microscopic crystals that have precipitated from seawater.

Some "minerals" are actually formed as a product of organic processes. For example, pearls and the shells of many organisms are actually organically formed calcite or aragonite. Some phosphate minerals and silica likewise have an organic origin.

COMMON MINERAL GROUPS

Minerals are classified by chemical composition and crystal structure. This seems to work well, for a good classification should separate what is being classified into well-defined groups while remaining flexible and enabling cross-linking of groups into larger units.

Native elements

Native elements are those that exist in nature in an uncombined state. For an element to exist in the native state it must remain relatively inert in the environment in which it occurs. Several are of considerable economic importance. Among the metals, native gold, silver, copper, and platinum and, among the nonmetals, diamond, graphite, and sulfur are economically important (Figure 5–12). Other elements that occur in the native state include iron, arsenic, bismuth, and occasionally liquid mercury. However, these elements are more abundant in the form of combined minerals.

Sulfides

Sulfides include minerals formed by one or more elements combining with sulfur (or occasionally arsenic, but these are properly termed arsenides). Many of the sulfides are very important economically. Common sulfides include pyrite, chalcopyrite, galena, and sphalerite (Figure 5–13). With the exception of pyrite, these minerals are the major sources of copper, lead, and zinc, respectively. Sulfides are not rock-forming minerals. Instead they occur in minor quantities dispersed throughout rocks, in veins, or as massive sulfide deposits formed by precipitation from very-hot-water (hydrothermal) solutions (100–400°C). Most of the common sulfides have a metallic luster and a dark gray to black streak.

Oxides

Oxide minerals consist of one or several elements combined with oxygen. These are not

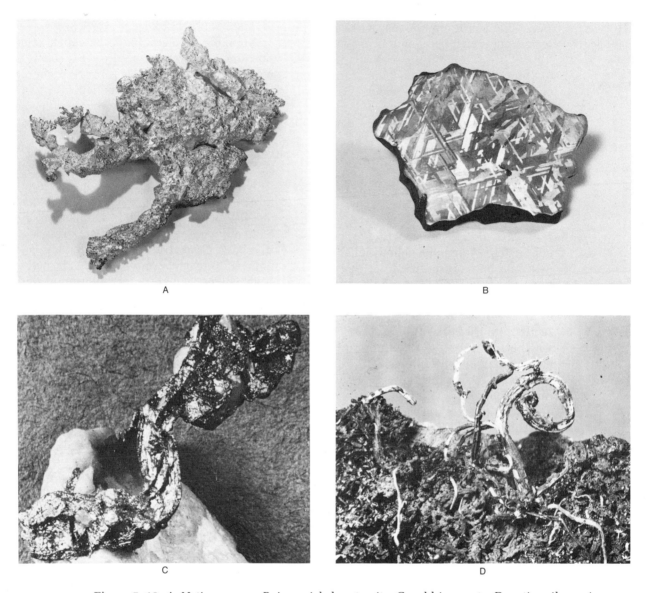

Figure 5–12. *A.* Native copper, *B.* iron-nickel meteorite, *C.* gold in quartz, *D.* native silver wire in quartz, (Courtesy of Ward's Natural Science Establishment.)

major rock-forming minerals, but several, such as hematite, may form by precipitation at surface termperatures and form widespread deposits. Hematite, magnetite, goethite, and corundum are common oxides that either have been or are presently economically important. Hematite and magnetite are major ores of iron. Ruby and sapphire are forms of corundum. Oxides form under a variety of conditions. Some form on or near the surface; others form at high temperatures and pressures as minor constituents of igneous and metamorphic rocks. Still others crystallize from hydrothermal solutions.

Figure 5–13. Crystals of galena on dolomite. (Specimen courtesy of Geology Department, Eastern Kentucky University.)

A

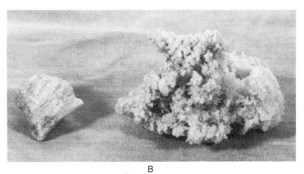

B

Figure 5–14. Crystals and cleavage fragments of calcite (A) and aragonite (B) Minerals of the calcite group have a contrasting crystal form and cleavage to those in the aragonite group.

Carbonates

Carbonates form by the combination of one or more elements with the carbonate (CO_3^{2-}) group. The most common and important carbonate minerals, calcite and dolomite, are major rock-forming minerals; they are the principal constituents of limestone and dolostone, respectively. Carbonates form at relatively low temperatures. The origin of the vast dolostone deposits is somewhat of an enigma. The mineral dolomite cannot be precipitated from normal seawater but may be synthesized in the laboratory at temperatures in excess of 300°C. Yet the extensive dolostone deposits are associated with other very-low-temperature (surface) marine sedimentary rocks, many of which contain abundant undeformed fossils.

The two groups of carbonates, are characterized by different crystal structures. The calcite group includes calcite, dolomite, and several others (Figure 5–14). All of these minerals have a rhombohedral cleavage, which resembles that of a cube that has been shoved over from the top. The aragonite group includes the mineral aragonite, a polymorph of calcite, and several others. This group does not possess the rhombohedral cleavage of the calcite group, but members of both groups react slightly to vigorously with acids.

Copper carbonates—malachite and azurite—are related to the other carbonates only by their compositions. Their respective bright green and blue colors and occurrence with other copper minerals make them easy to identify.

Halides

Halide minerals consist of metal ions, such as sodium or potassium, combined with a halogen, such as chlorine or fluorine. Several members of this group, including halite and sylvite, are held together by almost pure ionic bonding. They are identifiable by their clear to only slightly colored (light blue, gray, pink, white) state and their cubic (three directions at 90°) cleavage. Both have a salty taste, but sylvite is more bitter than halite. These two minerals and several others originate by the evaporation of seawater and form evaporite deposits.

Fluorite, another common halide mineral (Figure 5–15), has an octahedral cleavage (four directions). Unlike halite and sylvite, it is not soluble in water. It forms predominantly through hydrothermal activity.

Sulfates and phosphates

Sulfate and phosphate minerals are formed whenever metal ions such as calcium, lead, and several others combine with sulfate (SO_4^{2-}) or phosphate (PO_4^{3-}) groups. There are many sulfate and phosphate minerals, but the most common members of these two groups are gypsum and apatite respectively. Gypsum is a common constituent of evaporite deposits, because it precipitates on evaporation of seawater. Gypsum is identifiable by its hardness (softer than a fingernail), its colorless to white color, and its nongreasy feel. Apatite is a constituent of the commercially important sedimentary phosphate rock but also occurs in igneous rocks, sometimes in commercially recoverable concentrations. Apatite is recognized by its hardness (5), its vitreous luster, its hexagonal (six-sided) crystal habit, and, its bluish green or brownish pink color.

Silicates

Silicates are the most important group of minerals. Although they are relatively unimportant economically compared with oxides and sulfides, they form the great bulk of the rocks of the earth's crust and mantle. In addition, it is into silicate rocks that most economically important deposits are emplaced but partially only because they are so abundant.

There are many common silicate minerals. However, all silicate structures are based on the SiO_4^{4-} tetrahedral unit (Figure 5–16). Several subgroups arise within the silicate group, because the SiO_4^{4-} group can share

Figure 5–15. Crystals and cleavage fragments of fluorite. (Specimen courtesy of Geology Department, Eastern Kentucky University.)

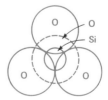

Figure 5–16. The SiO_4 tetrahedron.

oxygens on the corners and in a sense link to itself. The combinations that result include isolated SiO_4^{4-} groups, double tetrahedra, rings, single or double chains, sheets, and three-dimensional network structures. Positively charged ions fit into the structure of each silicate mineral to neutralize the charges on the SiO_4^{4-} groups. Table 5–5 summarizes the different silicate structural subgroups and contains one or more examples of common minerals in each.

The external structure of silicates frequently reflects their internal structure and bonding. Stubby olivine and garnet crystals reflect the isolated character of the SiO_4^{4-} group. Two prismatic cleavage directions at either 90° or 60° and 120° reflect the chain structure of the pyroxenes and amphiboles. Perfect cleavage in the micas reflects their sheet structure and the weak bonds between the sheets (Figure 5–17).

Silicate minerals originate in a variety of environments. Silicates make up almost all igneous rocks, most metamorphic rocks, and the majority of sedimentary rocks. Although quartz may form in all three major rock groups, most minerals that form in igneous rocks do not form in the sedimentary environment, and many minerals that form under conditions of metamorphism do not form under conditions of the igneous or sedimentary realms.

Appendix A contains descriptions of many common minerals.

Table 5–5. Structures and characteristics of silicate minerals.

Subgroup	Structural repeat unit	Common example(s)	Characteristics
Isolated tetrahedra		Olivine, garnet, zircon	Stubby crystals
Double tetrahedra		Epidote	Prismatic crystals
Rings (six-membered most stable, stacked like donuts)		Tourmaline, beryl	Prismatic crystals
Chains (single)		Pyroxenes	Prismatic crystals, 90° cleavages
Chains (double)	$\infty \leftarrow$ ... $\rightarrow \infty$	Amphiboles	Prismatic crystals, 60 and 120° cleavages
Sheets	$\infty \leftarrow$ ... $\rightarrow \infty$	Micas, clay minerals	Sheetlike cleavage fragments
3-D network	Too complex to represent here	Quartz, feldspars	Crystals somewhat prismatic, somewhat stubby

A

B

C

Figure 5–17. The internal structure of minerals is frequently reflected in a mineral's external form. *A*. Crystals of garnet (*left*) and zircon. *B*. Prismatic cleavage of amphibole (*left*) and pyroxene. *C*. Sheet structure exhibited by mica cleavage.

SUMMARY

A mineral is a naturally occurring crystalline inorganic material with certain physical and chemical properties that are either constant or vary within limits. Minerals are built of elements, which are composed of protons, neutrons, and electrons. These elements are bound together by either ionic, covalent, metallic, or other bond types or a combination of several bond types. The crystalline character of minerals is studied by x-ray diffraction to determine the exact nature of the structure that makes up the mineral. The atoms or ions in minerals are spaced a few angstroms apart and are arranged in a repetitious manner such that they form symmetrical patterns related to six crystal systems.

The compositions of some minerals may vary by isomorphic substitution. Minerals may be identified by determining certain physical properties, including color, streak, luster, density, hardness, cleavage or fracture, and magnetic properties. Most of these properties depend on the crystal structure and chemical composition. Native elements, sulfides, oxides, carbonates, halides, sulfates, phosphates, and silicates form the most common mineral groups. The silicates are most important, because they make up most of the crust and mantle of the earth.

Questions for thought and review

1. What is the difference between an atom and an ion, a mineral and a compound, a crystal and a crystalline solid, a molecule and an atom?
2. Briefly discuss the differences between isomorphism and polymorphism, and give examples of each.
3. How do ionic and covalent bonds differ from one another?
4. Why do crystal faces not form on all minerals when they crystallize?
5. What are the controlling factors of isomorphism?
6. Why is streak more useful than color in identifying certain minerals such as hematite?
7. Why are the physical properties of color, streak, luster, density, hardness, and so on called derived properties?
8. How is cleavage used to differentiate between amphiboles and pyroxenes?
9. Why do certain elements exist as native elements?
10. What is a hydrothermal solution?
11. Why are silicates the most important group of minerals?
12. Why do different subgroups exist within the silicate group?

Selected readings

Anders, Edward, 1965, Diamonds in Meteorites, *Scientific American*, vol. 213, no. 4, pp. 26–36.

Bragg, Lawrence, 1968, X-ray Crystallography, *Scientific American*, vol. 219, no. 1, pp. 58–70.

Bridgman, P. W., 1956, Synthetic Diamonds, *Scientific American*, vol. 193, no. 5, pp. 42–46.

Ernst, W. G., 1969, *Earth Materials*, Prentice-Hall, Englewood Cliffs, N. J.

Fullman, R. L., 1955, The Growth of Crystals, *Scientific American*, vol. 192, no. 3, pp. 74–80.

Hurlbut, C. S., Jr., 1963, *Dana's Minerals and How to Study Them*, 3rd ed., Wiley, New York.

Pearl, R. M., 1955, *How to Know the Minerals and Rocks*, McGraw-Hill, New York.

6 Igneous rocks

STUDENT OBJECTIVES

At the end of this chapter the student should:

1. understand what igneous rocks are and the historical aspects of the molten origin of igneous rocks

2. know the basis of classification of rocks and its usage in classifying igneous rocks

3. understand various igneous textural types and their origins

4. recognize common igneous rock types

5. understand Bowen's theory and the evidence for and against its use

6. be familiar with ideas that attempt to explain the origin of magma

7. recognize different types of plutons and differences among them

*All wish to be learned, but no one
is willing to pay the price.*

Juvenal

THE IMPORTANCE OF
IGNEOUS ROCKS

The importance of something cannot always be measured in terms of economics. Although igneous rocks are extremely important because they serve as host rocks for many economic minerals (for example, gold, copper, diamonds), they are also important for many other reasons. Igneous rocks are the most abundant within the earth's crust, and sedimentary rocks are most abundant at the surface of the continents.

Igneous rocks crystallize from magma. When magma is exposed to the surface through volcanic eruption, it is and has been a hazard to people for many centuries (see Chapter 7). Magma cooling beneath the surface has for several years served as a source of heat to create steam to generate electricity in Iceland and New Zealand. Today this source of energy (called *geothermal heat*) is being tapped in the Geysers area of northern California, and plans are being laid to use it to generate electricity in other areas of the United States, where cooling magma lies beneath the surface.

NEPTUNISTS, VULCANISTS,
AND PLUTONISTS

The molten origin of igneous rocks has not always been accepted. The ancient Greeks probably correctly interpreted the origin of volcanic rocks, because they lived in an active volcanic region and were sufficiently advanced thinkers. During the Middle Ages, much of their knowledge was lost. With the Renaissance came new beginnings of independent thought, and fresh ideas concerning the origin of igneous rocks were brought forth. Perhaps most notable of these were the ideas of Abraham G. Werner (1750–1817), to whom we alluded in Chapter 1. Reviewing his ideas, he espoused the idea that all rocks, including igneous rocks, precipitated from a universal ocean. Werner explained lava pouring from the mouths of

volcanoes as the result of the burning of subterranean coal beds and was restricted to certain areas. Werner ignored most of the evidence that would prove the molten origin for many of the rocks in his native Germany. Because he was such a renowned lecturer, his ideas were accepted for many years. The great eruption in 1783 in Iceland did not convince Werner that the dark rocks around extinct volcanoes were the same as the dark rocks formed from cooling lava. So it remained for others to disprove his ideas.

In 1751 Jean Guettard, who had pointed out the actual origin of the volcanic cones and lava flows in the Auvergne region in France, cited the baked soils beneath the lava flows as evidence for a molten origin (Figure 6–1). The publication of these observations, as we noted in Chapter 1, was received with no great enthusiasm, because scientists of the day felt it strange that there had been active volcanoes in the middle of France. In 1763 Nicholas Desmarest reviewed Guettard's evidence and actually traced several of the lava flows back to their sources in volcanic craters. Desmarest made a geologic map of this volcanic area and confirmed the origin of these rocks. He did not attempt, however, to convince skeptics of the validity of these discoveries. His simple axiom, "Go and see," was the means by which his discoveries were made in the first place. Observation remains today as the most powerful tool of the geologist. Several of Werner's students later came to the Auvergne and saw in dismay that Werner was wrong.

Those who followed the teachings of Werner were called Neptunists, and believers in the ideas of Guettard and Desmarest were called Vulcanists. The Vulcanists succeeded in proving the molten origin of volcanic rocks. During the same period James Hutton, who formulated the concept of uniformitarianism, observed that the granitic rocks of his native Scotland were not precipitates from Werner's universal

Figure 6-1. Part of the Auvergne region in France. (Courtesy of French Government Tourist Office.)

ocean. Hutton knew of the insolubility of quartz and other silicates in water and observed other relationships among granitic bodies that proved their intrusive nature. He also proved that granites are not the oldest rocks, because they are intrusive, and intruded rocks had to be there first. Those accepting the molten intrusive origin of granitic rocks were called Plutonists, and Hutton became a chief advocate of this school of thought.

So the molten origin of what we now call igneous rocks became known during the latter half of the eighteenth century. These ideas became widely accepted during the nineteenth century but not without waning resistance from the Neptunist school.

DESCRIBING AND STUDYING COMMON IGNEOUS ROCKS

Our basis for naming rocks in either of the three major classes—igneous, sedimentary, and metamorphic—is twofold: (1) *composition* and (2) *texture*. The composition of a rock can be expressed in terms of what minerals it contains. The texture of a rock refers to the sizes of mineral grains, their shapes, and the relations of one grain to another (whether they are all the same size or the same shape, or are of contrasting shape and/or size). In short, texture is the grain-to-grain relationship that exists in a rock. Thus two rocks with different compositions and the same texture would be given different names and vice versa.

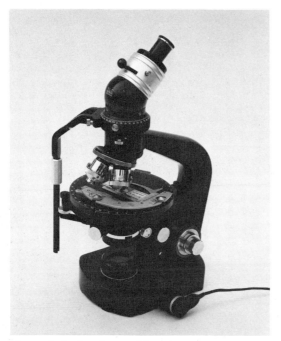

Figure 6–2. A petrographic microscope. (Courtesy of Wild Heerbrugg Instruments, Inc., Farmingdale, N.Y.)

Petrology is study of rocks. Petrologists are concerned with the origin and distribution of different rock types, and they employ many different field and laboratory techniques to carry out their studies. One tool that is an indispensable aid to the petrologist is the *petrographic microscope* (Figure 6–2). The petrographic microscope is an ordinary light microscope, like that used by a biologist, but has additional attachments and modifications. Perhaps the most important difference between the petrographic and biologic microscopes is that the petrographic microscope contains attachments for polarizing the light that passes through different lens elements. Observation of the manner in which light passes through minerals, principally as the colors produced by the optical properties of minerals in polarized light, aids the petrologist in identifying minerals and observing different textures and even effects of chemical reactions during for-

mation of the rock (Figure 6–3). The microscope also aids in observing very small crystals. Many minerals and textures that are not visible to the unaided eye may be observed and identified under the microscope.

Study of rocks with a petrographic microscope cannot be accomplished simply by placing a chip or rock on the stage and looking at it. The rock must be sawed, mounted on a glass slide and carefully ground to a standard thickness of 0.03 mm (thinner than paper). The product of this procedure is called a *thin section*. This renders many

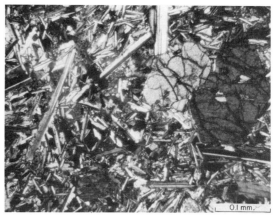

A

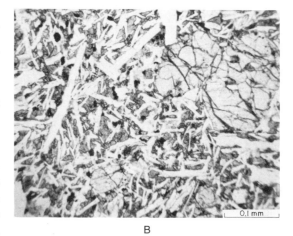

B

Figure 6–3. Photographs of part of a thin section of diabase in polarized light (*A*) and in plane light (*B*).

minerals that appear opaque in hand specimen translucent or transparent and enables the petrologist to investigate their optical properties.

Igneous minerals: the composition of igneous rocks

The common minerals in igneous rocks are relatively few and are all silicates. Table 6–1 is a summary of the common minerals in igneous rocks. A convenient way to classify minerals in igneous rocks is in terms of their relative abundance: as *essential* or *accessory* minerals. Essential minerals are those necessary (that *must* be present) to name the rock. The essential minerals of a granite are quartz and orthoclase. They must be there before the rock can be called a granite. Accessory minerals commonly make up less than 10% of the rock. They do not have to be there to name the rock. But if one accessory mineral appears rather obviously in the rock, the name of the accessory may be prefixed before the rock name, as in biotite granite. This serves to tell us more than that the rock is simply a granite. Accessories that function in the rock name are called *characterizing* accessories. Others not functioning in this way are termed *minor accessories* and are not commonly observed in a hand specimen.

Igneous textures

Different textures in igneous rocks are produced by the environment in which magma cooled. In general, there are two environmental realms for crystallization—on the surface of the earth and beneath it. Those rocks that crystallize on the earth's surface are collectively known as *extrusive* or *volcanic* rocks. Those crystallizing beneath the surface are called *intrusive* or *plutonic* rocks. Within each of these major groups lie several possibilities for modifications to produce different textural types.

Phaneritic texture. An igneous rock whose grains are all visible to the unaided eye and are all about the same size is said to have a *phaneritic texture* (Figure 6–4). Laboratory experiments have shown that this texture results from slow uniform cooling of magma so that the crystals attain a moderate size.

Aphanitic texture. An igneous rock made up of crystals too fine to be seen by the unaided eye has an *aphanitic texture* (Figure 6–5). Rocks possessing this texture crys-

Table 6–1. Igneous rock minerals

Felsic (light)	Intermediate	Mafic (dark)	Ultramafic
Essential minerals			
Quartz	Plagioclase	Plagioclase	Pyroxene
Orthoclase	Hornblende	Augite	Olivine
(K-feldspar)		(pyroxene)	
Plagioclase			
Accessory minerals			
Hornblende	Quartz	Biotite	Plagioclase
Biotite	Orthoclase	Olivine	Serpentine[1]
Muscovite	Biotite		Talc[1]

[1] These minerals do not crystallize along with other rock minerals. They are formed by alteration of some of the primary minerals, such as olivine.

Figure 6–4. Phaneritic textured igneous rock.

Figure 6–5. Aphanitic textured basalt.

tallized fairly rapidly either on or very close to the surface.

Porphyritic texture. Some igneous rocks contain both large and small crystals; these are said to have a *porphyritic texture* (Figure 6–6). Porphyritic texture arises whenever magma cools slowly at first, forming large crystals, and then more rapidly so that small crystals are formed. Two cooling rates may result when magma rests for some time deep within the earth and then moves toward the surface or is even extruded onto the surface. Early formed crystals are carried along with the liquid to the site of further cooling. The small crystals are called the *groundmass*. Almost any combination of grain sizes is possible, ranging from very coarse phaneritic phenocrysts and a coarse phaneritic groundmass to phaneritic phenocrysts and an aphanitic groundmass to phaneritic or aphanitic phenocrysts and a glassy (see below) groundmass.

Glassy texture. A glassy texture is produced when magma cools immediately by coming into contact with air or water so that it is quenched and heat is removed from it

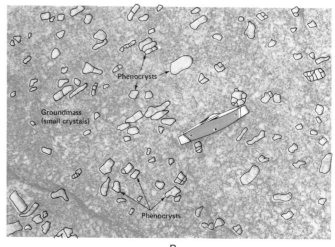

A   B

Figure 6–6. Porphyritic texture. Photograph in *A* is a granite porphyry containing phenocrysts of orthoclase in a groundmass of quartz, orthoclase, and biotite. Textural relations are illustrated in *B*.

Figure 6-7. Glassy texture illustrated with the rock obsidian. (Courtesy of Ward's Natural Science Establishment.)

Figure 6-8. Pumice.

very rapidly. Not all lava that comes into contact with the air becomes glassy, but most that enters water will be glassy, at least on the surface. One common glassy rock is *obsidian* (Figure 6-7). When magma releases larger amounts of gas and is also cooled rapidly, frothy glass called *pumice* results (Figure 6-8).

Figure 6-9. Fine-grained volcanic ash becomes tuff when it settles from the air and becomes consolidated.

Fragmental texture. When a volcano erupts explosively, magma may be congealed and then broken into fragments or even blown into the air with such violence that the largest particles are the size of dust (*volcanic ash*). An accumulation of fairly uniformly fine-grained volcanic ash is called *tuff* (Figure 6-9). A mixture of fine and coarser angular fragments is called *volcanic breccia* (Figure 6-10). Collectively these are known as *pyroclastic* or *fragmental* rocks and may contain quite a bit of glass fragments or shards.

Classification of igneous rocks

Several hundred varieties of igneous rocks are known. But only a few are really common. We will be concerned here with only the most abundant types that can be readily identified in hand specimen. A sim-

Figure 6-10. Volcanic breccia from Durham, North Carolina. (Specimen courtesy of J. E. Wright, Jr.)

plified classification of igneous rocks is presented in Table 6–2.

Igneous rocks may be broken down into several groups based on certain general compositional and color characteristics. *Felsic* rocks are light colored and may contain abundant visible quartz. They are rich in either or both orthoclase and sodium-rich plagioclase feldspar and are poor in dark minerals (see Table 6–1).

Intermediate rocks are darker than felsic rocks but lighter than mafic and ultramafic rocks. They do not, for the most part, contain abundant free quartz (usually not enough to see in a hand specimen) but are composed mostly of feldspar, commonly plagioclase, and one or two dark minerals, usually hornblende or biotite.

In *mafic* igneous rocks dark minerals become very abundant. Mafics are composed of dark (calcium-rich) plagioclase and pyroxene and may contain olivine.

Ultramafic igneous rocks are composed of a majority, commonly 80% or more, of the dark minerals olivine and/or pyroxene. Feldspar of any kind is rare in these rocks. When

it does occur, it is commonly a calcium-rich plagioclase.

Discussion

Granitic rocks are probably the most abundant plutonic rocks. The continental crust has a granitic composition. These rocks must contain visible quartz and dominant orthoclase to be called granites. Hand-specimen identification of a rhyolite should also be on the basis of visible quartz. If quartz is not visible in a light-colored aphanitic igneous rock, it should be called a *felsite*. Some andesites could probably be called felsite.

Obsidian may be black, reddish brown, or other colors. It is distinguished from basalt glass (also black) by the fact that thin slivers or edges of obsidian are translucent; those of basalt glass are opaque.

Diorite and andesite are the most abundant intermediate rocks. They contain the essential minerals plagioclase and hornblende. Andesite is distinguished from felsite and rhyolite by its slightly darker color and from basalt by its lighter color. Diorite

Table 6–2. Classification of igneous rocks

Texture	Composition				Mode of Emplacement
	Felsic	Intermediate	Mafic	Ultramafic	
Glassy	Obsidian Pumice	—	Basalt Glass	—	Extrusive
Aphanitic	Rhyolite[1] Felsite[2]	Andesite	Basalt	—	Extrusive
Aphanitic porphyritic	Rhyolite porphyry[1] Felsite porphyry[2]	Andesite porphyry	Basalt porphyry	—	Extrusive
Phaneritic	Granite	Diorite	Gabbro	Dunite Peridotite	Intrusive
Phaneritic porphyritic	Granite porphyry	Diorite porphyry	Gabbro porphyry	—	Intrusive
Fragmental	Tuff, volcanic breccia[3]				Extrusive

[1] Quartz visible.
[2] Quartz not visible.
[3] Fragmental rocks may have a composition ranging from felsic to mafic.

and andesite contain a higher percentage of dark minerals than do felsic rocks.

Basalt is the most abundant igneous rock type. Most of the ocean basins are floored by basalt. Lunar rocks are also mostly basalts or close relatives. Basalt and its intrusive equivalent, gabbro, must contain dominant plagioclase and pyroxene as essential minerals.

The ultramafic rock dunite is one composed mostly, 90% or more, of olivine. The remainder is usually pyroxene. Peridotite is a mixture of olivine and pyroxene, totaling greater than 90%, with accessory biotite and/or plagioclase. Extrusive equivalents of all but one type of ultramafic rock are rare. The single extrusive ultramafic that occurs fairly frequently on oceanic islands is a type of olivine-rich basalt called *picrite*. The mantle is thought to be composed mostly of peridotite.

CRYSTALLIZATION OF MAGMA

The end-product of crystallization of a magma—igneous rock—may be one of many rock types. These products may not always be the same, because magmas may have dif-ferent compositions, they may cool at different rates, and they undergo changes during cooling.

Every igneous rock type does not have a parent magma of that particular composition to produce it by simple cooling. N. L. Bowen studied the crystallization of artificial magmas for many years during the first half of the twentieth century. Early in his studies he recognized the importance of *fractional crystallization* in crystallization of different igneous rock types from a single magma. As the magmatic melt cools, different minerals crystallize at different temperatures.

Bowen studied many combinations of minerals in melts but discovered that through fractional crystallization most common igneous rock types could be produced by the crystallization of relatively few melts of different composition. He even postulated that basalt is the parent of all igneous rocks (this is called *Bowen's theory*) and formulated the *reaction series* (Figure 6–11), illustrating the order of crystallization of minerals that may react with the magma after crystallization to form others.

The reaction series is divisible into two subseries: a continuous and a discontinuous series. In *continuous series* reaction proceeds among members of the same group, the plagioclase feldspars. In the *discontinuous series* different minerals (with distinctly different structures) are involved at each stage of crystallization. After a mineral crystallizes, it tends to react with the remaining liquid again to form the next mineral below in the series.

Consider the sequence of crystallization from a high-temperature melt of basaltic composition. In the discontinuous series olivine will probably begin to crystallize first. Then, unless it is removed from contact with the magma, it will react with the magma to form a pyroxene, which will in turn react at lower temperature to form hornblende, an amphibole. Frequently the liquid is exhausted at this point. But conceivably some

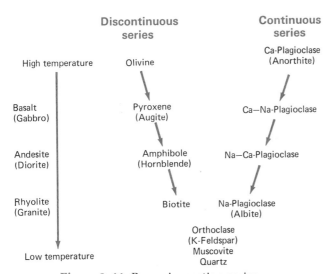

Figure 6–11. Bowen's reaction series.

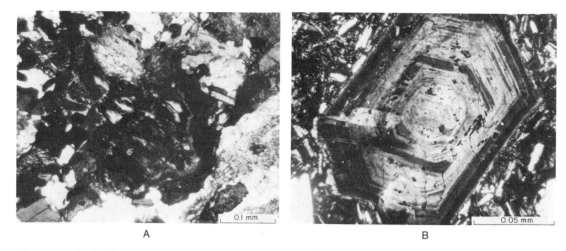

A   0.1 mm

B   0.05 mm

Figure 6–12. A. Thin section of a pyroxene grain with a reaction rim of hornblende. B. Zoned plagioclase crystal in a thin section.

of the hornblende could react with any remaining liquid to form biotite.

Paralleling the crystallization of olivine, pyroxene, and the other members of the discontinuous series is the crystallization of the continuous-reaction series. While olivine is crystallizing, a calcium-rich plagioclase feldspar may be forming. This will react as the temperature decreases to form more sodium-rich forms of plagioclase, provided it remains in contact with the magma. Again, the sodium plagioclase, albite, may not form because all liquid may be used up first.

The last three members of the reaction series, which are part of *neither* the continuous nor the discontinuous series will form only if liquid remains after those above have formed.

Can we actually see examples of this sequence in natural rocks? Although no examples of the complete series occur in a single igneous body, many igneous rocks contain olivine grains that are surrounded by a *reaction rim* of pyroxene. Some even have a second reaction rim of hornblende. Pyroxene grains have been observed with a rim of hornblende (Figure 6–12A) and even a second reaction rim of biotite. Likewise, in plagioclase grains in igneous rocks it is not

uncommon to observe *zoning* in individual crystals, in which the innermost (first-formed) layers are more calcium-rich and the outer parts are enriched in sodium (Figure 6–12B).

Several excellent examples of differentiation are visible in large igneous bodies scattered throughout the world. In the United States a large sill (p. 82) forms the Palisade of the Hudson River in New York (Figure 6–13). This body is over 1000 ft (300 m) thick and is well exposed along the Hudson to afford an excellent place for study.

The Palisade Sill was intruded as a molten mass that baked the sedimentary rocks above and below the sill. The cooler mass of sedimentary rocks flanking the igneous body caused some of the magma to be chilled and solidified very shortly after it was intruded. However, the remainder of the mass stayed molten for many years and cooled slowly so that crystals could form and react again with the liquid, as Bowen had predicted.

Not long after intrusion and chilling of the edges of the sill, heavier olivine crystals began to form and filter down through the liquid to accumulate as an olivine-rich layer along the bottom of the sill. Some pyroxene and plagioclase were also crystallizing then.

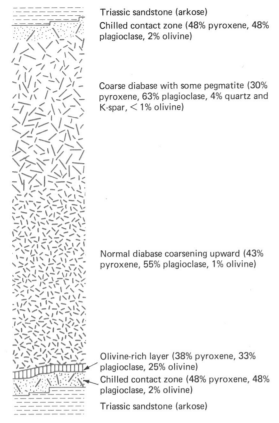

Triassic sandstone (arkose)

Chilled contact zone (48% pyroxene, 48% plagioclase, 2% olivine)

Coarse diabase with some pegmatite (30% pyroxene, 63% plagioclase, 4% quartz and K-spar, < 1% olivine)

Normal diabase coarsening upward (43% pyroxene, 55% plagioclase, 1% olivine)

Olivine-rich layer (38% pyroxene, 33% plagioclase, 25% olivine)

Chilled contact zone (48% pyroxene, 48% plagioclase, 2% olivine)

Triassic sandstone (arkose)

Figure 6–13. Cross section of the Palisade Sill along the Hudson River near New York City. (From Turner and Verhoogen, *Igneous and Metamorphic Petrology*, 2nd. ed., McGraw-Hill Book Co., New York, 1960).

But more of these minerals crystallized as the magma continued to cool. Toward the top of the sill the plagioclase becomes sodium-rich, and some quartz even appears in the rock mass. The sill also becomes coarser toward the top, which means that here was a concentration of the more volatile materials, such as water serving to lower the viscosity of the magma in increasing mobility of ions and promoting the growth of larger crystals. However, all the magma was solidified before Bowen's reaction series could be carried to completion. Despite this, the crystallization of the Palisade Sill is one of the best illustrations of Bowen's theory in nature.

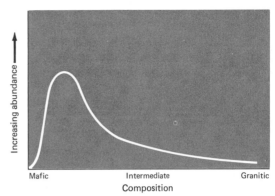

Figure 6–14. Abundances of the different groups of igneous rocks in the earth's crust, predicted from Bowen's theory.

Utilizing Bowen's theory and the reaction series we could predict that, starting with a basaltic magma, the most abundant igneous rocks should be mafic rocks (basalts) and should be followed in abundance by intermediate rocks (andesite and diorite); the felsic (granitic) rocks should be least abundant (Figure 6–14). However, when we look at the actual abundance of the different groups of rocks we find that basaltic and granitic rocks have about the same overall abundance and that intermediate rocks are rather rare (Figure 6–15). So, are Bowen's ideas to be abandoned? The Palisade Sill and many other igneous bodies scattered throughout the world lead us to believe that his ideas work. The

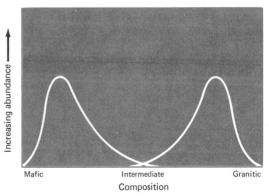

Figure 6–15. Actual abundances of the different igneous rock groups in the earth's crust.

abundance of data tells us it does not. The answer is probably that more than one mechanism has produced the diversity and observed abundances of igneous rocks.

WATER, VOLATILES, AND IGNEOUS MELTS

The minerals found in igneous rocks are stable at high temperatures and form from silicate melts. If we heat a piece of granite until it melts, it would have to be heated to about 1000°C. Basalt melts at 1100–1200°C. If we place another piece of granite in a highly pressurized container with some water, the granite's melting point may be lowered to 600–700°C. Thus the presence or absence of water in the system can have a significant effect on the freezing conditions of the magma and also on the minerals that will form.

During the late stages of crystallization of many magmas, substances with the lowest melting points remain in the liquid. In addition to water and the elements that would crystallize the minerals at the lower end of Bowen's reaction series, elements such as lithium, boron, fluorine, chlorine, and others tend to concentrate here, provided they are present in the original magma. As a result of the presence of water and these other substances, the viscosity of the magma is lowered considerably, and very large crystals tend to form because of the additional mobility of elements in the very fluid magma. The resultant very-coarse-grained rocks are called *pegmatites* (Figure 6–16) and are generally composed of K-feldspar, muscovite, and quartz. Those containing this assemblage are called *simple* pegmatites. Occasionally the abundance of other light elements in the final magma results in crystallization of certain minerals (e.g., fluorite or even rarer minerals). These are called *complex* pegmatites. The lithium deposits at Kings Mountain, North Carolina, are developed in complex pegmatites and are the largest such deposits in the United States.

Pegmatite veins range in size from a few inches (centimeters) in width to large masses a mile (kilometer) or more wide. Some pegmatites are so coarse that single crystals of feldspar have been large enough to mine. In some large pegmatites crystals the size of boxcars have been found.

Figure 6–16. Large crystals of K-feldspar, muscovite, and quartz in a pegmatite vein cutting a finer granitic rock mass, Georgia.

WHERE DOES MAGMA COME FROM?

Through the years various ideas have been advanced to explain the origin of magma, from ancient and medieval ideas based on superstition and fear to Werner's idea of coal burning beneath the surface to ideas based on sound scientific observation. The most obvious alternative to the problem with Bowen's theory is to have several molten masses of different composition. This may solve one problem, but it creates another, one of producing many different pots of magma of different composition. Also, the problem of the apparent overabundance of granitic rocks must be considered.

Part of the solution of these problems may

lie in the location of the igneous body, that is, where it was formed and its point of origin. The basaltic magmas that are erupted onto the ocean floor or on oceanic islands are probably formed in the mantle just beneath the crust (Figure 6–17). Likewise, basalts that are found on the continents may also have originated by melting of mantle material. But if basaltic magma must travel through continental material before it solidifies, its composition may be altered. This is one reason that continental basalts differ from those found on ocean floors. The ultramafic rocks dunite and peridotite are also probably derived from the mantle.

Granitic magma may arise from the melting of continental crust. It may also be produced as the end-product of the process of metamorphism, when rocks might be subjected to temperatures above their melting points, usually in the presence of water, which tends to lower melting points. This production of granitic magmas as a part of the metamorphic process is called *granitization* and is discussed more thoroughly in Chapter 10.

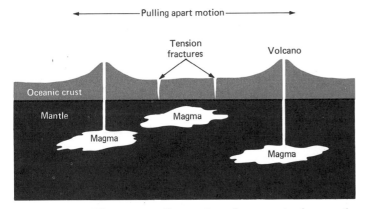

Figure 6–17. Generation of magma in the mantle directly beneath the crust. Tensional fracturing causes the pressure in the mantle to be reduced thereby lowering the melting point of mantle material and creating a magma body.

If basaltic magma may be derived from melting of mantle rocks, and if granitic magma originates in the continental crust, where do the intermediate rocks, like diorite or andesite fit into the picture? There are several possibilities. Intermediate magmas may be produced by differentiation of basaltic magmas. As more minerals rich in calcium and iron–magnesium (such as calcium plagioclase, olivine, and pyroxene) crystallize, they may leave behind a magma of more intermediate composition, rich in sodium, silica, and aluminum.

A magma of intermediate composition may be produced by contamination of a basaltic magma when it comes into contact with continental crust of a granitic composition. Granitic rocks are richer in sodium, silica, and aluminum, as well as potassium. And the end-product would depend on the degree of contamination. In the opposite sense, contamination of granitic magma by basaltic material could accomplish the same end. Many other possibilities arise, if we consider contamination of different kinds of magmas by sedimentary rocks of diverse compositions.

Finally, intermediate rocks could be early crystallization products of granitic magmas. Portions of this type of magma richer in iron, magnesium, and calcium may only be sufficiently mafic to produce diorite on crystallization.

Perhaps melting and/or contamination of existing magmas may occur to produce the diversity of igneous rocks. But how does melting occur in the upper mantle or within the continental crust? For the answers to this question we must turn to some of the most recent ideas and developments in geology.

The theory of plate tectonics (see Chapter 21) provides for the generation of magma along zones where oceanic crust is being pulled apart. The crust is fractured here, and the mantle directly beneath it has less than its normal high pressure. Under these conditions some of the upper mantle material

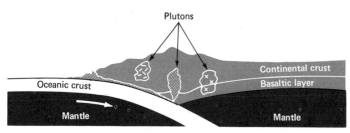

Figure 6–18. Generation of magma due to frictional heating of crustal and mantle materials where two oppositely directed segments of the outer skin of the Earth meet.

melts to produce a basaltic magma (Figure 6–17).

Magmas may also be produced when a mass of oceanic crust passes beneath the edge of a continent or beneath another mass of oceanic crust (Figure 6–18). Here diverse rock types and magmas arise because of the wide variety of possibilities for melting of different materials (sediments, old granitic crust, basaltic material, mantle material) and

Figure 6–19. Relationships between a dike and the invaded host rocks. (Courtesy of Ward's Natural Science Establishment.)

for contamination of each of these by one or more of the others. This is a situation of compression, and much of the heat for melting is derived from friction developed as one mass (of, say, oceanic crust) attempts to move past the other (continental crust). Radioactive heat may also contribute to the melting of rocks. So we can account for most igneous bodies by one or more mechanisms. As usual there are exceptions not explained, such as the igneous bodies found in North America in areas not thought to have undergone this type of activity. Some of these rocks are found in Missouri, Kansas, and Kentucky.

IGNEOUS BODIES

Magma crystallizes into masses of different sizes and shapes. These masses may be divided into two groups based on the environment from which they derive—either on the surface of the earth or under it. Those which crystallize on the earth, such as lava flows, are discussed in Chapter 7. We will confine our discussion here to intrusive igneous bodies called *plutons*. Plutons are of two types: concordant and discordant. Concordant plutons are intruded along and lie parallel to the layering in the *host* or *country rocks* (the rocks into which the body has been intruded). Discordant plutons have cut across the layering in the host rocks. In *tabular* plutons two dimensions greatly exceed the third, producing slablike bodies, which in some cases are concordant and in other cases discordant.

Concordant plutons

Of the several varieties of concordant plutons, the most common is the *sill* (Figure 6–19). A sill is a tabular pluton that may range in size from a few inches (centimeters) thick and a few square feet (square meters) in areal extent to sills the size of the Palisade Sill, which is more than 1000 ft (300 m) thick and is traceable over an area of many square

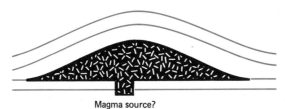

Figure 6–20. Cross-section of a laccolith.

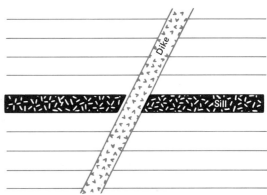

Figure 6–22. Relationships between a dike and a sill.

miles (square kilometers). Their composition may range from felsic to mafic.

Laccoliths are bell-shaped concordant plutons that have arched the layers of rock above the body by the force of the intrusion. These relatively large features may be several miles (kilometers) across (Figure 6–20).

A *lopolith* is a large bowl-shaped concordant pluton that assumes the form of a downfold (Figure 6–21). These plutons may be as much as 100 mi (160 km) across. It is not known for certain how magma was supplied to this feature, because the underside of a lopolith has never been observed. Some of the world's important metal deposits, such as the nickel–cobalt deposit near Sudbury, Ontario, occur in lopoliths.

Discordant plutons

The discordant tabular pluton is a *dike* (Figure 6–22). The difference between a dike and a sill is one of discordance or concordance. Dikes may also range in size from a few inches (centimeters) to hundreds of feet (meters) in width and may be traceable for many miles (kilometers).

More or less equidimensional discordant plutons having a surface extent of less than

Figure 6–21. Cross-section of a lopolith.

40 mi² (110 km²) are called *stocks*. These are thought to maintain the same shape or to become larger at greater depth. Those plutons whose surface extents exceed those of a stock are called *batholiths* (Figure 6–23). Batholiths form the cores of many of the large mountain ranges of the world. The Sierra Nevada are cored by a series of large stocks and batholiths that cover 300 mi (500 km) in length and about 50 mi (85 km) in width. The Coast Range batholithic intrusions are even larger. Most batholiths are composed of granitic rocks (Figure 6–24).

Batholiths and stocks have certain subordinate features associated with them. A *cupola* is a small discordant protrusion of the main magma body (Figure 6–25). It increases in size downward as it joins the main part of the stock or batholith. A *roof pendant* is a downward extension of country rock into a stock or batholith (Figure 6–25). A roof pendant is still connected to the main mass of country rock.

A portion of country rock that has become disconnected from the main mass sinks into the magma. If it is not completely dissolved by it but remains after the magma is congealed, it is called a *xenolith* (Figure 6–26) Xenoliths may occur not only in stocks and batholiths but also in other plutons as well. The process by which xenoliths are sepa-

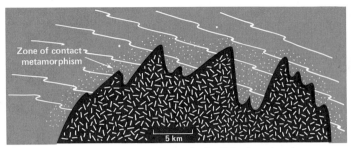

Figure 6-23. Large batholith showing its relationship to the country rocks and related features.

rated from the walls of a magma chamber is called *stoping*. This is accomplished by magma forcing its way into cracks in the wall rock and then surrounding and engulfing the loosened chunks. This is also one way in which magma may intrude and work its way toward the surface.

SUMMARY

Igneous rocks crystallize from magma — molten rock material. They are important not only as host rocks for many valuable minerals but also because they are the most abundant rocks in the Earth's crust. The magmatic origin of igneous rocks was not firmly established until the latter part of the eight-

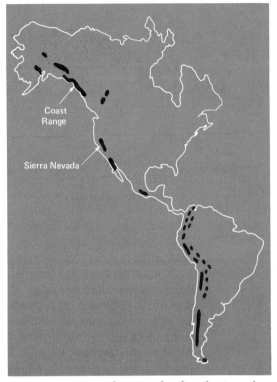

Figure 6-24. Map showing the distribution of batholith size intrusive bodies in North and South America.

eenth century when Desmarest proved the volcanic origin of basalt, and Hutton demonstrated the molten intrusive origin of granite.

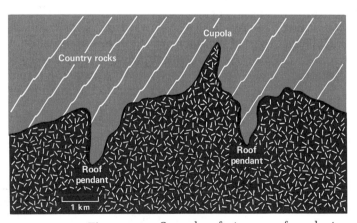

Figure 6-25. Secondary features, roof pendants, and cupolas.

Figure 6-26. Xenoliths of wallrock in a granitic body, South Carolina.

Rocks are classified based on composition and texture. Relatively few kinds of minerals make up the bulk of most common igneous rocks. All are silicates. Different textural types (phaneritic, aphanitic, porphyritic, glassy, fragmental) arise from different cooling rates and/or explosive volcanic activity. Igneous rocks may be divided into four general compositional groups: felsic, intermediate, mafic, and ultramafic. Each of these groups may contain several rock types with different textures.

The study of artificial rock melts led to the formulation of Bowen's reaction series and Bowen's theory. Much field and laboratory evidence supports these ideas, but the relative abundances of the different groups of igneous rocks do not. This suggests that igneous rocks may originate by means other than by differentiation of basaltic magma. Other sources of magma include contamination of magma by other rock types, melting of mantle material, melting of granitic rocks or sediments, and mixing of primary magmas.

A pluton is an igneous intrusive body. Two types of pluton exist: concordant (intruded parallel to layering in host rocks) and discordant (cut across layering in host rocks). Concordant types include the sill, laccolith and lopolith. Dike, stock, and batholith are examples of discordant plutons.

Questions for thought and review

1. Why was Werner not able to draw the same conclusions about the nature of basalt and granite that others of his time did?
2. On what bases are rocks classified?
3. What are the principal differences between a granite and a rhyolite? A rhyolite and a basalt? A granite and a diorite?
4. What group of minerals makes up the bulk of igneous rocks?
5. What is an essential mineral?
6. How does the continuous series differ from the discontinuous series?
7. How does a porphyritic texture originate?
8. List several means by which intermediate rocks may form.
9. How does the presence of volatiles affect the melting behavior of igneous rocks?
10. What is the principal difference between a dike and a sill?
11. Where do most mafic and ultramafic magmas probably originate? Granitic magmas?

Selected readings
Ernst, W. G., 1969, *Earth Materials*, Prentice-Hall, Englewood Cliffs, N.J.
Tuttle, O. F., 1955, The Origin of Granite, *Scientific American*, vol. 192, no. 4, pp. 77–82.

7 Volcanoes

STUDENT OBJECTIVES

At the conclusion of this chapter the student should:

1. know what a volcano is and the kinds of material it ejects

2. recognize various types of volcanoes and the nature of their activity

3. know about the distribution of volcanoes on the earth, both in the present and in the past

4. appreciate both the positive and negative aspects of the relationships between volcanoes and human beings

Then, with a loud roar, at 2 o'clock the great
convulsion came. Those who were in the open air
saw the huge black cloud rolling down the mountain
in globular, surging masses. Onward it rushed with
a loud rumbling noise and filled with lightnings.
Any who were in the open air perished at once.

Anderson and Flett
Eruption of La Soufriere, 1902

Figure 7–1. Mount Vesuvius. (U.S. Navy photograph.)

THE BIRTH OF VESUVIUS

When faint tremors and groanings within the earth were noticed by the 20,000 inhabitants of Pompeii on the morning of August 24, A.D. 79, few believed that a violent volcanic eruption was imminent. Perhaps an earthquake, yes, because the town was still rebuilding from an earthquake that shook it and nearby Herculaneum 16 years before. Small shocks and tremors had been commonplace since then. While it was true that Monte Somma to the north was a volcano, it had not erupted within the memory of anyone living and was thought to be extinct. Monte Somma's crater walls had been crumbling from the onslaught of erosion for centuries, and within the enlarged crater itself wolves and boars roamed in a thick tangle of brushwood and wild vines.

On that day in August this complacency was abruptly dispelled by a series of thunderous explosions. A column of smoke, gases, and steam rose into the air above the crater of Monte Somma and blew out part of its ancient walls. A dense rain of volcanic ash and glowing debris turned day into night over the towns of Pompeii, Herculaneum, and Stabiae. Vesuvius was born (Figure 7–1).

Most of the population of these towns escaped; many did not. A number sought safety within their villas, carrying food and water with them into lower rooms. They are still there, huddled in corners where they were suffocated by poisonous gases. A few in the act of flight died on the streets. Pliny the Elder, captain of the Roman fleet at Naples, engaged in rescue operations. In attempting to get a closer view of the eruption, he perished at the town of Stabiae. His nephew, Pliny the Younger, later described the tragedy in two letters to his friend Tacitus. These documents, still extant, provide us with details of what happened. In the several hours since the onset of the eruption, Pompeii became buried under 20 ft (6 m) of volcanic ash, with roofs of houses collapsing and finally disappearing in the inundation. Herculaneum, closer to the base of Vesuvius, received 50 ft (17 m) of water-saturated flows of volcanic ash. Ironically, the city of Pompeii was built originally on the site of an old lava flow.

In the centuries following this eruption, the location of these towns was forgotten. Their location remained unknown until excavations in the eighteenth century accidentally brought them to light. Since that time, systematic digging has uncovered 60% of Pompeii and a good portion of Herculaneum, resurrecting well-preserved art treasures and a magnificently detailed glimpse into Greco-Roman life of the first century (Figure 7–2). Paintings, mosaics, statues, temples, and private houses have been exhumed. On the mundane side, scribblings on walls (graffiti), chariot ruts in the streets, and ring marks from wine glasses on counters have been preserved. Loaves of bread with the baker's name engraved, medicinal pills, olives, almonds, and fish roe have also been found.

Vesuvius continued to be active in the intervening centuries since A.D. 79, erupting 50 times. A major outburst in 1631 sent torrents of lava and mud over several villages, and a conspicuous eruption occurred in 1944.

VOLCANISM

As we saw in Chapter 6, igneous activity takes place extensively beneath the surface of the earth. When igneous material reaches the surface, one of the most spectacular events in nature occurs: volcanic action. As the account of Vesuvius suggests, the eruption of volcanoes has a vital effect on human activities.

The term *volcanism* includes all activity associated with the expulsion of igneous material from the interior of the earth to the surface. This transferral is accomplished by means of a conduit or *fissure* that penetrates

Figure 7–2. Some of the remains of Pompeii after excavation. (Courtesy of Italian Government Travel Office.)

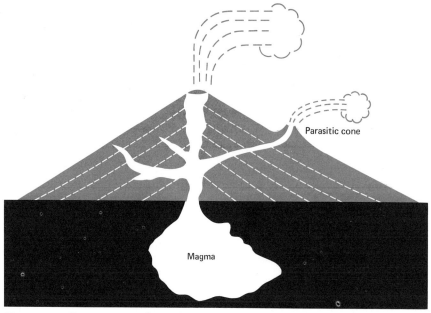

Figure 7–3. Cross section through a volcano shows possible structure.

deeply into the earth, tapping a supply of superheated fluids and gases. A *volcano* is the accumulated pile of igneous debris around the point of exit, usually called the *vent*. Volcanoes form when pressure buildup from below rends the overlying rocks. In some cases a slight upward bulging will appear in the region of a volcano prior to its eruption, as sensitive tiltmeters on Hawaii have shown. One main conduit usually feeds the volcano from below, but vents may be numerous. Offshoots from the main conduit may result in *parasitic cones* on the flanks of a volcano (Figure 7–3).

Expanding water vapor plays an important role in helping to propel magma up along fissures, especially in the more explosive volcanoes. Steam is plentiful in volcanic emissions. During a phase of eruption at Mount Etna in Sicily, it was estimated that 460 million gallons of water (as steam) escaped from one vent. On Hawaii, an average water content of nearly 80% has been found in magmatic gases issuing from volcanoes there (Table 7–1).

Table 7–1. Typical composition of magmatic gases in Hawaii.

Substance	Percentage
Water	79.31
Carbon dioxide	11.61
Sulfur dioxide	6.48
Nitrogen	1.29
Hydrogen	0.58
Carbon monoxide	0.37
Sulfur	0.24
Chlorine	0.05
Argon	0.04

SOURCE: J. F. White, ed., 1962, *Study of the Earth*, Prentice-Hall, Englewood Cliffs, N.J.

PRODUCTS OF VOLCANISM

Igneous material emitted by volcanoes can be classified into two groups: (1) *lava* and (2) *pyroclastics*. Lava is the fluid phase of volcanic activity. Pyroclastics (also called *tephra*) are various-sized, discrete particles of hot debris thrown out of a volcano.

Whether lava or pyroclastics are being ejected, the eruption is normally accompanied by the expulsion of water and other gases.

Lava

Lava usually expresses itself as elongate flows or, literally, rivers of molten rock, moving down and outward from the slopes of a volcano. In composition, lava is basaltic, andesitic, or rhyolitic (see Chapter 6); but basaltic flows are the most common. When congealed, many lavas still display flowage features. *Pahoehoe* or ropy-looking lava is one such type (Figure 7–4). Sometimes a lava flow cools and becomes sluggish and forms an outer crust. When the still-fluid lava beneath the crust starts to move again, it breaks the crust into jagged chunks and blocks. The final product of such a flow is called *aa* (pronounced ah-ah), as shown in Figure 7–5.

Basaltic lavas are known for their low viscosity and are capable of spreading out over wide distances from vents to form volcanoes with a low, broad profile. In the earth's distant past, crustal instability has produced fissures of regional extent from which many cubic miles (kilometers) of basalt have upwelled to form *basalt plateaus* thousands of square miles (square kilometers) in area. Rhyolitic lava is much more viscous. Because of this high viscosity it does not spread widely from its vent and is more restricted in areal extent.

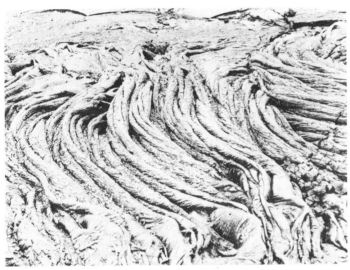

Figure 7–4. Congealed flow of pahoehoe or ropy lava. (Photo by John S. King.)

Pyroclastics

Pyroclastics are fragmentary materials that are ejected from volcanoes and fall in solid form to the ground. Some common types are shown in Table 7–2. The finest particles are called volcanic *ash.* Typically a light gray powdery substance representing the pulverization of lava during the eruptive process, ash can be swept by winds over vast distances before it settles. When volcanic ash forms solidified layers, the term *tuff*

Figure 7–5. Aa surface. (Courtesy of Oregon State Highway Division.)

Table 7–2. Pyroclastic materials ejected during volcanic eruption.

Type of material	Size range (diameter)
Blocks	Greater than 32 mm up to several meters
Bombs	4 mm up to several meters
Cinders	4–32 mm
Ash	Less than 4 mm

Figure 7–6. A volcanic bomb. (Courtesy of Jan Patterson.)

may exhibit contorted surface markings (Figure 7–6). *Blocks* are solid, angular pieces of debris of varying size, some weighing several tons. Blocks do not necessarily have to be of igneous material; blocks of limestone or other country rock lining the vent wall may be torn off and carried up along the fissure and expelled.

TYPES OF VOLCANOES AND CRATERS

Volcanoes can be classified according to the nature of their activity, that is, *quiescent* or *explosive*. Quiescent activity involves the relatively gentle extrusion of quantities of lava, usually basaltic in composition. Explosive activity is associated with the production of pyroclastics, including cinders, bombs, and blocks. The sudden release of pressures long contained shatters magmatic materials and catapults discrete particles over wide areas. A single volcano often displays both explosive and quiescent phases during its eruptive history and may change from quiet to explosive during a single eruption.

A more precise way to classify volcanoes is on the basis of their *form*, which bears a strong relationship to the kind of igneous material ejected from them. And this, in turn, may relate to the quiescent or explosive type of activity exhibited. Table 7–3 lists some common types of volcanoes and their characteristics.

applies. *Cinders* or *lapilli* range from the size of a pea to that of a golf ball and are irregular in shape. They consist of either solid or cellular lava fragments. *Pumice* is essentially a solidified volcanic froth with many watertight voids within it. Lapilli-sized lumps of pumice have fallen into the sea and, instead of sinking, have floated on the surface. So much of this debris floated on the sea near Krakatoa (following its eruption in 1883) that ships were not able to push through it.

Volcanic *bombs* consist of pieces of igneous material thrown out of a volcano while still in a soft, plastic condition. As the bombs rotate and solidify in flight, they may take on an elliptical or rounded shape and

Table 7–3. Common types of volcanoes.

Type	Form	Type of material	Type of activity	Examples
Shield	Broader than high	Basaltic lava	Generally quiescent	Mauna Loa, Mauna Kea
Strato-volcano (composite)	Often a symmetrical cone	Alternate lava and pyroclastics	Explosive, quiescent	Fujiyama, Vesuvius
Cinder cone	Steep-sided, conical	Chiefly pyroclastics, some lava	Mostly explosive	Paricutin

Figure 7-7. View of Crater Grande, Mexico. (U.S. Geological Survey.)

Shield volcanoes

The low, broad profile of the *shield volcano* is the result of great outpourings of highly fluid basalt, which spreads over wide areas. The fluidity of basalt prevents it from building up any volcanic cone with sides much steeper than 7°. Hawaii is a typical locality for the shield volcano. This group of islands is the result of eruptions of basaltic lava that originally poured out onto the sea floor and built up through about 3 mi (5 km) of water to reach sea level. Continued eruptions of basalt created the present land area of 6,424 mi² (10,360 km²), with such peaks as Mauna Loa and Mauna Kea rising to more than 13,600 ft (4,125 m) above sea level. With the exception of Iceland, Hawaii is the largest edifice of lava in the world; and Mauna Loa is the largest mountain in the world in terms of both height and volume (Figure 7-7).

The volcanoes of Hawaii are so characteristic of quiescent eruption that the term *Hawaiian type* is often used instead. Craters on Hawaii have in the past maintained lava lakes so large that waves of lava slosh against the crater walls (Figure 7-8). In such volcanoes pyroclastics are relatively minor, and explosive activity of the Vesuvian type is

Figure 7-8. Lava lake in crater of Pauahi, Hawaii. (Photo by John S. King.)

normally absent. But it should not be concluded that Hawaiian eruptions are harmless. Swiftly moving lava flows have overrun villages. Mild explosive phases occurred in 1790, when part of the Hawaiian army was wiped out, and later in 1924, when Kilauea spewed pyroclastics for ten days. Eruptions also took place in 1955 and in the early 1960s, providing excellent opportunities for scientists and tourists to observe and record their behavior.

Stratovolcanoes

A *stratovolcano*, the most common type of volcano, alternates between explosive expulsions of pyroclastics and relatively quiet extrusions of lava. The resulting volcanic cone is made up of alternating layers of each kind of material. (For this reason stratovolcanoes are known also as *composite* volcanoes.) Such volcanic action constructs the beautifully symmetrical cones exemplified by Mayon in the Philippines, Fujiyama in Japan (Figure 7–9), and Mount Shasta in the United States. Alternation of activity results when a quiescent lava-ejection phase con-

cludes and forms an effective seal of solidified lava within the conduit of the volcano, permitting gradual pressure buildup below and setting the stage for a violent blast of pyroclastics. The cycle then repeats itself. Stratovolcanoes such as Vesuvius gain a reputation for explosivity, because during such phases the greatest destruction occurs.

Cinder cones

A *cinder cone* is a conical hill of pyroclastics, most of which lie within the cinder-size range. These volcanoes tend to be explosive, but extrusions of lava are not unknown. Cinder cones are numerous, occur in all sizes, and tend to be steep sided. Paricutín, a cinder cone in Mexico, began in a cornfield on February 20, 1943, as a hole in the ground throwing out dense smoke. Within 24 hours a cone of cinders had risen to more than 100 ft (30 m), and the cornfield was no more. In ten days Paricutín was 400 ft (120 m) high. Lava started to flow, spreading out 3 mi (5 km) from the crater and invading two villages. After a year Paricutín was more than 1000 ft (300 m) high. It continued to erupt until 1953, when it lapsed into inactivity.

Calderas

The term *caldera* refers not so much to a particular type of volcano as it does to the crater that results. A caldera is essentially a very large crater that may form in different ways. Calderas can result from massive collapse or subsidence during volcanic activity, or later from erosion during dormancy. Some calderas may result from unusually violent explosions (Figure 7–10).

SOME SPECTACULAR EXAMPLES

Mt. Pelée

Certainly the people of St. Pierre on the island of Martinique in the West Indies had ample warning that Pelée was about to erupt. In the 2 weeks prior to May 8, 1902, Pelée an-

Figure 7–9. Fujiyama, Japan, is a stratovolcano. (Courtesy of Japan Air Lines.)

Figure 7–10. Crater Lake, Oregon, a caldera. (Courtesy of Oregon State Highway Division.)

nounced its revival (after 50 years of dormancy) with loud hissings of steam, boiling of lake waters within the crater, and discharges of flourlike ash that spread over the surrounding countryside to despoil crops and affect cattle. At night a strange glowing could be seen near the summit. As May 8 approached, the pent-up forces within the earth intensified, and detonations like distant artillery could be heard.

The population of St. Pierre was about 30,000. Vowing to avoid a panicky exodus, the governor posted troops on roads leading out of the city, and they turned many people back. The crater rim of Pelée was several hundred feet (meters) high but was breached on the side facing St. Pierre by the erosional activity of a stream, the Rivière Blanche,

which had carved a valley down the slope of Pelée to the edge of the town (Figure 7–11). Thus the gash in the crater was like the muzzle of a gun pointed at St. Pierre, 4 mi (6 km) away.

Eighteen ships rode at anchor in the harbor at St. Pierre at about 8:00 A.M. on May 8. Sailors aboard the *Roddam* heard a deafening explosion, and the ship lurched. A gigantic dark cloud appeared at the summit of Pelée and, glowing with increasing intensity, hurtled down and across the town and into the harbor. Ships burst into flames and sank. The *Roddam* managed to get under way and luckily escaped beyond the so-called "singe area." In a matter of minutes, 30,000 people had died, most of them either roasted alive or suffocated by incandescent dust. Only

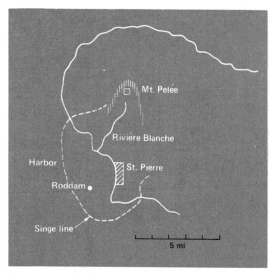

Figure 7–11. Sketch of St. Pierre and vicinity, Martinique.

one man, imprisoned below ground in a dungeon, survived. The fiery avalanche of gas, ashes, and cinders that engulfed and destroyed St. Pierre later was named a *nuée ardente*, or glowing cloud. The eruption of Pelée was not accompanied by significant lava flows, although mudflows were common.

Krakatoa

Krakatoa is situated in the Sunda Strait between Java and Sumatra, a region of numerous volcanoes. In 1883 it recorded one of the most violent eruptions in history. Like Vesuvius, Krakatoa had been the scene of volcanic activity in the remote past. The remnants of a large caldera lay submerged beneath the sea. Resumption of volcanism centuries before had produced three small coalesced cones protruding above sea level. Beginning in May 1883 and throughout the summer months, these three cones emitted smoke and steam. More vents appeared below the water's surface. If any people lived on the island, they left.

On August 26, loud explosions were heard 100 mi (160 km) away, and dense clouds of ash and pumice shot 17 mi (27 km) into the air. Along the coast of Java and Sumatra, darkness fell as volcanic clouds shut out the sun. This darkness lasted two and a half days. Torrential ash-laden rains added to the turmoil. On August 27, Krakatoa reached its peak of explosivity. The sounds of a series of detonations were heard in Australia, 3000 mi (4840 km) away. At the same time, volcanic debris was thrown many miles (kilometers) into the sky. The finer particles, riding on stratospheric winds, encircled the earth and took two years to settle. It is estimated that 4–5 mi³ (6–8 km³) of rock debris were blown into the air during the paroxysms that pulverized Krakatoa, with an energy release matching that of the most powerful hydrogen bomb.

Had Krakatoa erupted in a densely populated area, millions of lives might have been lost. As it was, the reverberations of Krakatoa unleashed a *tsunami*, a large sea wave that attained a height of 120 ft (38 m) from base to crest as it crashed against the coasts of Java and Sumatra and swept 36,000 people to their deaths. The force of the tsunami can be appreciated by the fact that it carried a large ship 1½ mi (2½ km) inland and stranded it there 30 ft (10 m) above sea level. Rocks weighing up to 50 tons were transported even further. Krakatoa is not dead. A new cone rose above the sea in 1928 and was named Anak Krakatoa (child of Krakatoa). By 1953 the child had grown to a height of 360 ft (108 m).

Surtsey

Surtsey appeared in the North Atlantic south of Iceland on November 14, 1963. By the second day it was 33 ft (10 m) high. Scientists and other observers kept the new volcano under close watch. The basaltic intrusions had built up from the ocean floor 425 ft (128 m) below. Loud explosions accompanied a towering cloud of pyroclastic debris that rose thousands of feet (meters) into the air. The cloud could be seen from

Figure 7–12. The Surtsey volcano during an eruption in August 1966. (Courtesy of William A. Keith and the Icelandic Consulate.)

the capital city of Reykjavik, 75 mi (120 km) to the north. By the end of a week the infant island had grown to 200 ft (60 m) in height and 2000 ft (600 m) in length (Figure 7–12). The abrupt contact of red-hot magma with cold seawater caused a continuous display of explosions. Blocks up to 3 ft (1 m) in diameter were hurled hundreds of feet (meters) into the air, to fall whistling like bombs.

Between November and April the effusives from Surtsey were pyroclastic. But the survival of the new island was questionable, because the loose particles of ash and cinders would not be able to withstand the pounding of ocean waves, and the island would be washed away. In April, however, flows of basalt created a solid rock barrier to protect the island from excessive wave erosion. Lava continued to extrude until, in June 1967, Surtsey had grown to a height of 570 ft (171 m).

VOLCANOES IN SPACE AND TIME

Volcanoes are not scattered at random over the earth. They are more abundant in certain areas of the world. The zone of greatest volcanic activity (called the *ring of fire*) encircles the Pacific Ocean basin. In the Mediterranean region, another area of conspicuous volcanism, the earliest observations of volcanoes were made by Greeks and Romans. The question arises, Why only certain zones of volcanic activity? The question cannot be answered with utmost certainty, but these areas coincide with those of greatest earthquake activity. Advocates of plate tectonics theory believe that these are areas of plate contact where energy release takes place. This subject will be considered in more detail in Chapter 20.

Volcanoes are not unique to the present. Geologists, probing among the rocks of past geologic periods, find abundant evidence of continued volcanism throughout earth history. For example, in Pennsylvania and Virginia, extensive thin layers of *bentonite*, an altered volcanic ash, attest to protracted volcanic activity in the eastern United States hundreds of millions of years ago. In Canada hundreds of cubic miles (cubic kilometers) are present. In fact, virtually every period in earth history yields evidence of volcanism somewhere in the world.

A GOOD WORD ABOUT VOLCANOES

It would seem that, so far in this chapter, we have portrayed the volcano as a villain: erupting and sending thousands to their deaths by roasting, gassing, or—equally insensitively—by spawning tsunamis to drown them. However, volcanoes do have positive aspects. From the scientist's point of view, close study of active volcanoes such as those on Hawaii afford us an appreciation of the forces of nature and a knowledge of the internal condition of the earth, laying the groundwork for possible predictions of volcanic eruptions.

If this is small comfort, then consider that it has been through volcanic eruption, according to most scientists, that the hydrosphere and much of the atmosphere came into existence. Volcanism has resulted in the creation of new land (such as Surtsey) available for settlement by many life forms, including humans (such as Hawaii). Volcanic islands may have served as steppingstones for migrations of life to continents and island areas. Another benefit has been the creation of fertile volcanic soils for raising coffee beans in South America, grapes in the Mediterranean area (that make exceptionally good wines), and pineapples in Hawaii.

SUMMARY

Volcanoes can erupt explosively at the earth's surface, taking a heavy toll in lives and property. A volcano is the accumulation of igneous material around a vent, or exit from the interior. Volcanism is a general term that refers to all the activity resulting from the transferral of igneous products to the surface. These igneous products include lava, which produces relatively quiet eruptions, and pyroclastics, which include various-sized fragments of hot debris ejected in an explosive manner.

Three types of volcanoes include: (1) shield volcanoes, having a low, broad profile and emitting fluid basalts like the volcanoes of Hawaii; (2) stratovolcanoes, which alternate between quiescent and explosive activity and build up a composite cone of alternating lavas and pyroclastics reflecting this activity; and (3) cinder cones, composed chiefly of pyroclastic material such as Paricutín in Mexico. A caldera is an unusually large crater produced by collapse, protracted erosion, or exceptional explosivity.

Mount Pelée, Krakatoa, and Surtsey are vivid examples of volcanic eruption. Such volcanic activity has been going on extensively and continuously since the earliest days of the earth, but today the greatest concentration of volcanic activity is in the ring of fire around the borders of the Pacific. Volcanoes are not all bad, despite their threat to humankind. Through volcanism the hydrosphere and atmosphere probably have been produced. Volcanoes provide new land for settlement and may have served as stepping-stones in the migration of plants and animals. Rich productive soils also result from volcanic debris.

Questions for thought and review
1. What is a volcano?
2. What is the role of water in a volcanic eruption?
3. Might you find a shield volcano composed of rhyolite? If not, why not?
4. Why can some volcanic ejecta float in water?
5. During the eruption of Krakatoa it was estimated (afterwards) that 4–5 mi^3 (10–11 km^3) of rock debris were blown into the air. Could you speculate on how scientists arrived at this estimate?

Selected readings
Bullard, F. M., 1962, *Volcanoes in History, in Theory, and in Eruption*, University of Texas Press, Austin.

Maiuri, Amedeo, 1958, Pompeii, *Scientific American*, vol. 198, no. 4, pp. 68–82.

Williams, Howel, 1951, Volcanoes, *Scientific American*, vol. 185, no. 5, pp. 45–53.

8 Modern environments and sediments

STUDENT OBJECTIVES

At the conclusion of this chapter the student should:

1. know that sediments forming today are part of the rock cycle and will become sedimentary rocks at some time in the future

2. understand how systematic study of modern sediments helps us to understand better how ancient sedimentary rocks were formed

3. know that physical, chemical, and biological factors in any environment strongly influence the sediment that will result

4. recognize why geologists actively investigate modern environments and the controlling factors not only to understand sediment formation but for economic and ecological motives as well

Some circumstantial evidence is very strong,
as when you find a trout in the milk.

Thoreau

A PART OF THE ROCK CYCLE

In the discussion of the rock cycle in Chapter 4 we pointed out that any type of preexisting rock could become exposed and eroded at the earth's surface. Erosional debris, both in the form of solids and dissolved substances, may then be transported to a depositional site to settle out or to be precipitated. Under favorable circumstances, these materials in time would become sedimentary rocks.

Sedimentary materials can be observed today in a variety of environments: sands along beaches, pebbles and cobbles in a stream channel, the odorous muck of a swamp, or limy muds dredged up by an oceanographic research vessel. These sediments have been in transition since their release from the parent rocks (Figure 8–1).

Figure 8–1. Corer being brought aboard research vessel. The sediment it contains is a part of the rock cycle. (Photo by S. D. Heron, Jr.)

The route taken by eroded materials may be complex. Stream sediments, for example, may be picked up and deposited several times along a river course before they reach the sea. These same sediments then may form part of a strandline or perhaps a sand dune before removal to deeper water. During occupancy of an environment, the sediments will be affected and modified by the conditions that prevail in that environment.

REASONS FOR STUDY

Most practicing geologists are involved in the study of sediments and sedimentary rocks. Many of them specialize in modern sediments forming at the present time or in the recent geologic past. This makes sense if we recall that, in accordance with the principle of uniformitarianism (Chapter 1), ancient sedimentary rocks probably were formed under conditions similar to those of today (Figure 8–2). The more we know and understand conditions of modern sedimentary formation, the better able we will be to unravel the events in the earth's history represented by ancient sedimentary rocks.

Another motive is economic—almost all petroleum and natural gas is formed in sedimentary rocks. Not only do oil and gas originate there from organic components, but they migrate through them and become entrapped as pools or reservoirs in sedimentary rocks (Figure 8–3). As one oil company executive put it, "We operate with the conviction that, the more geology we know, the more oil we'll find." It is not surprising, then, to find major oil companies employing staffs of scientific personnel whose investigative efforts include the geology of modern sediments.

There are obvious ecological aspects in the study of recent sediments. Throughout the history of their transportation and deposition, most sediments are handled by water in streams, rivers, lakes, estuaries, and oceans. Any understanding we gain concerning the behavior of natural materials in these

Figure 8–2. Underwater scene of modern reef in the Bahamas. Similar reefs grew in the past. (U.S. Navy photograph.)

Figure 8–3. Specimen of limestone obtained by drilling shows residual oil in porous openings. (Courtesy of Exxon Company, Inc.)

waterways can be useful in understanding the behavior of pollutants, resulting from human activity, that are entrained with the sediments. If, for example, a study of the pattern of distribution of bottom sediments and currents is made in a large lake, these data will be of use in controlling a pollutant such as mercury that may have been introduced into the lake.

CLASSIFICATION OF MODERN ENVIRONMENTS

Where we see sediments accumulating today, the specific environmental setting in which they are forming probably can be assigned to one of the environments listed in Table 8–1. Geologists generally agree on the threefold categorization of continental, marine, and mixed (such as lagoonal and deltaic) environments. Some of these depositional environments are of considerably more importance than others in terms of volume and areal extent of sediments. For example, cave deposits (spelean) are more restricted than stream deposits.

Table 8–1. Classification of sedimentary environments.

I. Continental Environments
   A. Subaqueous
      1. Stream or river (fluvial)
      2. Lake (lacustrine)
      3. Swamp (paludal)
      4. Cave (spelean)
   B. Subaerial
      1. desert
      2. glacial

II. Marine Environments
   A. Continental shelf (neritic)
   B. Continental slope (bathyal)
   C. Deep-ocean basins (abyssal)

III. Mixed Environments
   A. Deltaic
   B. Lagoonal
   C. Estuarine
   D. Littoral

ENVIRONMENTAL FACTORS

Even a casual observer can see that in certain environments types of sediment are distinctly different from those found in other environments. Swamp deposits, for example, consisting of clay and silt in a setting of luxuriant vegetation, contrast sharply with the fairly coarse, clean sands of an ocean shoreline. Why? Table 8–2 lists a few of the important *environmental factors* that influence sediment development.

Consider the swamp. Streams entering a swamp are slow, and their currents typically are capable of carrying in only fine particles of clay and silt. Rapidly growing vegetation (reeds, grasses, mangroves) choke the waterways, decrease water depth, and further restrict current activity. Heat (climate) accelerates chemical reactions and organic decay. In time, layers of sediment, consisting of silt and clay intermingled with dark organic matter from decaying plants, are built up. These layers of swamp sediment could not form if we were able to lower drastically the regional temperature and to introduce stronger circulating currents. Heavy vegetation would not grow, chemical reactions would be slowed, and much silt and clay would be swept away by stronger streams.

Consider, on the other hand, the beach. A variety of particles (sand, silt, clay) is delivered to the beach. Constant to-and-fro wave action quickly sorts out the easily transportable clay and silt and removes it, leaving a residuum of sand particles that are less readily moved. Vegetation finds it difficult to take root in or near the surf zone, even if the climate is warm. Mechanical energy is dominant. Were we to remove all wave action, the shoreline would look quite different. A placid pond with little or no wave action gives evidence of this. The shoreline of a pond is typically one of encroachment by vegetation, with a mixture of various-sized sediments.

In similar fashion we can analyze each of the environments listed in Table 8–1 and try to determine which factors are important in determining the sedimentary makeup of that environment. Although we have stressed energy and biological factors in our examples, it should be borne in mind that each environment is complex and that perhaps dozens of other factors, some of them subtle, contribute toward the ultimate characteristics of a sediment (e.g., depth and temperature of water, salinity, pH, amount and composition of materials available, abundance of burrowing organisms such as worms and so on).

In the next chapter, we will note certain *primary features* found in ancient sedimentary rocks. These features, such as tracks and trails of animals, ripple marks, and raindrop imprints, formed at the same time as the sediment. When found in solid rock, these features tell us about environmental conditions at the time of deposition. In a number of cases, inferences are obvious. For example, raindrop impressions indicate that the sediment was exposed to air for a while. However, the discovery of a complex assemblage of fossil marine organisms in a rock may prompt the investigator to study modern counterparts of these animals to determine the conditions under which they (the fossils) might have lived (Figure 8–4). Each organism or group of organisms prefers a certain range of water depth, salinity, temperature, and other conditions. It is probable that the fossil predecessors may have been similarly inclined.

Table 8–2. Comparison of some of the environmental factors found in swamp and beach sediments.

Swamp	Beach
High thermal energy	Strong wave energy
Abundant vegetation	Little vegetation
Sluggish currents	Vigorous water circulation
Little oxygen in water	Water well oxygenated
	Much abrasion of particles
High decay rate	

Figure 8–4. Limestone slab shows several types of benthonic animals. (Courtesy of Wards Natural Science Establishment.)

MODERN SEDIMENT INVESTIGATIONS

Contemporary sands

Earlier we noted the striking difference between swamp and beach sediments. Most people, geologists or not, can look at two samples of sediment from these environments and say, "They are different." Suppose, however, that an average person is shown a handful of sand from a beach, a dune, and a river bottom. To the casual eye these sands look quite similar (Figure 8–5).

For many years geologists have sought ways to distinguish among such sediments and to apply the results to ancient rocks.

One approach is to analyze statistically the size distribution of the particles that make up the sand. Not all particles in a natural sand are of the same diameter, and the distribution of the sizes is not so much by chance as it is by the transporting agent and the conditions that prevail in the depositional environment. To find out the distribution of sizes, the geologist takes a weighed amount of representative sample and passes it through a nest of screens with decreasing size of mesh openings (Figure 8–6). Each particle will be caught on a screen with openings too small for the particle to pass. The finest particles (some silt and clay) will pass through all the screens to a pan at the bottom. Thus each screen will catch a portion, of a certain size range, of the total sample (Figure 8–7). Each size fraction can then be weighed and expressed as a percentage of the total sample.

Equipped with this information, the geologist can use various formulas to determine such characteristics as average diameter within the particle population in much the same way that a census bureau calculates the height of an average person within a population. When the samples are plotted on

Figure 8–5. Beach, dune, and river sand (*left to right*). Special analysis is needed to distinguish them.

Figure 8–6. This machine, called a ro-tap, separates sands into several size fractions.

Figure 8–7. Separated size fractions of a natural sand.

graphs, distinctions may be made between groups of samples taken from known environments. For example, in Figure 8–8, beach and dune samples occupy separate parts of the graph when average diameter (the mean) is plotted against skewness. Skewness is a measure of the bunching of particles at either the coarse or the fine end of a particle size distribution. Beach sands tend to be skewed (or bunched up) toward the coarse end and dune sands skewed toward the fine end of a particle size distribution (Figure 8–8).

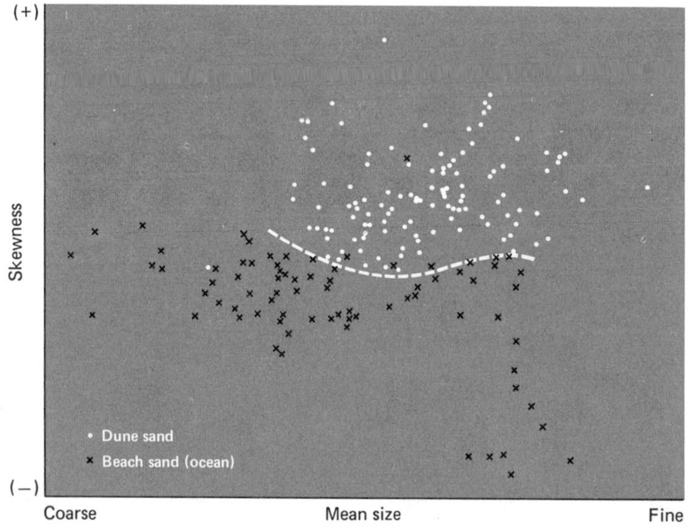

These and other statistical devices have been applied to sediments in a variety of environments. It should not be assumed that success is achieved in all cases. In some studies, individual samples cannot be clearly distinguished among beach, dune, river, or other environments. For this reason other particle attributes such as degree of particle rounding (Figure 8–9) or types of minerals present have been investigated. In the former case, studies show that particles become better rounded in the downstream direction of modern rivers (Figure 8–10). Also, windblown sands tend to be even better rounded than those handled by flowing water.

Figure 8–8. Graphical distinction between beach and dune sands. (Modified from Gerald Friedman, 1961, Distinction between dune, beach, and river sands from their textural characteristics. *Journal of Sedimentary Petrology.* Vol. 31)

Evaporite deposits

Layered sequences of sedimentary rock composed chiefly of the minerals halite and gypsum commonly occur in the geologic rec-

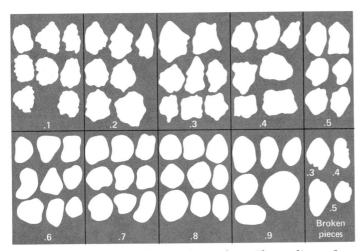

Figure 8–9. Roundness chart. The outlines of sediment particles are compared with these silhouettes in order to interpret transportational history of the sediment.

more than 300 million years ago. It would be difficult to imagine this were it not for such modern situations as Great Salt Lake in Utah and the Dead Sea in the Middle East, where similar evaporite sequences are being formed today (Figure 8–12).

A number of detailed studies of evaporite deposition around the world have been made. For example, Robert Morris and Parke Dickey investigated a shallow estuary, the Bocana de Virrila, along the arid northwest coast of Peru (Figure 8–13). They recognized three depositional environments within the estuary: (1) normal marine, (2) penesaline, and (3) saline. These subenvironments were differentiated not only geographically but also in terms of water composition and mineral associations. With distance from the mouth of the estuary, the water becomes shallower and warmer, and ion concentrations change as minerals are precipitated. The saline environment, located where the estuary forks, contains abundant halite and gypsum, with crystals of the latter growing to ½ in. (1¼ cm). The water is pink, apparently because of red algae, one of the few forms of life that can live in supersaline waters. Studies such as this provide details and insight into the circumstances of evaporite deposition in the past.

ord and are often hundreds of millions of years old (Figure 8–11). These rock types are called *evaporites*, because they form as precipitates from restricted bodies of water (usually seawater) in hot, arid climates.

Extensive deposits of evaporite minerals in Michigan and western New York point to a widespread desert in the northeastern United States during Late Silurian Time

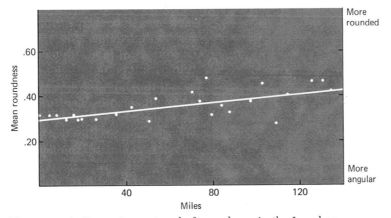

Figure 8–10. Downstream trend of roundness in the Lynches River, South Carolina.

Figure 8–11. Exposure of evaporite beds. (Courtesy of J. D. Martinez.)

Figure 8–12. Great Salt Lake, Utah. The greater density of the water makes objects more buoyant. (Courtesy of Salt Lake Arca Chamber of Commerce.)

Oceanic sediments

Most ancient sediments are of marine origin; therefore we must look for clues concerning their conditions of formation in modern oceans. The mysteries of the oceans have been a focus of attention for humankind ever since early peoples used the oceans for trade routes, for transportation, and as sources of food. Better ships and navigation permitted more extended scientific voyages throughout the seventeenth, eighteenth, and nineteenth centuries. The famous *Challenger* expedition (1872) was a 3½-year voyage around the world to observe the physics, chemistry, biology, and geology of the oceans. The data gathered by the *Challenger* are still referred to by oceanographers. During the twentieth century numerous expeditions by several countries have amassed a staggering amount of infor-

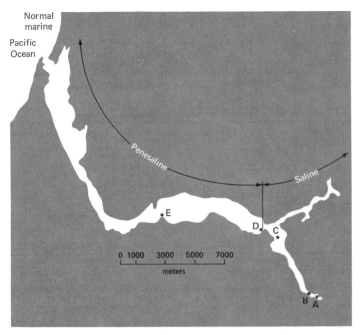

Figure 8–13. Depositional sediments of west coast of Peru and Bocana de Virrila estuary. (After Robert Morris and Parke Dickey, 1957, "Modern Evaporate Deposition in Peru," American Association of Petroleum Geologists Bulletin. Vol. 41) Points *A–E* are sample locations.

mation, and data are still accumulating.

The data obtained from ocean-bottom samples show gross sedimentary distribution and composition. In the Atlantic Ocean calcium carbonate in bottom sediments is concentrated generally away from continental land masses and in shallow water (Figure 8–14). In seas bordering land masses noncalcareous land-derived sediments become mixed with calcareous sediments, reducing the total calcium carbonate content of the sediment. These basic ideas can be applied to studies of limestone, a common sedimentary rock, which in most cases formed as calcareous muds on sea bottoms in the geologic past. For example, a trend of decreasing calcium carbonate content in a limestone might indicate the proximity of a land mass. Purer limestones would form further offshore.

The *trace elements*, or minor constituents, of oceanic sediments are being studied with increasing frequency. Some trace elements, such as cobalt, show distinct patterns of distribution in the Atlantic Ocean (Figure 8–15). Higher concentrations of cobalt are associated with finer sediments and relatively shallow water, whereas low concentrations occur in coarser sediments of the abyssal plain. The application of these and other trace element studies to ancient rocks has yet to be fully explored.

The importance of trace element inves-

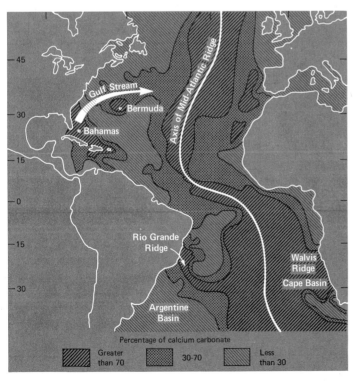

Figure 8–14. Calcium carbonate distribution in sediments of the Atlantic Ocean. (After K. K. Turekian, 1968, *The Oceans*, Prentice-Hall, Englewood Cliffs, N.J.)

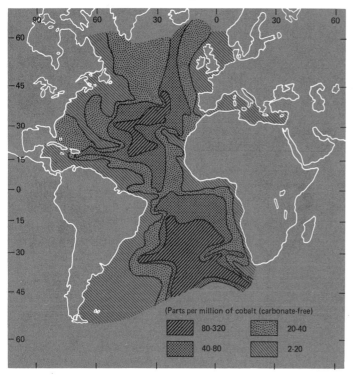

Figure 8–15. Distribution of cobalt in the Atlantic Ocean. (After K. K. Turekian, 1968, *The Oceans*, Prentice-Hall, Englewood Cliffs, N.J.)

(Parts per million of cobalt (carbonate-free)

- 80-320
- 40-80
- 20-40
- 2-20

were formed. In addition, there are economic and ecological reasons for these investigations.

Modern sedimentary environments can be broadly classified as continental, marine, and mixed. The latter category includes lagoons and deltas. In a specific environment, numerous physical, chemical, and biological factors will influence the ultimate character of the accumulating sediment. These factors are exemplified by water depth, salinity, wave energy, and kinds and numbers of burrowing organisms, although there are many others.

Modern sediment investigations, seeking tools to apply to ancient sediments, have probed most contemporary environments. These include detailed study of size distribution and shape of sand-sized particles and arid regions where salt and gypsum deposits are forming, such as in Peru. The oceans are a special focus of study, because most ancient sediments were formed in the marine environment. The occurrence of calcium carbonate and trace elements in the Atlantic Ocean and the inferences drawn from their distribution are only a few examples of the wide scope of oceanographic research being carried out today.

Questions for thought and review

1. What kinds of preexisting rocks contribute to the formation of modern sediments?
2. Give three reasons why modern sediments are studied.
3. What kind of sediments might form in a desert? In a lagoon?
4. What are some important environmental factors that would influence the sediment types mentioned in question 3?
5. Look up the term "littoral." What are some features of a littoral environment?
6. Do you think you could distinguish between swamp and glacial deposits? How?
7. What are primary features? List three

tigations to ecological problems of the oceans is underscored by the fact that significant increases of tetraethyl lead, from automobile exhausts, and mercury, probably from insecticides and pesticides, have been measured in the surface waters of the oceans. In the Mediterranean Sea, increases of lead on the order of 10–100 times normal have been observed.

SUMMARY

That part of the rock cycle involving erosion, transportation, and deposition of sediments can be observed operating at the earth's surface today. Many geologists study these modern sediments, because they provide insight into how sedimentary rocks that might form above water level (sub-aerially).

8. Why is the geologist who studies different sands like a census bureau statistician?

9. What is skewness?

10. Why might the water composition of Great Salt Lake be different from the waters of the Bocana de Virrila?

11. What is a trace element? Name several not mentioned in this chapter.

12. What kinds of pollutants, other than lead and mercury, might find their way into the oceans?

13. Once a pollutant gets into the oceans, do you think it will be easy or difficult to remove? Why?

Selected readings

Krumbein, W. C., and Sloss, L. L., 1970, *Stratigraphy and Sedimentation*, 2nd ed., Freeman, San Francisco.

Shepard, F. P., 1973, *Submarine Geology*, 3rd ed., Harper & Row, New York.

Turekian, K. K., 1968, *The Oceans*, Prentice-Hall, Englewood Cliffs, N.J.

Vetter, R. C., ed., 1973, *Oceanography: The Last Frontier*, Basic Books, New York.

9 Sedimentary rocks

STUDENT OBJECTIVES

At the end of this chapter the student should:

1. understand the processes by which sediments are converted into sedimentary rocks

2. be aware of the variety of sedimentary rocks and how they are classified

3. know how geologists use primary features and other aspects of sedimentary rock units to interpret environmental conditions of the earth's past

*Thus the sedimentary rocks of many dates
. . . constitute a vast and authoritative
book of facts. In reading this book we
do not have to allow for any bias on the
part of the author.*

Playfair

LITHIFICATION AND DIAGENESIS

In the last chapter we noted that a large number of geologists are engaged in the study of modern sedimentary environments, because it is within these environments that future sedimentary rocks are created. An understanding of how this is happening today gives the geologist, following Hutton's concept of uniformitarian, a better grasp of the conditions that obtained during the formation of sedimentary rocks millions of years ago.

Sedimentary rocks form at the earth's surface when unconsolidated materials derived from preexisting rocks are transported into a basin or precipitated within a basin, there to be buried progressively and then to be compressed or crystallized into a consolidated rock.

Lithification describes the processes that convert an unconsolidated sediment into a consolidated sediment. Once a layer of sediment is deposited, the weight of additional layers on top of it causes *desiccation*, or the expulsion of excess water in the sediment. At the same time *compaction*, or the crowding of individual particles together, takes place. The result is a reduction in the volume of the rock. It is estimated that bodies of sand formed under water in the marine environment will be reduced to one-half their original volume by desiccation and compaction. With finer silts and clays, the amount of volume reduction is even more. A soupy mud on the ocean floor may be compressed eventually to one-fifth or less of its original volume because of the squeezing out of water.

As desiccation and compaction take place, *cementation* may occur also. Individual grains crowded together may become bonded by calcareous or silica cement carried in and deposited by solutions percolating through the spaces or *interstices* between the grains. A sedimentary rock therefore may become indurated to varying degrees.

Diagenesis involves all of the processes of lithification but is a more all-embracing concept. It includes any change that may occur after the sediment is deposited but excludes metamorphism and weathering. For example, diagenesis includes any chemical reorganization that takes place. Some of the original mineral constituents of the rock may dissolve and may be redeposited as overgrowths on nearby grains, or new minerals may grow in the interstices of the rock. Such new minerals are called *authigenic*. Common authigenic minerals include mica, chlorite, and feldspar.

DISTRIBUTION OF SEDIMENTARY ROCKS

At the surface of the earth the most common rock types are sedimentary. To assume, however, that these rocks comprise the bulk of the earth's crust would be incorrect. Sedimentary rocks cover about 90% of the surface land areas, masking the larger bulk (90–95%) of igneous and metamorphic rocks that lie below. The reason for this is obvious: Sedimentary rocks originate at the surface and are usually found there; other major rock types usually originate at depth and are found there (with the exception of volcanic extrusives). Despite their subordinate volume, sedimentary rocks offer intriguing clues to past physical events on the earth, and their fossil content provides the means of assessing the evolution of life on earth from early beginnings.

Sedimentary rocks form a relatively thin veneer over the crust. The thickest accumulations exceed 40,000 ft (12,133 m) in places where crustal downwarping has maintained basins of deposition for long periods of time. But in most places the crust has been fairly stable, and sediments have accumulated to a few thousand feet (meters) or less. Sediments were laid down originally as continuous, horizontal layers. Field study reveals, however, that these layers may as-

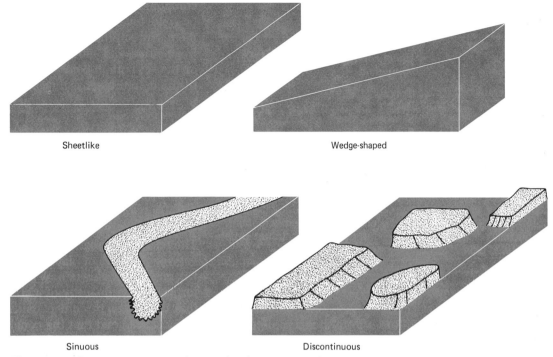

Figure 9–1. Common geometric forms of sedimentary rock units.

sume varied large-scale geometric forms. Some are thin and sheetlike, others are wedge-shaped or prismatic, and still others are sinuous or snakelike as a result of deposition in ancient stream channels. In a sense all sedimentary rocks are lenslike, thickening and thinning and eventually pinching out (Figure 9–1).

The most abundant sedimentary rocks were formed in a marine environment, especially one of shallow water. Shale is the most dominant type and accounts for more than one-half the volume of sedimentary rocks, according to most estimates. The next most common rock types are sandstone and limestone. Other types are minor in terms of volume, but it should be noted that even minor amounts of such sedimentary rocks as evaporites (see Chapter 8) are very important in helping to evaluate certain geologic problems.

CLASSIFICATION OF SEDIMENTARY ROCKS

Basis for classification

Most sedimentary rocks are amenable to classification on the basis of texture and composition just as igneous rocks are. The terminology of sedimentary rocks also follows a scheme whereby the properties stressed are obvious and fundamental to the rock and provide some implication, at least, of how the rock originated. Table 9–1 is a classification of common sedimentary rocks. The rocks are divided into two groups. The first group, *land-derived clastics*, includes those rocks made up of particles or fragments formed as a result of erosion of preexisting rocks. (The word *clastic* means broken.) These rocks are named and arranged according to the size of the particles (*texture*). The composition of particles also is

Table 9–1. Classification of common sedimentary rocks.

	Land-derived clastics	
Rock name	Unconsolidated equivalent	Size of particles[1]
Conglomerate	Gravel	> 2 mm
Breccia	Rubble	> 2 mm
Sandstone	Sand	$\frac{1}{16}$–2 mm
Siltstone	Silt	$\frac{1}{256}$–$\frac{1}{16}$ mm
Shale	Clay and/or silt	< $\frac{1}{16}$ mm
	Biochemical rocks and precipitates	
Rock name	Composition	
Limestone	Calcium Carbonate	
Dolomite	Calcium magnesium carbonate	
Evaporites (halite, gypsum)	Sodium chloride, calcium sulfate	
Coal	Complex of organic compounds	

[1] Geologists usually do not use units of measurement in the English system to describe sedimentary particles.

necessary in order to form a complete name for the rock. Most geologists handle this problem by using a mineral or rock name as an adjective to describe the composition. Thus "quartzose conglomerate" would convey the idea of a rock made up of quartz particles that exceed 2 mm in diameter (and would also distinguish it from a limestone conglomerate in which fragments consist of limestone).

Biochemical rocks form as a result of organic activity; limestones are the most notable. Colonial marine organisms such as corals and certain algae extract calcium carbonate ($CaCO_3$) from seawater to create many limestones. Some limestones are made up almost entirely of skeletal remains of lime-secreting organisms. Coal, another type of biochemical rock, is a complex of organic compounds formed from plant remains. Halite and gypsum form *precipitates* on the bottoms of evaporating seas when concentrations of sodium chloride ($NaCl$) and calcium sulphate ($CaSO_4$) reach their saturation points.

Sedimentary rocks are complex and are sometimes difficult to classify, because they are often mixtures of clastic and nonclastic components. A sandstone, for example, may be held together by calcium carbonate cement, a precipitate. Many limestones may be fragmented and moved around by currents on the sea bottom before they finally become lithified. In this sense they are clastic rocks.

Land-derived clastics

Conglomerates consist of *framework* and *matrix* (Figure 9–2). The framework is the assemblage of rounded fragments (*clasts*) larger than 2 mm in diameter. Most commonly the framework is composed of quartz pebbles. The matrix consists of the finer particles in which the framework is imbedded. The matrix may be sand, or silty sand with varying amounts of clay. The amount of matrix may give a clue to how the conglomerate formed. If the quantity of matrix is so small that clasts are in contact with each other, chances are good that the conglomerate was water-laid. With an abundance of matrix, with clasts generally out of contact with each other, deposition may have been by mudflow or possibly by melting ice.

Figure 9–2. Conglomerate. (Courtesy of Jan Patterson.)

A conglomerate generally is found close to its source. This is particularly true if the conglomerate is made up of something other than quartz or other siliceous particles. Limestone or granite conglomerates, for example, are relatively unstable and usually would be unable to endure the rigors of prolonged weathering and transport. So stable quartz is the dominant constituent of most conglomerates. The often excellent rounding of pebbles and cobbles displayed by conglomerates does not necessarily indicate long-distance transport. Field observation and laboratory experiments show that larger angular fragments are rounded quickly after only a few miles of sliding and rolling. Some conglomerates are *lag deposits;* that is, the currents were unable to move larger particles and preferentially swept away finer, more easily moved particles. *Breccia* is simi-

lar to conglomerate, except that clasts are angular rather than rounded (Figure 9–3).

Sandstones consist of sand-sized particles that are consolidated to form a rock. The word *sand* in itself conveys only size and not composition of the particles. Geologists, however, use several terms that transmit both the idea of sand size and a particular composition. Four names can describe almost any sandstone encountered:

Arkose is a sandstone that consists chiefly of pink or gray feldspar (chiefly microcline) and quartz. In this coarse sandstone, particles tend to be angular. Arkose is formed usually as a consequence of the weathering of granite. Some geologists might even define an arkose as a reconstituted granite.

Graywacke is a sandstone that consists of assorted rock fragments and unstable minerals bonded together by a matrix of clay- and silt-sized particles that constitutes more than 15% of the rock. These rocks are gray or dark colored.

Subgraywacke is the most common type of sandstone. It is similar to graywacke but contains more quartz (as much as 30% or more) and has less matrix.

Orthoquartzite is a "clean" (little clay and silt), well-sorted (grains are about the same size) sandstone that contains greater than 95% quartz particles (Figure 9–4).

Figure 9–3. Breccia. (Courtesy of Jan Patterson.)

Figure 9–4. Orthoquartzite. (Courtesy of Jan Patterson.)

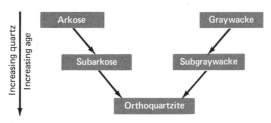

Figure 9–5. The maturity concept. Quartz content of sandstones is often a measure of maturity.

Figure 9–5 illustrates the *maturity concept*. The minerals and rock fragments that make up a sandstone are not all equal in their ability to withstand weathering and chemical attack. Of all the common minerals, quartz is the most stable under surface conditions; all others are of varying but usually less stability. This means that a given sandstone will become more enriched in quartz as it passes through two or more cycles of erosion and deposition in the rock cycle. Thus arkose can become modified to subarkose and graywacke to subgraywacke and eventually to orthoquarzite, if all of the original particles were able to remain close together during repeated entrainment.

Most geologists use the term *shale* to describe a thinly laminated rock made up of fine particles of silt and clay. In addition, a shale also possesses the property of *fissility*, that is, the ability to be split in thin, parallel layers. This property is attributable to the alignment of platy minerals such as mica

Figure 9–6. Shale, showing fissility. (Courtesy of Jan Patterson).

(Figure 9–6). Some fine-grained rocks, however, are not fissile; they tend to be blocky and massive. In such cases the following terms can be applied:

Rock name	Particles
siltstone	more silt than clay
claystone	more clay than silt
mudstone	not specified

Although the general particle size of shales and other fine-grained sedimentary rocks is apparent, the composition is less obvious because of the minute size of the grains. Chemical analysis, expressed as oxides, of numerous fine-grained rocks of various ages reveals that silica (SiO_2) makes up about 60% of the rock and alumina (Al_2O_3) another 15%; the remaining 25% consists of several minor ingredients including iron, sodium, magnesium, and potassium compounds. Common minerals involved are clay minerals, mica, quartz, and feldspar. Some of these constituents form as a result of chemical reorganization after deposition. These new minerals (e.g., feldspar) have not been transported and are authigenic.

The geological significance of shales and other fine grained rocks is still largely a moot question. Composed mainly of clay minerals, they have been traditionally regarded as the stable end-products of chemical decay, whose particles have traveled long distances, finally settling out in quiet, relatively deep water. This concept is probably too simple, and the problem and meaning of shales represent a challenge to further geologic inquiry.

Biochemical rocks and precipitates

Limestone is a layered sedimentary rock composed mainly of calcium carbonate ($CaCO_3$), usually in the form of the mineral calcite. In gross appearance limestones may be massive, blocky, indurated, and colored various shades of gray. Close examination in

Figure 9-7. Limestone. (Courtesy of S. D. Heron, Jr.)

hand specimens frequently show a crystalline aspect and fossil fragments (Figure 9-7). The oceans are the most common source of calcium carbonate. There precipitation takes place on the bottom in warm, shallow waters. This precipitation, as noted earlier, seems to result chiefly from the activity of lime-secreting organisms. Many creatures of the oceans, especially bottom dwellers such as corals, molluscs, bryozoans, and some algae, have the ability to extract calcium carbonate from the water and to construct their shells or skeletons from this material. Their remains often comprise significant portions of limestones. Sometimes entire reefs are preserved intact just as they grew in life. More often the remains are mixed fragments of a variety of organisms. Their fragmental nature indicates that, following death, skeletal parts were broken up and scattered by marine currents.

It should not be construed that limestones are simply organic accumulations and little else. Limestones are complex. Although some limestones are almost chemically pure calcium carbonate ($CaCO_3$), fine silt and clay may drift in from various sources and be-

come mixed with the lime prior to lithification. Some limestones also contain quartz sand. Such materials constitute an *insoluble residue*; calcium carbonate ($CaCO_3$) can always be dissolved by acid to leave an insoluble material that can be studied separately. Geologists informally refer to limestones with appreciable quantities of insolubles (say, 20–30%) as "dirty limestones." It is generally believed that such limestones indicate formation in near-shore waters, whereas purer limestones form further offshore.

Dolomites are closely related to limestones. They have similar appearance and characteristics, differing in their composition by the addition of magnesium to form the mineral dolomite. Some geologists prefer to call them *dolostones.* Even experienced geologists may have difficulty in distinguishing between limestone and dolomite. But a drop of acid on the rock is usually diagnostic. Limestone reacts vigorously with the acid and fizzes; dolomite reacts very slowly, sometimes almost imperceptibly.

The question of the origin of dolomites has been a subject of much discussion and investigation among geologists. The controversy relates to the concept of uniformitarianism. Because geologists cannot find dolomites forming on the sea floor today, they are led to believe that *primary* dolomites resulting from direct precipitation of calcium-magnesium carbonate onto the sea floor cannot occur. This has led to the idea of *secondary* dolomites. Magnesium, an important ingredient of dolomite, is present in seawater. It was visualized that, some time after limy muds were deposited on the sea floor and were buried under later accumulations, the magnesium-bearing seawater slowly circulated through limy sediments and introduced the necessary magnesium by replacement. This process is spoken of as *dolomitization.* Cores taken from the deep-sea bottom during oceanographic investigations seem to bear out this idea, because there is an enrichment in magnesium in the lower parts of these cores. For reasons not clear, most limy sediments do not undergo dolomitization and become limestones instead.

Evaporites are direct chemical precipitates from seawater, but they form only when certain environmental conditions prevail. These conditions are: (1) warmth and aridity, (2) a body of seawater cut off at least temporarily from the open sea, and (3) progressive evaporation and saturation of dissolved substances in the reduced volume of water. Evaporites are essentially *monomineralic* (consisting of one mineral) but may contain minor amounts of other minerals. Widespread occurrences of evaporite deposits in Texas and Michigan indicate that during the time that these deposits formed (Permian Period in Texas; Silurian Period in Michigan) these states were dominated by warm desert conditions (see Chapter 8).

Coal, an essential fuel, is found in nature as layers or seams ranging from a few inches (centimeters) to several feet (meters) thick. It is mined underground or by stripping off

layers of overburden where it occurs close to the surface (Figure 9–8). There is little doubt about the origin of coal, because we can see all gradations in the formation of coal from original plant material. In the past, abundant vegetation in widespread swamps underwent death and burial below water and mud. This inhibited normal decay processes (oxidation), and the vegetation was converted to coal. In this gradual process many varieties of coal formed. Each variety is essentially a stage in carbon enrichment.

The first stage in the transformation to coal is *peat*, a soft, brown, fibrous material used as a fuel in many parts of the world. *Lignite* or brown coal contains recognizable plant remains and is intermediate between peat and *bituminous* coal. Bituminous coal is extensively used and is what most people know as coal. It is more enriched in carbon than is lignite and burns readily. *Anthracite* is the highest grade of coal. It is hard, is high in carbon, and is difficult to ignite. Once it starts to burn, however, it gives off considerable heat with relatively little residual ash. Anthracite is more properly considered a metamorphic rock. Coals are found in many states and in rocks of many periods, reflecting the extensive swamps that once existed. Coal seams found in Pennsylvania and elsewhere are often underlain by a layer of clay known as *underclay*. This layer may represent the original soil in which the plants grew. The United States has enormous untapped coal reserves. As oil supplies dwindle, more coal will be used to replace them.

Figure 9–8. Exposure of coal seam. (M. R. Campbell, U.S. Geological Survey.)

INTERPRETATION OF SEDIMENTARY ROCKS

Primary features

During the process of sediment accumulation, while sediment is still soft and unconsolidated, certain features can be formed within the sediment or on the upper

Table 9–2. Simplified listing of some primary features of sedimentary rocks.

Internal features	Bedding plane features
Stratification	Fossil remains
Cross-bedding	Ripple marks
Graded bedding	Mud cracks
Fossil remains	Raindrop imprints

or lower surfaces of individual layers. These *primary features*, formed at the same time as the sedimentary rock, provide important clues concerning the environment of deposition of the rock. An abbreviated listing of primary features is shown in Table 9–2.

The most conspicuous feature within a sequence of sedimentary rocks is *stratification* or layered appearance. An individual layer is called a *bed* or *stratum* (plural, *strata*). The layered effect is produced in a number of ways as sediment accumulates. It can result from an interruption in the supply of material being brought in (nondeposition), from a change in the type of material deposited, or from other ways.

Cross-stratification or *cross-bedding* is a feature most commonly found in sandstones, although some conglomerates and limestones may exhibit it, too. It consists of thin-bedded strata lying at some angle (often between 18 and 24°) to the bedding of the larger rock unit in which it is contained (Figure 9–9). Cross-beds form by sediment deposited on a forward slope of a small delta, a dune, or other slope on which sediment might come to rest. Cross-beds have been produced in laboratory flume experiments that attempt to reproduce natural processes.

We know from flume experiments that cross-beds incline in the same direction as

Figure 9–9. Cross-bedding in sandstone. The lower beds (inclined to the left) were deposited by currents flowing from right to left. (Courtesy of S. D. Heron, Jr.)

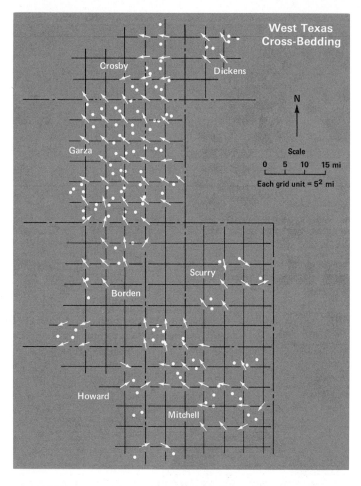

West Texas Cross-Bedding

N

Scale
0   5   10   15 mi

Each grid unit = 5² mi

Crosby   Dickens

Garza

Scurry

Borden

Howard

Mitchell

Figure 9–10. Arrows indicate the dip direction of cross-beds in the Dockum Sandstone in Garza County, Texas. The sediment supply was to the southeast and east. (After C. Cazeau, 1960, *Journal of Sedimentary Petrology.*)

that of the depositing current, and herein lies their significance for the geologist. Suppose a geologist is studying a cross-bedded sandstone originally formed by stream activity. If a compass bearing is taken of the dip direction of the cross-beds at many outcrops scattered over the region of exposure, a pattern of directions may emerge; such a pattern is shown in Figure 9–10. We can infer in this case that the source of sediment supply was generally to the southeast and that there must have been higher ground in that direction, because the source of sediment would be opposite to the direction of inclination of the cross-beds. Studies such as these help to depict the *paleogeography* of a region. Paleogeography expresses the probable position of lands and seas at certain times in the earth's geologic past. Cross-bedding can also develop by wind action and in the marine environment by longshore currents. In the latter case, though, interpretation of source direction would be different from in the example shown in Figure 9–10. In marine sediments longshore currents would produce cross-stratification inclined parallel to the shoreline; this would be 90° (at a right angle) to the landmass supplying the sediment to the coast (Figure 9–11).

Graded bedding is usually a result of continuous deposition from a current of declining velocity. The coarsest sediment will lie toward the base of such a deposit with in-

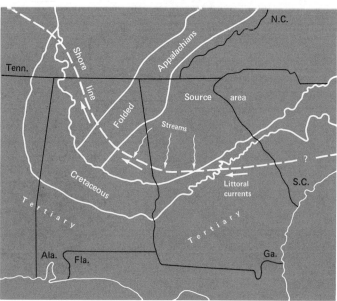

N.C.

Tenn.

Appalachians

Shore line

Folded

Source   area

Streams

Cretaceous

Littoral currents

S.C.

Tertiary

Tertiary

Ala.   Fla.

Ga.

Figure 9–11. Shoreline reconstruction in the Mississippi-Alabama area for the Cretaceous Period. Cross-beds represented by the arrows parallel to the shore were formed by longshore currents. The dip direction lies 90° from the assumed northerly source. (After W. F. Tanner, 1955, *American Association of Petroleum Geologists Bulletin.*)

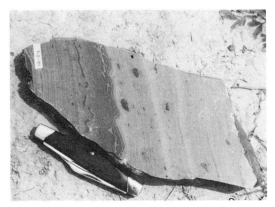

Figure 9–12. Graded bedding in a turbidite. (Courtesy of J. E. Wright, Jr.)

creasingly finer particles toward the top of the sedimentary unit. Many such deposits are interpreted as *turbidites,* because graded bedding often occurs in sediments deposited on the sea floor from turbidity currents. These sediment-laden currents are capable of rapid downslope movement when triggered by earthquakes, severe storms, or other disturbances (Figure 9–12).

Bedding plane features provide some interesting sidelights in interpreting ancient environments, although they are restricted in their usefulness simply because they are not sufficiently common. *Ripple marks* generally signify shallow water, but they have been photographed at depths well beyond diving range. Particular types of ripple mark can tell us whether wave action or unidirectional currents have been at work. Symmetrical ripple marks are produced by the to-and-fro movement of waves such as along a shoreline. Current ripples, however, will be asymmetrical (Figure 9–13). *Mud cracks* and *raindrop impressions* suggest not only shallow water but also periodic exposure to air, as in a tidal flat (Figure 9–14).

Fossils, the most important primary feature found in sedimentary rocks, occur either internally or on bedding plane surfaces. Fossils represent the remains, direct or indirect, of former life now entombed in rock. We say "indirect" because, although a certain fossil such as an animal track or burrow is not the animal itself, it gives evidence that the animal was there. Like most bedding plane features, animal tracks are not particularly abundant. An interesting exception is the abundance of dinosaur tracks in rocks of Triassic age in Massachusetts, Connecticut, and other eastern states (Figure 9–15). These tracks reveal that the dinosaurs in this locality were bipedal and walked like ostriches

Figure 9–13. Ripple marks in sandstone.

Figure 9–14. Rain drop imprints in modern sediment. Such impressions are found also in rocks. (Courtesy of S. D. Heron, Jr.)

Figure 9–15. Dinosaur tracks. (Courtesy of Field Museum of Natural History.)

rather than hopped like kangaroos. In one place it can be seen where a dinosaur sat down, leaving the impression of his "behind" and a long tail.

Fossils occur most abundantly in marine beds such as limestones, sandstones, and shales. The process of fossilization usually requires hard parts, such as a skeleton, and rapid burial. These conditions are best met on the ocean floor. Many marine organisms are *benthonic*, that is, bottom dwellers. They live attached to the sea bottom or burrowed into it. So in a sense they already have one foot in the grave. Examples of such bottom dwellers are corals, clams, and snails. When buried, a creature with hard parts does not remain unchanged.

Replacement is the most common type of preservation. In this process the original shell or bone is dissolved, often by groundwaters, and simultaneously replaced by other substances, especially calcium carbonate ($CaCO_3$). In this volume-for-volume substitution all of the original external details are faithfully preserved. Replacement materials may also be silica, pyrite, or carbonized remains. Other modes of ultimate preservation include *intact preservation*, as refrigerated or mummified remains and buried bones in tar sands; *casts and molds*, in which, for example, worm borings or animal burrows, are filled in with solid material; and *impressions*, such as footprints and tracks.

The significance of fossil assemblages is that: (1) we can use them to correlate widely separated rock sequences, (2) they provide us with information about past environmental conditions, and (3) they show in gross aspect the evolution of life on earth through geologic time.

Facies

The term *facies* refers to the sum total of characteristics that make up a sedimentary rock. As noted in Chapter 8, the influence of environmental factors determines the nature of the sediment. It is not surprising, then, to find within a sedimentary unit, laid down during a specific interval of time in the past, lateral variation in the character of the rock. These *facies changes* are not accidental. As environmental factors varied from place to place, so did the characteristics of the sediment.

Two basic types of facies include *lithofacies* and *biofacies*. Within each type the rock character and fossil assemblage, respectively, remain consistent. When traced laterally in the field, a sandstone may grade into a shale or perhaps into a sandy limestone. These are separate lithofacies, reflecting variation in environmental circumstances during the time of sediment formation. Alteration in the type of fossils may or may not be accompanied by a change in lithofacies, but often it is. Variation in fossil assemblage (biofacies) may reflect, among several possibilities, transition from shallow to deep water or from brackish to normal saline waters. Of course, when fossils are not affected by variation in lithofacies, they establish time correlations.

Many geologists extend usage of the term *facies* to the name of the environment it is thought to represent. Thus geologists speak of deep-water facies, shallow marine facies, lagoonal facies, and others. Recognition of facies and their environmental significance leads to more accurate and meaningful interpretation of the paleogeography represented by a rock unit.

SUMMARY

Sedimentary rocks form at the earth's surface when unconsolidated materials derived from preexisting rocks are transported into a basin or precipitated within a basin of deposition, there to be buried progressively and compressed or crystallized into a consolidated rock. The processes that convert a sediment into rock are known collectively as lithification, involving volume reduction by desiccation and compaction as well as cementation.

Sedimentary rocks are not so abundant in the earth's crust as are igneous and metamorphic rocks, but they form a thin veneer over the crust and are exposed extensively at the surface. The most common sedimentary rocks are of marine origin. Shales, sandstones, and limestones are dominant rock types.

Sedimentary rocks are classified on the basis of texture and composition into two broad groups: (1) land-derived clastics and (2) biochemical rocks and precipitates. The first category is subdivided on the basis of size of particles into conglomerates, sandstones, shales, and other fine-grained rocks. The second group includes limestones, dolomites, evaporites, and coal as common representatives.

Sedimentary rocks represent ancient environments that once existed on the earth and therefore are useful in reconstructing the earth's physical history. Primary features such as ripple marks, cross-bedding, and animal tracks and burrows are formed at the same time as the sediment and provide insight into the nature of the environment. Fossils are another primary feature contained within sedimentary rocks that contribute data on the evolution of life on the earth.

Questions for thought and review

1. How do lithification and diagenesis differ?
2. What kind of cements can bind a sedimentary rock together? Which cement is the most durable?
3. What is the basis for classification of sedimentary rocks?
4. Why are sedimentary rocks more difficult to classify than other rocks? Can you think of alternative ways to classify sedimentary rocks other than the classification we have presented here?
5. What is meant by the maturity concept?
6. Is it possible for limestones to form without organic assistance? If so, under what conditions?
7. In an evaporating sea, which would be precipitated first, halite or gypsum?
8. How would you describe the process of dolomitization?
9. Of what elements is coal formed? Which elements are removed or reduced in the conversion from peat to bituminous coal?
10. Symmetrical ripple marks can form along shorelines because of to-and-fro motion of waves. Do you think that asymmetrical ripple marks could form in this environment? If so, why?
11. List as many characteristics as you can think of that would define the lithofacies of an arkosic sandstone.

Selected readings

Bascom, Willard, 1964, *Waves and Beaches*, Doubleday, Garden City, N.Y.

Flint, R. F., 1973, *The Earth and Its History*, Norton, New York.

LaPorte, L. F., 1968, *Ancient Environments*, Prentice-Hall, Englewood Cliffs, N.J.

Matthews, W. H., 1971, *Invitation to Geology*, Natural History Press, Garden City, N.Y.

10 Metamorphism and metamorphic rocks

STUDENT OBJECTIVES

After reading this chapter the student should:

1. know what metamorphism is

2. be able to distinguish important kinds of metamorphism

3. be able to recognize major rock types and how they are classified

4. know something about grades or intensities of metamorphism

5. have some familiarity with the question of granitization

To behold is not necessarily to observe.

Humboldt

WHAT IS METAMORPHISM?

Most of us have seen snow. Consider an area of the earth where a snowstorm is taking place. In this area the snow may remain on the ground for some time. As the blanket of snow thickens, the beautiful hexagonal flakes become pressed together under the pressure of their own weight. The snow crystals may then recrystallize into granular-looking ice as the pressure builds up. Eventually the snow at the bottom of the pile becomes welded into a single interlocking mass of solid ice.

In this example the snowflakes have undergone a *change of form* to ice. The word *metamorphism* means "change of form" and we can speak of the snow as having been *metamorphosed* into ice by means of pressure. Preexisting rocks may also become metamorphosed into something else. The "something else" comprises one of the three great groups of rocks; these are known as *metamorphic rocks.*

Most geologists would admit that ice formation is a very special kind of metamorphic

Figure 10–1. Severely contorted gneiss produced by regional metamorphism in the Shuswaps complex, British Columbia.

process, because it takes place on the surface of the earth. Metamorphic rocks, on the other hand, form mainly beneath the earth at elevated temperatures and pressures (Figure 10–1). In other words, metamorphism is primarily an *internal process.*

Like the change from snow to ice, metamorphism is a *solid-state* process. This does not mean the rocks must be rigid. On the contrary, under the very great pressures and high temperatures to which these rocks are subjected, they are often in a plastic state and can be deformed (Figure 10–1). During recrystallization characteristic new textures and structures emerge. The appearance of the newly formed metamorphic rock may be markedly different from that of the parent rock (primarily igneous or sedimentary). While metamorphism is taking place, there is ordinarily no change in the overall bulk chemical composition of the rock mass being recrystallized. However, if water or gases enter or leave the rock mass, they may remove or deposit matter. If the composition of minerals making up the rock is thus changed, we call this *metasomatism.* This replacement process often accompanies metamorphism.

Recrystallization during metamorphism is another example of nature's effort to establish equilibrium (see Chapter 4). The imposition on rocks of pressures and temperatures different from those under which they were originally formed creates disequilibrium. The response by recrystallization is an attempt to reestablish equilibrium.

The study of metamorphic rocks presents considerable challenge to the geologist for a number of reasons. First, metamorphic processes take place within the earth and are not directly observable. In comparison we can directly observe stages in the development of igneous or sedimentary rocks as in the case of volcanic eruptions or sediment transport by streams. Second, metamorphic rocks tend to be very old by virtue of the fact that, the older is the rock, the greater are the chances

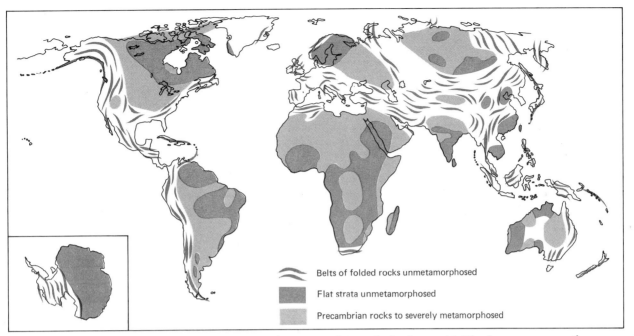

Figure 10–2. Surface distribution of metamorphic rocks in the world. The larger areas are the shields while the smaller elongate striped areas are the cores of the major younger (post-Precambrian) mountain systems. (After Umbgrove, 1947).

for it to be metamorphosed. This is a complicating factor, because one or more event is superimposed on another. Some large areas consist of rocks that have been subjected to repeated metamorphic events. Thus antiquity tends to mask the original nature of these rocks. Third, geologists have been disposed until recently to neglect study of metamorphics. Not until the turn of the century or later were attempts made to understand their complex nature. In some areas metamorphic rocks are still shown on geologic maps as "metamorphic complex" or "metamorphic rocks, undivided." Today we are learning a great deal about the earth's early history by studying these rocks in the great continental shield areas and in the cores of mountain belts (Figure 10–2).

Metamorphic rocks serve as an important source of building materials. Marble, a metamorphosed limestone, is widely used in building façades, tombstones, and public rest rooms; it is a favored material of sculptors. Marble deposits in Greece enabled Greek sculpture to reach such perfection. Slate is also important as building stone, roofing material, ornamental stone, and material for many blackboards. Metamorphic rocks also make up some of the largest and most beautiful mountains in the world (see Chapter 22).

KINDS OF METAMORPHISM

As we have seen, metamorphism is a process whereby rocks are placed into a pressure–temperature environment that is not the same as that in which they were formed. The most common of these environments are regional, contact, and cataclastic metamorphism (Figure 10–3).

Regional metamorphism

Recrystallization of rocks at considerable depths under temperature conditions ranging from 300°C to 600–700°C and pressures

129

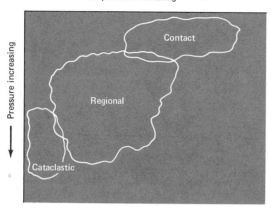

Temperature increasing ⟶

Pressure increasing

Contact

Regional

Cataclastic

Figure 10–3. Diagram showing the relative pressure-temperature realms of the different kinds of metamorphism.

of thousands of atmospheres is termed *regional metamorphism*. In every major mountain system in the world, including the Alps, Himalayas, Andes, Appalachians, and North American Cordillera, are areas where one can pass gradually from rocks that have not been affected by metamorphism into areas where the rocks have been heated to

temperatures near or even above their melting points. We see them exposed today, because they have been carried to the surface and exposed by millions of years of erosion. Much of the pressure applied to these rocks is directed and is in the form of shearing stresses (such as that applied to a deck of cards when you spread it across a table). As a result, regional metamorphic rocks commonly have a layered aspect produced by recrystallization and parallel reorientation of minerals in positions where they experience the least pressure. This layered aspect of regional metamorphic rocks is called *foliation* and is characteristic of most rocks of this class (Figure 10–4).

Contact metamorphism

When a hot igneous body intrudes cooler rocks, the intruded rocks are likely to be recrystallized as the cooling magma loses its heat to its surroundings. This is called *contact metamorphism*. The zone of contact metamorphism surrounding an igneous body is called the *contact aureole*, actually

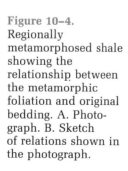

Figure 10–4. Regionally metamorphosed shale showing the relationship between the metamorphic foliation and original bedding. A. Photograph. B. Sketch of relations shown in the photograph.

A

Figure 10–5. Contact metamorphism around an igneous pluton.

meaning halo (Figure 10–5). The aureole may range from a few inches (centimeters) to several miles (kilometers) wide. Contact metamorphism may also occur beneath lava flows on the surface.

Different rocks exhibit different susceptibility to contact metamorphism. As in regional metamorphism, shales show recrystallization effects best, but impure limestones likewise exhibit metamorphic effects better than many other rock types (Figure 10–6). Generally contact metamorphic rocks possess a less obvious foliation or no foliation at all, because little directional pressure is applied by this process.

Cataclastic metamorphism

Movement along faults (see Chapter 19) at depths of several miles (kilometers) in the earth results in crushing and grinding of rocks along the fault rather than breaking them cleanly. Some cataclastic zones exposed by erosion are 1 mi (1.67 km) or more in width. Rocks affected in this way are subjected to intense pressure and heat derived from friction. As a result the rocks are ground up. This is *cataclastic metamorphism*.

ORIGIN OF METAMORPHIC ROCKS

Where do metamorphic rocks form? In only a very few places can we actually see metamorphic rocks being formed on or near the surface. One such place is under a cooling lava flow as the heat of the flow bakes the material beneath. But studies of metamor-

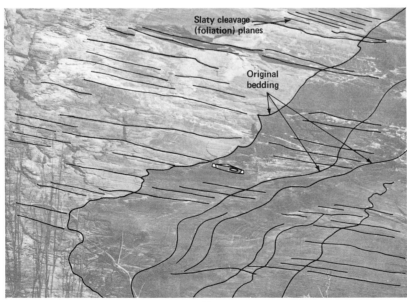

B

A

B

Figure 10–6. A. Small-scale contact metamorphism of marble along a granite contact. The contact zone is composed of calcite, pyroxene, amphiboles, and other silicate minerals. B. Sketch showing the relations shown in the photograph. (Specimen from Ware Shoals, S.C., D. S. Snipes, Clemson University).

phic minerals in the laboratory tell us that regional metamorphic processes occur under pressures of thousands of atmospheres and temperatures above 200–300°C. Certainly there is no place on the earth's surface where these conditions prevail.

Because both temperature and pressure increase with depth in the earth, the logical place to look for the conditions of metamorphism is into the earth's crust. However, mines about 10,000 ft (3,000 m) deep in South Africa and drill holes about 25,000–30,000 ft (7,500–9,000 m) deep in Texas and Oklahoma have not reached temperatures and pressures of metamorphism. If we consider the rate of increase of temperature with depth, the *geothermal gradient*, this rate, about 50°F/mi (20°C/km) requires depths greater than either mines or drill holes have reached: greater than 12 mi (20 km). It is unlikely that we will drill to these depths soon.

We see metamorphic rocks in the cores of mountain ranges like the Alps, Appalachians, and Himalayas. These rocks formed at great depths and have been exposed by erosion as each mountain system was uplifted. We know that these are not older metamorphic rocks that have been caught up in the process of mountain building, because the metamorphic ages, determined by radiometric dating (Chapter 4), are the same as the age of formation of the mountain system. So there must be a relationship between the formation of metamorphic rocks and mountain systems.

Some of the ideas derived from plate tectonics theory (see Chapter 21) provide a means by which we can explain metamorphism. Frictional heat and pressure produced by two converging masses of the outer part of the earth cause the rocks near the zone where they meet to recrystallize as regional metamorphic rocks (Figure 10–7). Others in higher-temperature realms may melt to form igneous bodies (see Figure 6–18). These may find their way into the upper crust, where they cool and produce contact metamorphic aureoles. Pressure in some shallower regions causes material to break and move about. These breaks are large faults; crushing along these faults at depth results in cataclastic metamorphism. However, this is not the only way in which faults form (see Chapter 20), nor for that matter igneous bodies (Chapter 6).

MINERALS IN METAMORPHIC ROCKS

We have made the point that, when a preexisting rock becomes metamorphosed, its appearance is often drastically altered. This is attributable not only to changes in texture and structure but also to changes in the minerals that make up the rock mass. Not all minerals respond alike in a new metamorphic environment. Some can "take it" and remain pretty much as they were. Others cannot, and change by rearranging their internal atomic structure or by reacting with other substances to form entirely new minerals. Table 10–1 lists minerals that often survive metamorphism and new minerals that might form.

The principal original mineral in most sandstones is quartz. It survives by recrystallizing and remains as quartz. The major constituents of most carbonate rocks, calcite and dolomite, likewise survive, unless the rock contains impurities (quartz, clays). If impurities are present, reaction of some calcite or dolomite may take place to form new minerals. Three examples of metamorphic reactions are given below:

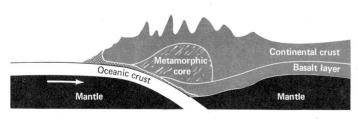

Figure 10–7. Metamorphic processes may be derived from heat and pressure arising from two colliding plates.

$$CaMg(CO_3)_2 + 2SiO_2 = CaMgSi_2O_6 + 2CO_2$$

dolomite   quartz   diopside
(a pyroxene)

$$CaMg(CO_3)_2 + 4SiO_2 + H_2O =$$

dolomite   quartz   water

$$Mg_3Si_4O_{10}(OH)_2 + 3CaCO_3 + 3CO_2$$

talc   calcite

$$CaCO_3 + SiO_2 = CaSiO_3 + CO_2$$

calcite   quartz   wollastonite
(a pyroxene-
like mineral)

Several of the minerals in shales, because of their composition and fine particle size, react readily to form new minerals. Others, such as quartz, simply recrystallize or participate in reactions as additionally needed constituents. Various clay minerals in shales recrystallize to muscovite and/or chlorite at low temperatures and moderate pressures. Muscovite and chlorite may react with each other as temperature increases to produce the new minerals biotite and/or garnet. As temperature and pressure increase further, minerals such as kyanite and sillimanite may form from muscovite and quartz.

Minerals of igneous rocks have crystallized from high temperature melts and thus behave differently under metamorphic conditions. Like those in sedimentary rocks, some minerals, such as quartz, biotite, and

Table 10–1. Minerals in metamorphic rocks

Persistent minerals[1]	New minerals[2]
Quartz	Quartz
Calcite	Plagioclase
Dolomite	Orthoclase
Plagioclase	Chlorite
Orthoclase	Serpentine
	Talc
	Muscovite
	Biotite
	Garnet
	Amphiboles
	Pyroxenes
	Kyanite
	Sillimanite
	Staurolite

[1] Minerals that commonly survive from other environments.
[2] Minerals that commonly form in the metamorphic environment.

plagioclase feldspar, may persist unchanged into a metamorphic environment. However, many minerals, such as olivine and certain pyroxenes, break down and form talc, serpentine, amphiboles, and other hydrous silicates.

CLASSIFICATION OF METAMORPHIC ROCKS

Geologists have found that classifying metamorphic rocks is a simpler task than

Table 10–2. Classification of metamorphic rocks

	Rock type	Precursor	Metamorphic process
Foliated	Slate	Shale	Regional, contact
	Phyllite	Volcanic rocks	Regional
	Schist		Regional
	Gneiss	Impure sandstone	Regional
		Plutonic igneous rocks	
	Amphibolite	Mafic igneous rocks	Regional
		Impure carbonate rocks	
Nonfoliated	Mylonite	Siliceous rocks	Cataclastic
	Phyllonite	Mica- and feldspar-rich rocks	Cataclastic
	Quartzite[1]	Sandstone	Regional, contact
	Marble[1]	Limestone, dolostone	Regional, contact
	Hornfels	Shale	Contact

[1] May be foliated.

classifying either igneous or sedimentary types. They can be readily divided into two broad categories: (1) those that display foliation and (2) those that do not. Table 10–2 shows this twofold division with major rock types under each heading.

Once the geologist has decided whether a rock is foliated or nonfoliated, he further distinguishes the rock by noting changes in grain size or other textural differences. Important minerals are also taken into account. In naming metamorphic rocks, it is common practice to use the name of an abundant or conspicuous mineral as a modifier, as, for example, "mica schist," "garnet mica schist," or "hornblende gneiss."

Foliated rocks

Slate–phyllite–schist sequence. This sequence results from the progressive regional metamorphism of shale and other fine-grained rocks (mudstone, some volcanic rocks). They all possess a marked foliation. They are typically grouped together, because slate will be transformed to phyllite and phyllite in turn converted to schist under continuing regional metamorphism. Consider the changes that might take place in, say, a shale–mudstone sequence as it becomes subjected to increasing metamorphism. This will also serve to demonstrate the salient features of slate, phyllite, and schist.

Slate. The initial change that takes place in the formation of a slate is one of reorientation of its mineral grains with only minor recrystallization. The platy clay–mica grains, which give a shale its thinly laminated character, rotate until the applied pressure is somewhat relieved (Figure 10–8). This does not change the appearance of the rock body very much. But when the rock is struck, it no longer breaks parallel to its original sedimentary bedding. The mass of rock now breaks parallel to the new orientation of

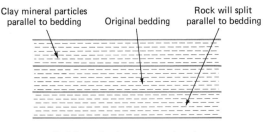

Undeformed and unmetamorphosed

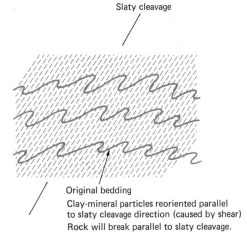

Deformed and slightly metamorphosed

Figure 10–8. Relationship between grain orientation and the development of foliation (slaty cleavage) in slate.

the clay–mica grains. This breakage direction is called *slaty cleavage*. Although superficially similar to shale, slate is harder, has a noticeable sheen on slaty cleavage surfaces, and will ring when struck sharply (Figure 10–9).

Phyllite. Phyllite is a foliated metamorphic rock in which the micas have recrystallized to a greater extent than those in a slate (Figure 10–10). It forms by recrystallization of slate. Foliation surfaces in phyllite have a more shiny appearance than those of a slate. The clay–mica grains of slates have recrystallized to muscovite, biotite, or chlorite but remain fine grained, only marginally visible to the unaided eye. Small

A

B

Figure 10–9. A. Hand specimen of a slate.
B. Thin section of slate. Compare these
photographs with those of Figures 10–10 and
10–11.

A

B

Figure 10–10. A. Hand specimen of a phyllite.
B. Thin section of phyllite. Note that the grains
in this rock are coarser than those in the slate in
Figure 10–9 yet finer than those in the schist in
Figure 10–11.

porphyroblasts (larger crystals surrounded
by smaller grains in a metamorphic rock) of
garnet, biotite, and other minerals may occur
in phyllite.

Schist. Schist is a foliated metamorphic

rock in which micas and most other minerals
are completely recrystallized (Figure 10–11).
It is produced by regional metamorphism of
phyllite, and all major constituents of the
rock should be visible to the unaided eye.

A

Figure 10–12. Photograph of a typical banded gneiss, North Carolina. Dark bands rich in biotite are separated by quartz and feldspar-rich layers.

B

Figure 10–11. A. A schist containing porphyroblasts of garnet (round, dark grains) and kyanite (blades). B. Thin section of muscovite-garnet-kyanite schist. Note how coarse this rock is.

Porphyroblasts of garnet, kyanite, staurolite and other minerals are common in schists. Instead of clays the micas muscovite and biotite, and talc or chlorite may be the main constituent of this rock type.

Gneiss. Gneiss is a foliated metamorphic rock in which different minerals may be segregated into light- and dark-colored bands (Figure 10–12). Gneisses generally contain more feldspar and quartz and fewer micas than schists; consequently they do not contain as obvious a foliation. Gneisses result from high-grade regional metamorphism of graywacke, arkose, and other impure coarse-grained sedimentary rocks as well as plutonic igneous rocks such as granite, diorite, syenite, and others, and occasionally even volcanic rocks. A gneiss containing lenticular or eye-shaped porphyroblasts is called an *augen gneiss* (Figure 10–13). (Augen is German for "eye.")

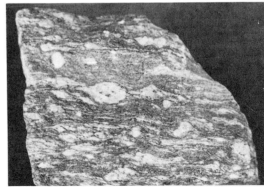

Figure 10–13. An augen gneiss. The augen (eyes) are composed of feldspar.

Amphibolite. Regional metamorphism of mafic igneous rocks (basalt, gabbro, diabase) may produce the rock *amphibolite* (Figure 10–14). Amphibolites may or may not possess marked foliation. They are composed principally of the amphibole hornblende, but other minerals such as biotite,

plagioclase, calcite, and quartz may be important constituents.

Mylonite. A faintly to strongly banded fine-grained rock formed by cataclastic metamorphism of other rocks is called *mylonite* (Figure 10–15). Mylonite very closely resem-

A

Figure 10–14. Regional metamorphism of mafic igneous rocks produces the rock amphibolite. A. Hand specimen of amphibolite. B. Thin section of amphibolite under the petrographic microscope.

B   0.1 mm

bles the sedimentary rock chert because of its usually high quartz content. It may contain large crystals or fragments of the coarser material from which it was derived.

Phyllonite. Phyllonite is a strongly foliated mica-rich rock formed by cataclastic metamorphism of coarser feldspathic or micaceous rocks (Figure 10–16). Micas are created from feldspars in feldspathic rocks by crushing and reaction with water (not a purely cataclastic process). Phyllonites closely resemble phyllites but are not so common.

Figure 10–15. A. Hand specimen of mylonite. B. Thin section of mylonite. The fine grains in this rock are due to crushing of a coarser rock.

Figure 10–16. A. Hand specimen of a phyllonite. B. Photomicrograph of a phyllonite.

Nonfoliated rocks

Quartzite. The product of regional or contact metamorphism of sandstones is quartzite (Figure 10–17). Sand grains are recrystallized and become "welded" together to produce a very hard rock. Not infrequently the original sandstone contains some clay or detrital mica that produces some foliation in the rock during regional metamorphism.

A

B

0.1 mm

Figure 10–17. A. A quartzite. Note how even grained this rock appears. B. Photomicrograph of the same rock. The quartz grains are welded together. A few grains of muscovite impart a faint foliation to the rock.

Hornfels. The baking of shale or other fine-grained clastic sedimentary rocks during contact metamorphism produces *hornfels* (Figure 10–19). Hornfels is commonly a gray fine-grained, almost aphanitic, nonfoliated rock type that frequently contains porphyroblasts of biotite or other minerals. These rocks are commonly composed of micas, quartz, and feldspar.

A

A

B

Figure 10–18. A. A coarse marble. B. Thin section of marble.

Marble. Marbles are produced by regional or contact metamorphism of limestone and dolostone (Figure 10–18). They may be fine to coarse grained; some regionally metamorphosed marbles may be foliated. Although dark-colored marbles exist, most are of light color. Crystals of garnet, micas, amphiboles, pyroxenes, talc, and/or olivine appear during metamorphism of impure limestone or dolostone. Some of these crystals grow to a size of several inches (centimeters).

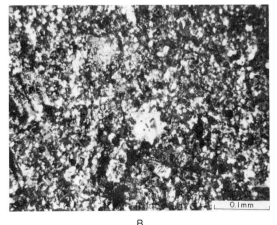

B

Figure 10–19. A. A hornfels. B. Thin section of hornfels.

METAMORPHIC ISOGRADS AND FACIES

George Barrow, a geologist in the latter part of the nineteenth century, focused his professional attention on the Scottish Highlands, a hilly region of somber, windswept beauty in the British Isles. Long ago this region was a vast sea where mud and silt accumulated, compacted, and solidified into shale. In time, but still shrouded in antiquity, the sea vanished, and the area and its shales were subjected to metamorphism.

Barrow, mindful of the fact that metamorphism could vary in its degree of intensity, clambered amidst the glens and crags of the Highlands, seeking data. The nagging question in his mind was, Granted, we probably have zones of low, moderate, and high grades of metamorphism, but how do we recognize them? Where are the clues? With diligent effort, he found the clues.

Knowing full well that metamorphism is a recrystallization process that produces new minerals, Barrow reasoned that, if the composition of Scottish Highlands rocks had remained the same (which they had), then new, diagnostic metamorphic minerals ought to appear predictably as temperature and pressure increased. His field work eventually bore this out. In other words, each degree of metamorphism was announced by the appearance of a new mineral.

Following in Barrow's footsteps, another geologist, C. E. Tilley, confirmed and amplified on his work. The new minerals associated with each degree of metamorphic intensity are known today as *index minerals*. Here are the conclusions of Barrow and Tilley:

The first mineral to appear as temperature and pressure increase is (1) *chlorite*. Then (2) *biotite* appears and is followed by (3) *garnet*. Then minerals (4) *staurolite* (an iron-aluminum silicate), (5) *kyanite* (Al_2SiO_5), and (6) *sillimanite* (a higher-temperature polymorph of Al_2SiO_5) follow in succession. Sillimanite generally represents the highest grade of regional metamorphism reached in most metamorphic zones. This series of minerals today bears Barrow's name as the standard high-pressure–high-temperature metamorphic series—called the *Barrovian series*. As a result of this work we can today draw lines (*isograds*) on a map through zones of equal metamorphic intensity and quickly grasp the metamorphic history of a region (Figure 10–20).

Interestingly, if there had been no shales in the areas studied, Barrow would not have seen these well-developed index minerals, for they appear only in silicate rocks that are rich in aluminum and iron and contain some magnesium. The shales fit this best. Sandstone, limestone, and other sedimentary and igneous rock types yield other minerals, but usually not the broad spectrum and variety produced by metamorphism of shales.

How do we determine metamorphic grade when we have no metamorphosed shales and none of Barrow's index minerals? The Finnish geologist Pennti Eskola came to the rescue and proposed that metamorphic rocks represent equilibrium systems. He maintained that minerals that make up a rock body were formed under a particular set of pressure–temperature conditions. Eskola called such a mineral assemblage a *metamorphic facies* and defined several facies that exist regardless of the composition of the rocks. The only requirement is that of equilibrium under a set of pressure–temperature conditions.

Isograds and metamorphic facies may also be recognized in contact metamorphic aureoles. Because temperature decreases away from the cooling magma body, the contact aureole may vary in thickness from a few inches (centimeters) to a few miles (kilometers). Contact metamorphism is typically a high-temperature–low-pressure process. As a result many minerals form that are different from those that develop during regional metamorphism. Therefore a different set of isograd minerals, as well as dif-

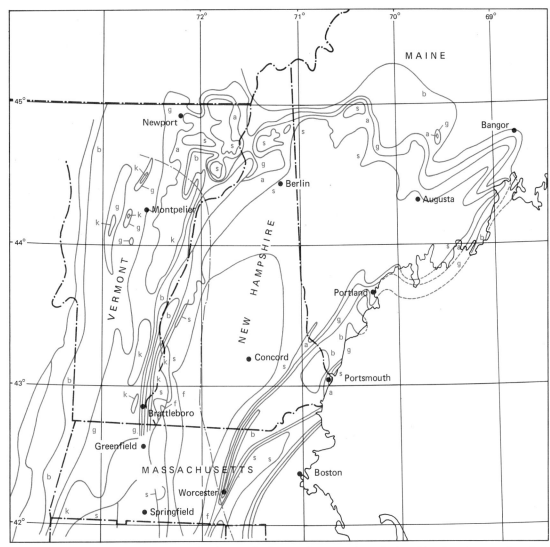

Figure 10–20. Part of the New England Appalachians showing the relationships between regional metamorphic isograds. b-biotite, g-garnet, a-andalusite, s-sillimanite. (From B. A. Morgan, 1972, "Metamorphic Map of the Appalachians," U.S. Geological Survey.)

ferent metamorphic facies, may be observed in the contact aureole (Figure 10–21).

TIME AND METAMORPHISM

How much time is necessary for limestone to be changed into marble or shale into schist? In the laboratory many minerals can be made to recrystallize over a period of a few weeks or less. But in nature heat and pressure may be applied for periods of millions of years. The result is complete recrystallization of large volumes of rock material and the attainment of a state of equilibrium in most cases.

Are all rocks metamorphosed at the same rate? The answer is, No. The composition of a rock partially determines the way in which

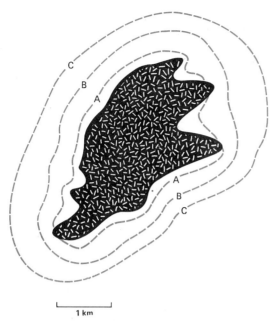

Figure 10–21. An igneous pluton and the intruded country rocks showing the contact metamorphic isograds developed around the pluton.

a rock will be affected by metamorphism. The type of metamorphic process affecting the rocks also affects the rate at which metamorphism will take place. A rock composed of calcite and quartz may not react at all (except for recrystallization of calcite) under low-grade regional metamorphic conditions.

But in high-grade regional or contact metamorphism several new minerals may form.

Grain size may also have a very important effect on the rate of metamorphism and the degree to which it affects a mass of rock. The smaller the individual grains in the rock, the more rapid the rate of recrystallization and/or the more complete the effects of metamorphism. Why? It is related to the increase in the total surface area of grains as the size of individual particles is reduced. Consider a simple calculation. What is the surface area of a cube that is 1 cm on each edge (Figure 10–22)?

There are six sides on a cube, so:

1 cm $\times$ 1 cm $\times$ 6 sides = 6 cm^2 of surface area

Cut the cube so that cubes ½ cm on each edge result. Now what is the total surface area of all the cubes? We now have eight cubes of ½-cm size, so:

½ cm $\times$ ½ cm $\times$ 6 sides/cube $\times$ 8 cubes = 12 cm^2

Thus by cutting a cube into cubes of half of the original size, the total surface area is doubled. If we cut the ½-cm cubes into ¼-cm cubes, the surface area is doubled again, but the volume is still the same as the original cube with which we began. 1 cm^2.

Rocks with the greatest surface area generally react faster, because more material can

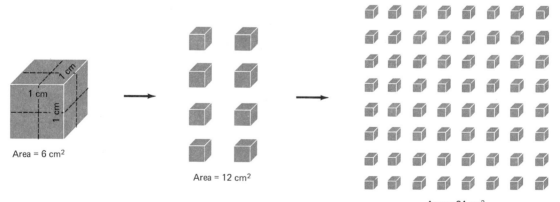

Figure 10–22. A 1-cm cube when cut into eight 1/2-cm cubes doubles its surface area. It is doubled again when the 1/2-cm cubes are cut into 1/4-cm cubes.

be in contact for reaction to take place. A single crystal of calcite measuring 1 cm on an edge would react much more slowly than the same volume of powdered calcite when mixed with a reactant.

GRANITE: AN IGNEOUS OR A METAMORPHIC ROCK?

Granite was introduced in Chapter 6 as an intrusive igneous rock composed mostly of orthoclase feldspar and quartz. This is no doubt true of many granites, but many granites have been produced on a large scale by high-grade regional metamorphism of sedimentary and volcanic rocks. They are produced on a small scale by metasomatism along some igneous contacts. Metamorphic granites represent the near completion of the rock cycle (see Chapter 4).

Many square miles (square kilometers) in shield areas and cores of mountain belts are composed of granitic gneiss. Also, large granitic batholiths are observable in many places in the world (see Figure 6–24). If these were intruded by pushing aside country rocks, what happened to all those rocks that were displaced? A controversy arose in the

early part of this century over the origin of granite, similar to the one that existed in the late eighteenth century, which established the igneous origin of basalt and at least of some granites.

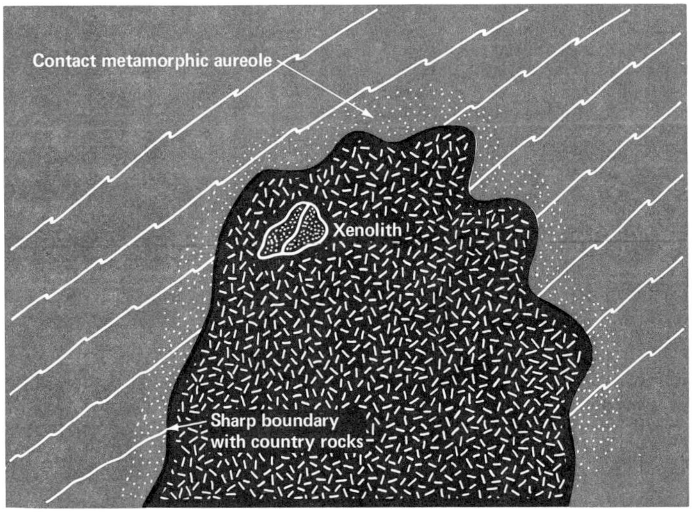

How can we distinguish between igneous and metamorphic granites? When James Hutton proved the intrusive origin of some granites in Scotland, he used many of the same means that we use today to determine that granite is intrusive (Figure 10–23). He noted the following:

1. sharp, discordant contacts with country rocks
2. baking (contact metamorphism) of country rocks
3. xenoliths of country rock in the intrusive body

Granites formed by high-grade regional metamorphism might have some of the following characteristics (Figure 10–24):

1. Diffuse, poorly defined boundaries exist between the granitic body and adjacent rocks.
2. Nongranitic rocks are commonly high-grade metamorphic rocks also, so sharp contact metamorphic zones are not usually present.
3. Original layering (sedimentary bedding or metamorphic foliation) may be preserved in the granite; it may be possible to trace original bedding or foliation in adjacent rocks into the granite body.

The process of transforming nongranitic rocks into granites is a metasomatic one, because the composition of the original material must in most cases be altered either slightly or considerably. This process is called *granitization*. Many of the rocks formed during granitization have many characteristics common to both igneous and metamorphic rocks. They are therefore called *mixed rocks* or *migmatites*.

What transforms high-grade gneisses into migmatites? Heat is important, but the pre-

Figure 10–23. Characteristics of an intrusive granite.

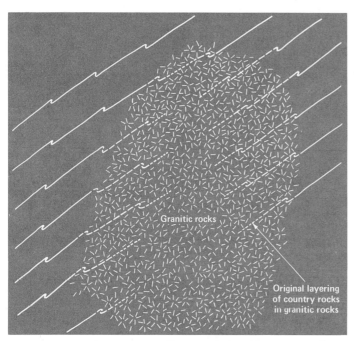

Figure 10–24. Characteristics of a metamorphic granite.

sence of water is probably of equal if not greater importance. In the absence of water, higher-grade metamorphic minerals may form at temperatures of 600–700°C. But if there is sufficient water in the rock mass, it may melt partially, and granitization may take place at temperatures in this range. A magma may be generated at still higher temperatures with ample water.

SUMMARY

Metamorphism is an internal process in which rocks are recrystallized under pressures and temperatures different from those under which they formed. Metamorphism is a solid-state process. Regional metamorphism occurs under elevated temperatures and high pressure. Regional metamorphic rocks are commonly foliated. Contact metamorphism occurs at high temperatures next to cooling igneous bodies and produces nonfoliated rocks. Cataclastic metamorphism consists of crushing and grinding of rocks along faults deep in the earth at low temperatures and high pressures. A number of minerals survive from sedimentary and igneous environments into an environment of metamorphism, whereas many others recrystallize to form minerals that are more stable here.

Plate tectonics theory provides a means by which we may explain the origin of the principal kinds of metamorphic processes. Heat and pressure produced by one outer segment of the earth pushing against another may result in regional metamorphism in some areas and melting in others. Molten material may move to cooler regions of the crust to solidify and produce contact metamorphism in country rocks. Fracturing of part of one segment may produce faults that at depth may crush and grind rocks to produce cataclastic metamorphic rocks.

Classification of metamorphic rocks is primarily textural, but composition is also important. Two broad categories of metamorphic rocks may be selected: foliated and nonfoliated rocks.

Grade of metamorphism may be determined by recognizing index minerals and by constructing isograds. Grade may also be derived by studying the entire mineral assemblage and determining to which metamorphic facies these rocks belong.

Different minerals behave differently under varying metamorphic conditions; some recrystallize faster than others. Grain size is another important factor in determining rate of recrystallization.

Granites may be products of either igneous or metamorphic processes. The metamorphic process of granite formation is called granitization.

Questions for thought and review
1. How would you distinguish an igneous or a sedimentary rock from a metamorphic rock?
2. How would you distinguish between a

regional and a contact metamorphic rock?

3. Where on the earth would you have the greatest likelihood of seeing metamorphic rocks at the surface?

4. What are the principal differences between the three metamorphic processes discussed in this chapter?

5. Why are some minerals persistent while others recrystallize during metamorphism?

6. What factors strongly influence the rate of metamorphism?

7. What are the principal differences between a phyllite and a slate? A gneiss and a schist?

8. What limitations are imposed on laboratory investigations that seek answers to how metamorphic rocks are formed?

9. How can you distinguish an igneous from a metamorphic granite?

10. Why do we believe that there is a connection between the origin of metamorphic rocks and mountain systems?

Selected readings

Ernst, W. G., 1969, *Earth Materials*, Prentice-Hall, Englewood Cliffs, N.J.

Mason, Brian, 1966, *Principles of Geochemistry*, 3rd ed., Wiley, New York.

11 Weathering and weathering products

STUDENT OBJECTIVES

At the completion of this chapter, the
student should:

1. have an understanding of weathering
   and its processes

2. understand the differences between
   major kinds of weathering and principal
   factors that control the kind that
   dominates in a given area

3. see the relationships between
   differential weathering, climate, and
   landscape development

4. be able to relate the differences and
   similarities between exfoliation and
   spheroidal weathering

5. know what a soil is, its constituents,
   and the factors that affect its formation

6. understand various soil classifications

7. be more aware of the importance of
   maintaining our soil resources

Whole mountain chains of the geological
yesterday have disappeared from view, and,
as with the ancient cities of the East, we
read their histories only in their ruins.

Merrill

Where would you go to find a sample of fresh bedrock material in the area where you live? To a quarry? To a highway cut? Perhaps in your own back yard? Many of us would have difficulty finding fresh, unweathered rock at all. The reason for this is that surface rocks have been changed by surface processes and conditions into products that are more stable (Figure 11–1). Here again is the stability/equilibrium relationship that was discussed in Chapter 4.

The process of soil formation is very important. Without soil we would have no agriculture and very little terrestrial life on our planet. Different materials weather at different rates. One of the best places to observe this is in a cemetery containing tombstones made of a variety of rock types that are several hundred years old (Figure 11–2). More resistant rocks maintain their lettering for many centuries, whereas those of less resistant material such as marble lose theirs.

Rocks that were formed at elevated temperatures and pressures or even at surface temperatures but in different environments, such as under marine conditions, will un-

Figure 11–2. Tombstones in a cemetery near Kings Mountain, North Carolina. Lettering on marble tombstones is illegible whereas that on tombstones made of granitic rocks can still be read.

dergo changes in response to surface atmospheric conditions so that their constituents are more stable. Various types of fracturing, abrasion, and expansion processes may affect rocks so that their composition may not be altered, but their physical shape or size may change. This is *mechanical weathering*. Also, surface conditions may dictate that the material decompose and the original rock minerals break down to form other minerals, with some elements going into solution. This is *chemical weathering*. Frequently a combination of mechanical and chemical processes operate in an area. In other areas one process may also be quite dominant over the other.

WEATHERING, EROSION, AND THE ROCK CYCLE

Weathering is a surficial mechanical or chemical process in which rock materials undergo a change toward greater stability. The weathering process is static and involves no large amount of movement. *Erosion* is a mechanical process in which loose material is removed and placed into the realm of transportation. As weathering

Figure 11–1. A well-developed soil, South Carolina.

precedes erosion in nature, it precedes it in the rock cycle (see Chapter 4). Weathering is an initial step in the surface processes of generation of sediment. Erosion is a dynamic process of removal of weathered materials, allowing them to be transported and later deposited as sediment. (The process of erosion will be discussed further in Chapter 12.)

KINDS OF WEATHERING

Weathering may take place as: (1) a predominantly mechanical process, in which materials being weathered do not undergo a change in composition; (2) a chemical process in which several of the constituents of the rock are broken down and change composition; or (3) a combined mechanical-chemical process in which one process aids the other.

Factors affecting and controlling weathering processes

Several factors control the processes of mechanical weathering, chemical weathering, or a combination of them in a given area. These include climate, degree of slope (topography), and availability of water and vegetative cover. *Time* of exposure to weathering processes also influences the degree to which a mass of rock will be weathered.

Climate is probably the most important factor influencing weathering processes. In areas of abundant rainfall, as in the eastern United States, chemical weathering will dominate. In more arid areas where rainfall is not so abundant or in areas like the Arctic where water remains frozen much of the year, mechanical weathering processes will dominate.

Slopes influence weathering processes to some extent. Areas of steeply developing cliffs will favor mechanical processes. Gently sloping areas that prevent moisture loss favor rapid chemical weathering processes.

Water is a powerful weathering agent,

both mechanical and chemical. When it is not available, as in an arid region, or when it has been rapidly removed in a humid or arid region, it cannot function in the weathering process. This will likely slow the overall process of weathering.

Vegetation cover or lack of it can influence weathering processes. Cover by vegetation may inhibit mechanical weathering except that accomplished by plant roots. But vegetative cover may enhance chemical weathering by helping to hold moisture and by adding organic acids to the soil.

Mechanical weathering

Mechanical weathering is mainly one of disintegration—a process of making smaller things of larger ones. This is accomplished by various means.

Probably the most widespread and important mechanical weathering process is *freeze–thaw*. When water changes to ice, it expands by 9% over its original volume (Figure 11–3). Water in a crack in a rock will generally freeze so that much of this expansion is directed as a force against the rock in the sides of the crack. This may amount to a sizable pushing-apart pressure that may exceed the strength of the rock and cause the crack to be enlarged. The *freeze–thaw* process depends on the temperature dropping below

Figure 11–3. Needle ice.

freezing and then rising again above the freezing point. This process works best when the freezing point is crossed frequently during the year. Perhaps the western mountains experience this better than any area in the United States. The temperate zone is best suited to maximize freeze–thaw, because it is neither too warm nor too cold to discourage the process.

Frost wedging is likely the most effective and widespread result of the freeze–thaw process. Water freezes in cracks in rocks, expands, and helps force the cracks apart (Figure 11–4).

Frost heaving is noticeable in the northern parts of the United States and Canada. It occurs when water collects and freezes a few inches (centimeters) below ground surface and thaws on the surface. Liquid surface water then percolates laterally and downward to freeze again and to add to the frozen mass, causing it to enlarge and expand. The soil material, or whatever is above, is thus heaved upward. If an ice lens is forming beneath part of a road, the road will become

cracked and broken there. This process is quite damaging in areas of poorly constructed roads.

Other mechanical weathering processes

Plant roots frequently grow into cracks in rocks and, as roots become larger, force the cracks apart. Many rocks are broken in this way, although this process is likely not so widespread as freeze–thaw.

One might think that *solar heating* then cooling at night might be an effective mechanical weathering process. On the contrary, experiments in the laboratory in which rocks have been heated and cooled repeatedly for the equivalent of many years have shown few signs of cracking or of undergoing any kind of disintegration. So this process can be discounted in most areas but is thought to be important in some places experiencing large day–night temperature variations. It has been noted that rocks heated by forest fires and then quenched by rain undergo spalling. Also, experiments have shown that the presence of water and salt greatly accelerates disintegration under changing temperature conditions.

Mechanical processes, such as glacial plucking and abrasion by wind and water, are considered by some to be mechanical weathering processes. Actually they are erosion processes involving removal and transport of material for varying distances. They will be discussed in Chapters 15 and 17.

Chemical weathering

Chemical weathering involves actual decomposition of the minerals that make up a rock. As we shall see, different minerals decompose at different rates. Also, the sizes of particles being weathered affect the rate of weathering. So mechanical weathering that reduces particle size can affect the rate of chemical weathering.

How does reducing the size of particles influence the rate of decomposition? It increases it. Why? It has to do with the effec-

Figure 11–4. Broken rock produced by frost wedging, Cariboo Mountains, British Columbia.

tive *surface area* exposed for chemical reaction. The same volume of smaller particles of the same material has a greater exposed surface area than does a smaller volume of larger particles. This greater surface area allows chemical reaction, in this case decomposition, to occur in more places at one time. (Refer to Chapter 10 in which we discussed the effect of surface area on rate and degree of metamorphism.)

Mineral stability in chemical weathering. Minerals break down under chemical weathering conditions, because they are not stable in the weathering environment. Some minerals break down more readily than others. Generally those igneous minerals that crystallized at the highest temperatures are least stable, whereas those that formed at lower temperatures are most stable. Crystal structures and compositions of minerals can also affect the rate of weathering. S. S. Goldich recognized this, and in 1938 he arranged the common minerals in order of their rate of decomposition (Table 11–1). Notice that this stability series parallels Bowen's reaction series; those minerals that are most stable form at conditions most nearly like those on the surface. Actually quartz can and does form under certain surface conditions.

Under chemical weathering conditions minerals in metamorphic rocks behave somewhat similarly to those in igneous rocks. Those that formed at higher tempera-

tures and pressures generally break down more easily. Yet there are notable exceptions to this rule. Garnets are more resistant than are feldspars to chemical weathering, as are kyanite and certain other high-temperature–high-pressure minerals. Some of these remain in the soil after most of the other rock minerals have decomposed. The reason for this added stability is that a great deal of energy was released in the formation of these minerals; so a great deal of energy would be required to decompose them.

Many of the minerals in sedimentary rocks have already been through the weathering cycle prior to deposition, but sedimentary rocks weather, too. In this case minerals that are being decomposed formed under conditions that are different from those that surround the rock mass today. They may have been deposited in a marine environment, in a lake, in a swamp, or elsewhere. All these conditions differ markedly from the *subaerial* conditions of the surface where materials are exposed to air and water at the same time.

Chemical decomposition of igneous and metamorphic rocks. The different minerals of igneous rocks respond differently to chemical weathering conditions. Quartz may dissolve slightly but generally remains unchanged. Feldspars decompose completely in the weathering scheme and form solid kaolinite and certain dissolved constituents. For example, an unbalanced reaction for the

Table 11–1. Stability of common igneous minerals under chemical weathering conditions.

Least stable	Olivine	Ca-plagioclase	High-temperature formation
	Pyroxene	Ca–Na plagioclase	
	Amphibole	Na–Ca plagioclase	
	Biotite	Na-plagioclase	
		K-feldspar, muscovite	
Most stable		Quartz	Lower-temperature formation

SOURCE: After S. S. Goldich, 1938, A study in rock weathering, *Journal of Geology*, pp. 17–58. Used by permission.

decomposition of potassium feldspar might look like this:

$$KAlSi_3O_8 + H_2O + CO_2 =$$
potassium
feldspar

$$Al_2Si_2O_5(OH)_4 + \quad SiO_2 \quad + \quad K^+ \quad + H_2O$$
kaolinite   silica   potassium
       (partially   ion (in
       dissolved)   solution)

The carbon dioxide is dissolved in water from the atmosphere. Its function is very important, because a little of the CO_2 reacts with water to form carbonic acid, H_2CO_3 (this also produces the bubbles in carbonated beverages). The groundwater becomes slightly acidic and thus more reactive with common rock minerals. The solid weathering product of the feldspars, kaolinite, is a clay mineral and remains to form an important constituent of soils.

The decomposition of the dark minerals of igneous rocks results in formation of other clay minerals, iron oxide-hydroxide, and certain other soluble materials; for example:

biotite + H_2O + CO_2 = clay mineral(s)
+ iron oxide-hydroxide + $[Na^+ + Mg^{2+} + K^+]$
                                   soluble ions
+ SiO_2 + H_2O
    silica
    (partially
    soluble)

Other dark minerals, such as olivine, pyroxene and amphiboles, behave similarly.

Chemical weathering of an igneous or metamorphic rock begins along fractures in the rock and then follows boundaries between individual grains. Once a grain has been weathered a bit all around, the process of breaking it down further is slowed, since additional decomposition must be accomplished through a weathered rim (Figure 11–5).

Chemical decomposition of sedimentary rocks. Sedimentary rocks weather differently according to their compositions. Sandstones tend to become disaggregated,

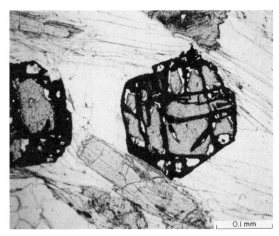

Figure 11–5. Photomicrograph of a partially decomposed metamorphic rock showing garnet crystals with fresh cores and weathered rims.

forming a sandy soil. Shales are composed of clay minerals; yet they may undergo further decomposition and form clayey soils. Carbonate rocks—limestone and dolostone—probably undergo the greatest changes of any sedimentary rocks because of the relative solubilities of the carbonate minerals in groundwater.

$$CaCO_3 + CO_2 + H_2O = Ca(HCO_3)_2$$
calcite   calcium
                          bicarbonate
                          (soluble)

Soils produced over limestones and dolostones are composed of clays, quartz, and chert, which are present as minor constituents in these rocks. When the carbonates are dissolved away, less soluble minerals remain behind as a residue.

DIFFERENTIAL WEATHERING AND LANDSCAPE DEVELOPMENT

The various ways in which minerals weather and the dominance of chemical over mechanical weathering, or vice versa, cause more resistant rocks to be etched into relief in the landscape, while those less resistant are weathered down and form low places.

This is termed *differential weathering*. If all the rocks of an area have similar resistances to the dominant weathering processes, the topography that is developed will bear little or no relationship to bedrock types and structures.

Differential weathering in a humid climate, where bedrock geology consists of limestone, shale, and sandstone that have been tilted or folded, will result in a ridge and valley topography such as that in parts of central Pennsylvania, western Maryland, western Virginia, eastern Tennessee, northeastern Georgia, and central Alabama (Figure 11–6). Here limestones are less resistant to the dominant chemical weathering process and form valleys; the more resistant sandstones form ridges. Shales form intermediate low hills. In areas of the western parts of the United States and Canada, where similar bedrock types occur in the same tilted or folded state, but where the climate is drier, mechanical processes dominate. There a ridge and valley topography may result, but the ridges are supported by limestones and valleys underlain by sandstones, because limestones are more resistant to mechanical weathering than are sandstones (Figure 11–7).

Numerous examples of differential weathering of igneous or metamorphic rock types could also be cited. It is important to remember that differential weathering of any group of rocks takes place because rocks vary

Figure 11–6. Ridge and valley topography near Harrisburg, Pennsylvania produced by differential weathering of folded sandstone, limestone and shale units in a humid climate. (Courtesy of John Shelton.)

Figure 11–7. Limestone ridges in Jasper National Park in the Canadian Rockies. Here the climate is less humid and mechanical weathering processes are more important. These mountains also have been glaciated.

in composition and texture. These variables, when coupled with the structural setting of the rocks and acted on by factors that produce the dominant weathering process in a region, will produce landscapes that are easily related to bedrock geology.

Flat-lying layers of resistant sedimentary or igneous rocks, such as lava flows, interbedded with nonresistant rocks may produce a topography of *buttes* and *mesas* (Figure 11–8). In parts of the western United States many buttes and mesas are capped by lava, limestone, or other resistant rocks. They also can be found in a few areas in the east and most commonly have sandstone caps. They are not so easily recognized here, because they are commonly covered with vegetation.

Figure 11–8. Buttes developed in flat-lying rocks due to a resistant cap and less resistant cap and less resistant rocks beneath, Hopi Buttes, Arizona. (Courtesy of John Shelton.)

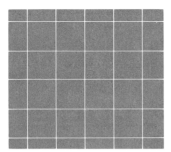

Intersecting fractures in unweathered rock produce sharp-edged rectangular blocks.

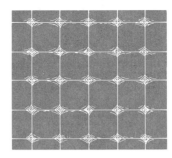

Weathering begins along fractures and rounds off sharp edges and corners.

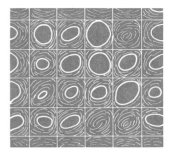

At an advanced stage many blocks contain no fresh material; others contain only a fresh spheroidal core.

Figure 11–9. *A.* Intersecting fractures in unweathered rock produce sharp-edged rectangular blocks. *B.* Weathering begins along fractures and rounds off sharp edges and corners. *C.* At an advanced stage many blocks contain no fresh material; others contain only a fresh spheroidal core.

SPHEROIDAL WEATHERING

Spheroidal weathering is a combined mechanical–chemical weathering process that produces rounded boulders. For this phenomenon to occur, the rock should contain two or three sets of intersecting fractures that are spaced about 4 in. (10 cm) or more apart (Figure 11–9). These fractures allow water to percolate into the rock mass and to begin the weathering process. Decomposition of minerals moves inward from the outside of the fracture-outlined blocks. Expansion of the rock mass and development of concentric layers take place as weathering proceeds. Initial layers develop parallel to the fracture surfaces. But, as more rock is decomposed, the corners of the blocks become rounded, and an isolated spherical core of unweathered rock remains (Figure 11–10). If carried to completion, the unweathered cores become decomposed and the spherical structure finally disappears. A partially decomposed spheroidal boulder can be broken to reveal the concentric layering of the weathered material (Figure 11–11).

Spheroidal weathering is present in very

Figure 11–10. Dike of mafic igneous rock which has been decomposed by spheroidal weathering, Georgia.

Figure 11–11. Spheroidal boulder broken to show the concentric layering of weathered and partially weathered zones surrounding a core of unweathered rock.

uniform, massive rocks, such as many igneous and thick-bedded sedimentary rocks. It is a rounding process produced by the partial chemical breakdown (generally hydration) of one or more minerals. This hydration results in an increase in volume of the weathered mineral grains; the weathered part of the rock must expand slightly to accommodate the greater volume.

EXFOLIATION

The process whereby the surfaces of massive rocks, such as granites, gneisses and some sedimentary rocks, peel away in thin layers is called *exfoliation*. Its results range from small areas of exposed rock *pavement* to large *exfoliation domes* like Half Dome in Yosemite National Park, Sugar Loaf in Rio de Janeiro, Stone Mountain in Georgia (Figure 11–12). It may be a purely mechanical process. Repeated expansion and contraction caused by the sun's heating and cooling an exposed mass of rock is thought by some to have produced exfoliation domes such as those in Yosemite. But rocks are very poor conductors of heat, so solar heating and cooling are a doubtful mechanism for exfoliation.

Expansion of lower portions of a mass or rock as erosion removes upper portions produces fractures called *sheeting joints* (Figure 11–13). They parallel the surface and become more closely spaced near there. Sheet-

A

Figure 11–12. A. Sugar Loaf (Pão de Açucar) and other exfoliation domes in Rio de Janeiro Brazil. (Courtesy of Pan American World Airways.) B. Forty Acre Rock, an exfoliation surface in South Carolina.

ing joints may lead to thin exfoliation sheets on the surface.

Exfoliation may also result from a combination chemical–mechanical weathering process, such as spheroidal weathering. Partial decomposition of some minerals on surface areas of an unfractured rock mass may produce expansion and exfoliation as these minerals are hydrated during chemical weathering.

SOILS: PRODUCTS OF WEATHERING

Soils represent surface accumulations of weathering products. They may be *residual soils*, which have developed from material directly underneath, or *transported soils*, which have been moved to their present sites.

Residual soils

Most of the world's soils are *residual soils*. Geologists and agronomists have made great efforts to understand the origin and development of these and other soils.

Rate of soil formation. One question that arose early in the study of the origin of soils considered how rapidly soils form. Residual soils vary greatly in thickness from place to place. They thin on steep slopes and vary in texture and composition with other factors, such as climate and amount of time that has elapsed since the soil began to form. Is there any place where we know exactly when a soil began to form? In the Midwest, glacial geologists have mapped precisely the extent of and rate of advance and retreat of Pleistocene ice sheets (see Chapter 15). Each of these ice sheets left a thick layer of fine-grained rock flour over the surface in many areas. Weathering of younger (8,000–10,000 years old) glacial material deposited in Ohio extends down 22–30 in. (60–90 cm), so the rate of weathering for that area is known to be about 0.002 in/year (0.008 cm/year). Older glacial material in Ohio and Indiana has been weathered in places to a depth of 25–30 ft (10 m). Volcanic ash has formed soils of appreciable thickness in about 100 years.

Soils in tropical or subtropical regions

B

Figure 11–13. Sheeting joints used to quarry massive granitic rock at Mt. Airy, North Carolina. (Courtesy of the North Carolina Granite Corp.)

that are more than 100 ft (30 m) thick have formed over long periods of time under intense chemical weathering conditions. It is difficult to estimate their rates of formation, but they must be at least on the order of hundreds of thousands of years.

Soil profiles. Soils may be divided vertically into different zones based on degree of weathering and composition of each zone. Each zone is called a *horizon*; soil classifications are based on the nature of the intermediate or B horizon.

The C horizon of a soil is the lowermost zone (Figure 11–14). It occurs directly above bedrock and contains partially decomposed fragments of underlying bedrock along with more completely decomposed material. In some areas bedrock structures may be traced uninterrupted into the overlying C horizon. When this phenomenon is well developed in a thick C horizon, the weathered mass preserving bedrock features is called a *saprolite* (Figure 11–15).

Overlying the C horizon is the B horizon. It is characterized by complete decomposition of all bedrock material and a more or less structureless appearance. No features from bedrock should be noticeable here. However, the B horizon may be enriched in some component, such as clays, particular oxides or hydroxides, or calcium carbonate. Plant roots in search of water may penetrate the B horizon and occasionally the C horizon.

Organic matter is concentrated in the up-

Figure 11–14. A well-developed soil profile showing the C, B, and A horizons.

Figure 11–15. Saprolite of interlayered light and dark gneisses preserving details of structures in the original bedrock, South Carolina.

permost A horizon. The components of the B horizon are usually present, but the surface accumulation of organic matter becomes part of the soil. It also contains the roots of small plants. The A horizon is seldom more than 19 in. (50 cm) thick. Leaching of water-soluble soil constituents is maximized here because of greater contact with the surface and percolating water and the presence of acids from decay of organic matter.

Early in the development of a soil only surface material will be present, perhaps only an organic enriched layer. In time more bedrock is weathered, and the C horizon becomes separable from the A horizon. The soil is said to be *immature* at this point. The B horizon develops later, as more material is completely broken down and accumulation of some component occurs. Soils containing a well-developed B horizon are called *mature* soils.

Transported soils

Soils that are not derived from underlying bedrock but have been moved to their present site are termed *transported soils* (Figure 11–16).

The most common and widespread transported soils are stream-deposited *alluvial soils*. These may vary considerably in composition, grain size, and thickness, depending on the nature of stream system that deposited them.

Other soils have been transported down

Figure 11–16. Alluvial material deposited upon weathered bedrock having a near vertical foliation, North Carolina.

slopes for varying distances by gravity. If a soil has been transported a few yards (meters) downslope, its structure and contact with the bedrock source would be lost.

Glacial soils involve transport for many miles (kilometers), and the soil material may bear no resemblance to the bedrock upon which it rests today (Figure 11–17). It may even be deposited on a residual soil that had formed prior to glaciation.

Soil types and classifications

Marbut's classification. An early scheme of classification was derived by C. F. Marbut, a soil scientist who greatly influenced the thinking of others in this field throughout the 1920s and early 1930s, when he was chief of the U.S. Soil Conservation Service. His classification of soils was first published in 1927 and was revised after his death in 1936. Marbut's classification is based on characteristics of the B horizon. It therefore is a classification of mature soils only, de-

spite the fact that Marbut recognized that soils in many places are immature. Because immature animals and plants are not considered in their classification, Marbut believed

Figure 11–17. A glacial soil resting upon bedrock. This soil is derived from material brought far from its present site and is not related to the bedrock beneath. (Courtesy of Ward's Natural Science Establishment.)

that immature soils should not be considered either. The fallacy in his logic is, of course, the fact that soils are not living things. Yet the major subdivisions of his classifications are worth considering.

One of Marbut's major divisions is the *pedalfer soil*. (Pedalfer means rich in aluminum and iron.) Pedalfer soils have a B horizon that contains an abundance of clays and iron oxides and hydroxides. Aluminum enrichment occurs in the clays. Such soils generally develop in humid climates with good vegetative cover, such as that of the eastern United States.

Marbut called soils that contain a concentration of calcium carbonate *pedocal soils*. White or tan calcium carbonate nodules—caliche—may be seen in highway cuts and natural exposures of soil profiles in areas where these soils occur. Pedocal soils form in temperate zones where temperatures are fairly high, rainfall low, and vegetative cover sparse.

A north–south line drawn through east Texas, Oklahoma, Kansas, Nebraska and the Dakotas roughly separates pedalfer soils to the east from pedocal soils to the west. This is the line of 25 in. (63 cm) of annual rainfall (Figure 11–18).

Climate and soils. Climate is a major factor in controlling the types of soils that develop in different areas, even if the bedrock type in each area is the same. This relationship was first discovered by Russian soil scientist V. V. Dokachaiev. In 1886 he proposed the first soil classification based solely on climate. Alluvial and wind-deposited soils were considered to be nonsoils. He later revised his classification to include them but called them abnormal soils.

Soil scientists of the U.S. Soil Conserva-

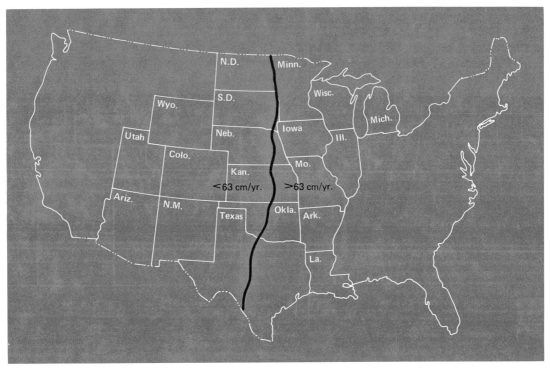

Figure 11–18. Map of the United States showing the approximate location of 20 in/yr (63 cm/yr) rainfall line.

tion Service devised a soil classification that was climate related but included soils that cross climatic boundaries as well as other soils that develop within particular climatic zones (Table 11–2). It remained in use from 1938 until 1965. The group of *zonal soils* is related to climatic zones (Figure 11–19). Those which develop under special conditions within a climatic zone are *intrazonal soils*. *Azonal soils* are not climate related and can develop in more than one climatic zone. Transported soils fall into this group.

Many of the names of the great soil groups seem self-explanatory. Even though they are to some degree, each great soil group possesses certain characteristics that are not inherent in a descriptive name. For example, the group of brown soils of arid regions are colored brown on the surface, but they form in temperate to cool arid regions. They contain a lighter colored subsurface horizon that generally contains a limy (calcium carbonate) zone. Several of the names of other great soil groups are not so descriptive to the nonsoil scientist. *Sierozem soils* have a gray A horizon and are calcareous at shallow

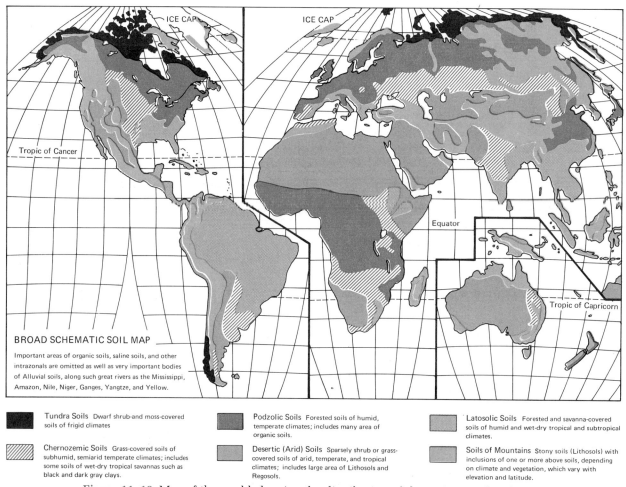

BROAD SCHEMATIC SOIL MAP

Important areas of organic soils, saline soils, and other intrazonals are omitted as well as very important bodies of Alluvial soils, along such great rivers as the Mississippi, Amazon, Nile, Niger, Ganges, Yangtze, and Yellow.

Tundra Soils Dwarf shrub-and moss-covered soils of frigid climates

Chernozemic Soils Grass-covered soils of subhumid, semiarid temperate climates; includes some soils of wet-dry tropical savannas such as black and dark gray clays.

Podzolic Soils Forested soils of humid, temperate climates; includes many area of organic soils.

Desertic (Arid) Soils Sparsely shrub or grass-covered soils of arid, temperate, and tropical climates; includes large area of Lithosols and Regosols.

Latosolic Soils Forested and savanna-covered soils of humid and wet-dry tropical and subtropical climates.

Soils of Mountains Stony soils (Lithosols) with inclusions of one or more above soils, depending on climate and vegetation, which vary with elevation and latitude.

Figure 11–19. Map of the world showing the distribution of the major zonal soil types. (U.S. Department of Agriculture.)

Table 11–2. Classification of soils into higher categories

Order	Suborder	Great soil groups
Zonal soils	1. Cold zone soils	Tundra soils
	2. Light-colored soils of arid regions	Desert soils Red desert soils Sierozem Brown soils Reddish brown soils
	3. Dark-colored soils of semiarid, subhumid, and humid grasslands	Chestnut soils Reddish chestnut soils Chernozem soils Prairie soils Reddish prairie soils
	4. Soils of the forest–grassland transition	Degraded chernozem Noncalcic brown soils
	5. Light-colored podzolized soils of the timbered regions	Podzol soils Gray podzolic soils Brown podzolic soils Gray-brown podzolic soils Red-yellow podzolic soils
	6. Lateritic soils of forested warm-temperate and tropical regions	Reddish brown lateritic soils Yellowish brown lateritic soils Laterite soils
Intrazonal soils	1. Halomorphic soils of imperfectly drained arid regions and littoral deposits	Saline soils Alkaline soils
	2. Hydromorphic soils of marshes, swamps, seep areas, and flats	Humic gley soils Alpine meadow soils Bog soils Half-bog soils Low-humic gley soils Groundwater podzol soils Groundwater laterite soils
	3. Calcimorphic soils	Brown forest soils Rendzina soils
Azonal soils		Lithosols Regosols (includes dry sands) Alluvial soils

SOURCE: After Soil Survey Staff, 1960, *Soil Classification, A Comprehensive System*, 7th Approximation, U.S. Department of Agriculture, Soil Conservation Service.

depths. They form in temperate to cool arid climates under a cover of desert plants, short grass, and brush. *Chernozem soils* contain a black organic and calcium-rich A horizon underlain by a lighter transition zone, which is above a zone of calcium carbonate accumulation. Chernozem soils form in cool subhumid climates under a grass cover.

Many of the rich farmland soils of the Midwest are of this type. *Chestnut soils* are similar to chernozems, but their A horizon is brown rather than black. *Podzol* soils form in forested temperate regions and generally contain a zone from which many clays and iron have been leached. They may have a prominent gray or other colored A horizon.

Laterite soils are red iron- and aluminum-rich soils that form beneath humid tropical and subtropical forests. Very little soluble material in the form of calcium and magnesium is present. Likewise silica has been leached from these soils (silica is leached under alkaline conditions, whereas calcium and magnesium are leached under acidic conditions). Acidic groundwater conditions produced during rainy seasons of tropical regions and more alkaline groundwater conditions of dry seasons may be responsible for the formation of laterites. Iron has been leached from some laterites to leave a lighter colored aluminum-rich residue. If the aluminum content is high enough, the rock *bauxite* results (Figure 11–20). Most bauxite, the principal source of aluminum comes from Jamaica and countries in northern South America. Laterites or lateritic soils are found in the southeastern United States, but leaching has not been intense enough to form large commercial bauxite deposits. Laterites are widespread throughout the tropics. But although they will support a rainforest cover, they make very poor agricultural soils.

Gley soils are formed under strong reducing conditions such as those found in water-saturated boggy or swampy areas. Iron and manganese may be dissolved and reprecipitated to produce a mottled appearance in the soil.

Rendzina soils are marly and friable at the surface, but beneath are light gray to pale yellow calcareous parent horizons. These soils develop in humid and semiarid climates beneath a cover of grass or mixed grass and forest. These soils are common in Texas and Oklahoma.

Lithosols are "soils" that contain no soil characteristics but are really unweathered or imperfectly decomposed bedrock. In these even the C horizon is missing.

Glacial drift, beach sands, dune sands, and any other material that is unconsolidated and is without a definite genetic horizon will produce a *regosol*.

Modern soil classification. The present soil classification used by agronomists throughout the United States is closely related to that originally set up by Dokachaiev and to the great soil groups. It was devised in 1960 and was adopted by the U.S. Department of Agriculture in 1965. The classification is based on the physical properties of different soil horizons. Great-group soils with similar profiles will be placed into the same order in the present classification. Soils that have similar origins and similar properties are placed into different orders. Table 11–3 contains the major orders within the present classification, their meaning, and some of the equivalents in the great soil groups.

Our dependence on soils

We are heavily dependent on soils for

Figure 11–20. Bauxite. (Courtesy of Ward's Natural Science Establishment.)

food. This dependence began thousands of years ago when early peoples learned to grow crops and to domesticate a few animals to feed their families. Humankind has wasted the soil resources of the world so that today, in a time when world population is at its greatest point and increasing more rapidly every year, soils become ever more important. Many innovations in agricultural technology serve to protect and replenish the soils in this country. But primitive agricultural techniques still are used over much of the world.

The United States is fortunate to have the right combination of climate, good soils, and abundant supporting resources, such as

Table 11–3. Present soil classification and equivalents in the great soil groups.

Soil order	Approximate meaning	Great soil group equivalents
Entisols	Soils showing no profile development (recent soils)	Azonal soils and some humic gley soils
Vertisols	Swelling (cracking) clayey soils	Swelling clays (grumusols)
Inceptisols	Weakly developed soils (young soils)	Some brown forest, low-humic gley and humic gley soils
Aridisols	Light-colored soils of desert shrub areas and desert grasslands	Desert soils, red desert soils, sierozem, some brown and reddish brown desert soils, associated alkaline soils
Mollisols	Dark-colored soils of grasslands (prairies) (soft soils)	Chestnut, chernozems, rend-zina, some brown and brown forest soils, associated alkaline and humic gley soils
Spodosols	Soils with an "ashy" horizon	Podzols, brown podzolic soils, groundwater podzols
Alfisols	Light-colored soils of humid-temperature areas (pedalfers)	Gray-brown podzolic, gray wooded soils, noncalcic brown soils, degraded chernozem
Ultisols	Light-colored soils of warm, temperate forest lands	Red-yellow podzolic soils, reddish brown lateritic soils of the United States
Oxisols	Highly weathered soils of the tropics	Laterite soils, latosols
Histosols	Areas dominated by organic soils	Bog soils

SOURCE: Compiled from Soil Survey Staff, 1960, *Soil Classification, a Comprehensive System*, 7th Approximation, U.S. Department of Agriculture, Soil Conservation Service; and Buol, S. W., ed., 1973, *Soils of the Southern States and Puerto Rico*, Southern Cooperative Series Bulletin 174, U.S. Department of Agriculture.

phosphates for making fertilizers. The yields of our lands have increased each year. Yet we are losing our soils to erosive processes, and our resources for replenishing them are finite (see Chapter 23). It takes many years for an A horizon to be restored once it has been removed. The B horizon is certainly not well adapted for growing things. So we should make every attempt to conserve our soils and not allow them to be rendered unusable.

Soils and economic deposits

Weathering processes have in some instances concentrated valuable materials in soils. Most of our aluminum comes from tropical laterite soils. If rocks from which this aluminum-rich soil were mined, the concentration of this metal would be so low, and the processing would require so much energy, that it would not be worth the effort of mining. Without the processes that concentrated the aluminum, it would not be such an important metal.

Aluminum is not the only metal concentrated in this way. Some copper deposits have been found by locating weathered residue on the surface of copper-iron sulfides that reside at depth. The surface material generally is rich in the iron hydroxide limonite and is called a *gossan* (iron hat). A few copper minerals may also be found in the gossan, but beneath is a leached zone in soils or near surface rock and then a zone in which secondary copper minerals have concentrated (Figure 11–21). Many rich copper deposits have formed in this way. Weathering and leaching by groundwater start the process; then secondary concentration occurs below the water table (see Chapter 14).

Weathering processes may also serve to release certain valuable minerals that are resistant to weathering by breaking down the surrounding rock material. Gold is very stable at the surface and has been released in many areas to accumulate in soils and

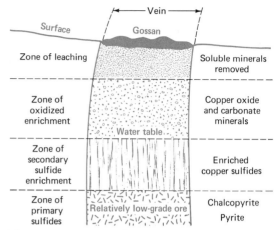

Figure 11–21. Secondary concentration of copper minerals due to weathering and enrichment below the water table. (After L. G. Berry and Brian Mason, *Mineralogy: Concepts, Descriptions, Determinations.* W. H. Freeman Co. Copyright © 1959.)

stream deposits. Platinum and diamonds may also accumulate in this way. A concentration of a valuable material in a stream or beach deposit is called a *placer.* Deposits of materials that may be mined from soils are called *residual deposits.*

SUMMARY

Weathering is a surface chemical or mechanical process whereby materials undergo changes toward greater stability. Climate, slope, water, and vegetation determine the rate of weathering and also whether chemical or mechanical weathering processes predominate.

Mechanical weathering involves reduction in the sizes of grains. Freeze–thaw is probably the most widespread mechanical weathering agent. Frost wedging, frost heaving, plant roots, and solar heating may be important locally.

Chemical weathering involves actual breakdown of the minerals in a rock. Those minerals that form at lower temperatures are more stable under chemical weathering conditions than those that form at higher temperatures. Mechanical weathering may aid chemical weathering by decreasing grain size, thereby exposing more surface area for chemical decomposition. Carbonic acid in groundwater solutions also aids the decomposition process.

Differential weathering involves different rates of weathering of minerals and rocks. It may take different forms from one climatic zone to another.

Spheroidal weathering is a combined mechanical–chemical process. A uniform rock mass that contains fractures that intersect to produce blocks may be rounded by a combined decomposition–expansion process to produce round boulders.

Exfoliation is a dominantly mechanical process in which surface layers of a uniform mass of rock peel off, leaving an exposed pavement or dome. Unloading and/or solar heating and cooling have been proposed to explain this phenomenon.

Soils are surface accumulations of the products of weathering. Most soils are residual, but some have been transported. Rate of soil formation varies with the same factors that control weathering rates of minerals.

Several zones or horizons may be recognized in soils that are well developed: a lower C horizon of incompletely decomposed material overlies the bedrock, a B horizon of completely decomposed material and material derived from leaching from above, and an A horizon that is enriched in organic matter but impoverished in other constituents. Soils that have a well developed profile are mature soils.

Soil classifications in prominent usage have evolved from those based primarily on climatic bases to a classification based on physical characteristics of the B horizon back again to climatic bases. The modern classification is based on climatic factors and also on the physical properties of the different soil horizons.

Our dependence on soils for food increases each year. This dependence underscores the need for effective conservation of this resource.

Certain valuable materials, such as aluminum, are mined from soils and have been concentrated by the weathering process. Weathering may also release certain minerals to be transported and concentrated elsewhere.

Questions for thought and review

1. What is the principal function of mechanical weathering? Of chemical weathering?
2. How do weathering and erosion differ?
3. How does water function in the process of mechanical weathering? In chemical weathering?
4. What kinds of minerals are generally most stable under chemical weathering conditions?
5. Why in general do rocks weather?
6. Why do sedimentary rocks, which were formed on or near the surface, weather?
7. What is differential weathering?
8. How do spheroidal weathering and exfoliation differ?
9. In what kinds of rocks does spheroidal weathering occur?
10. How fast or slow do soils form?
11. How does a mature soil form?
12. What is the basis of modern soil classifications?
13. What is a zonal soil?
14. How are laterite soils thought to form?
15. Briefly outline the economic importance of soils.

Selected readings

Hausenbuiller, R. L., 1972, *Soil Science, Principles and Practice*, Brown, Dubuque, Iowa.

Hunt, C. B., 1972, *Geology of Soils, Their Evolution, Classification and Uses*, Freeman, San Francisco.

McNeil, Mary, 1964, Lateritic Soils, *Scientific American*, vol. 211, no. 5, pp. 96–102.

12 Mass movement

STUDENT OBJECTIVES

At the conclusion of this chapter, the student should:

1. understand what mass movement processes are and factors that influence their occurrence

2. be able to differentiate between those factors that influence mass movement and those that initiate movement

3. know the differences between and causes of slow and rapid types of movement

4. understand the differences between various types of rotational, translational, and block types of movement

5. be able to suggest ways in which mass movement processes as environmental hazards might be better controlled

Everything changes but change.

Zangwill

Have you ever seen a rock or lump of soil become detached from its position on a slope and roll or slide to the bottom? What causes this rock or lump of soil to move downward? If you thought that it was the force of gravity, you were right. Gravity is the force that makes rivers run to the sea, glacial ice flow from the heights, and chunks of dislodged rock or soil tumble downhill. We refer to these processes that involve downslope movement of material as *mass movement*. Stream and glacial processes are excluded from mass movement processes and will be considered separately.

Processes of mass movement are processes seeking a state of equilibrium. Unstable masses of rock or loose material resting high above their surroundings are likely at some time to move downward to a more stable position (Figure 12–1). Processes of movement may be slow or rapid, but the same result is achieved in the end: a state of rest or equilibrium.

Human activity can affect any natural surficial process. Mass movement processes are

no exception. These processes frequently are effected unintentionally with costly results. Later in this chapter we shall examine several examples of human intervention into natural systems with disastrous results; then we shall look at possible ways of avoiding or controlling costly mass movement processes.

FACTORS INFLUENCING MASS MOVEMENTS

In some situations in nature the occurrence of any mass movement process will be extremely unlikely. The purpose of this section is to outline those geologic, climatic, and vegetative factors that do make mass movement probable. That some form of mass movement, whether slow or rapid, may be likely does not mean necessarily that it will occur. However, knowledge of areas where there is movement, particularly rapid types, can be information of vital importance in land use planning.

Slope

The most obvious factor influencing the possibility of mass movements is the steepness of the *slope*. Certainly the steeper the slope, the greater the danger of mass movements. However, when other factors are involved, even rapid types of movement can happen on deceptively gentle slopes. For example, the underwater flow that occurred at the Lake of Zug, Switzerland, in 1888 moved over a slope of less than 3°. And in arctic and subarctic regions massive transport of material can take place on slopes of only 1°.

Water

The presence, absence, or amount of *water* present can have a major effect on mass movement processes. Water has a tendency to separate and lubricate particles and fracture surfaces. When a mass of material is saturated with water, a buoyancy effect is

Figure 12–1. Small scale mass movement illustrated by slumping of a mass of soil down a steep slope (Photo by Jan Patterson).

added to the separating and lubricating effects of water. If pressure is applied to a saturated mass, water tends to support the pressure, and the mass may move as a fluid, not a solid material. Such pressure may arise simply from the weight of the saturated mass. However, when a mass is undersaturated, the cohesive surface tension of water tends to bind the mass together.

With the presence of water is the added possibility of freeze–thaw effects. The expansion of water during freezing may loosen material further, opening fractures and increasing the probability of movement. The rate at which water accumulates in an area also may have an effect. Rain falling at a uniform rate over a long period of time will have considerably less effect that a sudden accession of water by rainfall, such as a cloudburst.

Vegetation

The type, density of growth, or absence of *vegetation* may provide circumstances under which mass movement processes might be initiated. Roots of certain types of vegetation tend to stabilize slopes and to prevent sliding in many cases. In some, though, growth of a thick mass of vegetation could eventually result in overloading of the slope and actually increasing the chances of mass movement. Generally speaking, an overall lack of vegetation promotes erosion and mass movement. This is true particularly when the surface is made up of loose soil or rock material that tends to hold moisture.

Bedrock structure

The inclination, or dip, of layering or foliation in *bedrock structure* can have an effect on slope equilibrium conditions. The chances of slope adjustment are greater if the dip is parallel to the surface. This situation provides a weakness in the form of the foliation, fractures, or bedding planes. This factor combined with one or more additional factors, such as saturation by water and occurrence of certain favorable rock types, would make mass movement in some form very likely. Joints, faults, or fractures paralleling the surface may produce the same situation.

Active faults may create a condition of oversteepened slopes by moving and exposing a portion of the fault plane. They may also place material in an unstable condition, where none existed previously. (Refer to Chapter 18 for a complete discussion of joints, faults, and other structures.)

Rock type

Certain *rock types* singly or in combination may increase chances of mass movement in an area. For example, a sandstone may not tend to move under the influence of gravity, but if it is underlain by an impervious shale, it may move, particularly if water becomes trapped in it and if it is highly fractured or weathered.

Limestone is generally a strong, stable rock type. However, it tends to fracture. If these fractures become opened by freeze–thaw, the likelihood of movement is increased. Massive igneous rocks, such as granite, may fracture into large tabular masses parallel to the surface. As these sheets break up, gravity will move these products downslope. Certain clay minerals can promote movement by their ability to expand and absorb water and to become plastic. With a layer of absorptive clay comes the increased possibility of movement.

Weathering

Solid unweathered rock material is less susceptible to mass movement than is weathered rock. Moreover, some products of *weathering* are more stable to mass movement than others. Some soils do not hold water, while others do; and, as moisture builds up, they will become unstable.

Conclusion

Any of the factors listed above may influence mass movement processes. Generally several factors combine to produce eventual movement. For example, steep slopes, dip of bedrock parallel to the surface, impermeable clay shale beneath a weathered porous sandstone, and heavy precipitation all combined to produce the Gros Ventre, Wyoming, landslide in 1925.

INITIATORS OF MOVEMENT

The various factors discussed in the previous section may exist at a given locality, even in the proper combination, but mass movement may not occur. Something else is needed: a factor to set the process in motion. Most initiators of movement apply to rapid types of movement, but in some cases they may serve to initiate slow types of mass movement as well.

Earthquakes

Shock waves associated with earthquakes trigger landslides over an area whose size is directly related to the magnitude of the earthquake. Material that is in an unstable state because of one or more of the factors discussed above may be set in motion by an earthquake. Damage by the secondary effect of landslides has been in some cases more severe than that from the shock waves of the earthquake itself. This was perhaps the case with the Alaska earthquake in 1964.

Changes in slopes

A slope may be in a reasonably stable condition, but erosion of its lower portions or undercutting of the slope by streams or by man may steepen the slope so that mass movement is initiated. Other natural processes, such as freeze–thaw, ocean waves, and in a very few cases wind, may also oversteepen slopes.

Water

Precipitation in large amounts over a short period of time may initiate mass move-

ment. Seasonal precipitation may also enhance movement processes, especially if there is almost no precipitation during part of the year, and most of the annual precipitation falls during a few weeks or months. As an example, mudflows are not uncommon during the rainy winter months in southern California.

Changes in properties of materials

A mass of relatively unstable water-saturated soil or weathered bedrock may be stabilized by freezing during the winter. When it thaws, it may become unstable and move under the influence of gravity.

Certain clays, when water saturated and affected by some outside force, such as earthquake waves, behave more like a liquid than a plastic solid. When these clays are present as layers or make up a large proportion of a mass of unconsolidated material, they can produce rapid movement where slow movement might have been expected.

Human activities in shaping the surface of the earth, growing crops, removing vegetation, and mining frequently upset systems that already possess several factors that would favor movement. Oversteepening during highway construction, either by excavation or by building fills, produces slopes that will not stand. On steep slopes, clearing of vegetation may result in exposure of unstable material that was previously held in place by that vegetation. Both surface and underground mining operations may change surface conditions so that mass movements may be set in motion. Mine dumps on the surface, blasting, and settling of subsurface openings can effect mass movement.

Reservoir construction can also increase the possibility of mass movement by raising the water table (Chapter 14) in the area around the impoundment. This frequently results in saturation of weathered material with water that previously allowed percolation and drainage. The end result may be mass movement in some form.

MASS MOVEMENT PROCESSES

We can divide mass movement processes into two major groups: slow and rapid types of movement. Slow types of movement move so slowly that their motion is imperceptible to the eye. Indirect means, such as the offset of a line of driven stakes, must be used to measure their movement. Slow types of mass movement move at rates varying from 3 ft (1 m) per several years to several feet (meters) per day. Rapid types of mass movement travel down slopes at speeds ranging from 0.01 ft/sec (0.3 cm/sec) to more than 100 ft/sec (30 m/sec). D. J. Varnes has proposed a classification of mass movement processes. A simplified version is presented in Figure 12–2.

Slow types of mass movement

Creep. The slow downslope movement of soil or weathered or loose rock material is called *creep.* An extremely slow process, creep may be detected by the apparent downslope "bending" of rock layers as they pass from unweathered (unbent) material into the soil (Figure 12–3), by trees whose trunks have been tilted downhill but the trees continue to grow straight upward, and by fenceposts that have been tilted downhill. Animal paths in pastures may become distorted because of creep, and the animals may aid the process of downslope movement by continued use of the trails.

Masses of loose material (called *talus*) at

TYPE OF MOVEMENT	TYPE OF MATERIAL			
	BEDROCK	GRANULAR SOILS PLASTIC		
FALLS	ROCKFALL	SOILFALL		
SLIDES FEW UNITS	ROTATIONAL SLUMP	PLANAR BLOCK-GLIDE	SLUMP, BLOCK-GLIDE	
MANY UNITS		ROCKSLIDE	DEBRIS SLIDE	FAILURE BY LATERAL SPREADING
FLOWS DRY	ALL UNCONSOLIDATED			
	ROCK FRAGMENTS	SAND OR SILT	MIXED	MOSTLY PLASTIC
	ROCKFALL- AVALANCHE	SAND RUN	LOESS FLOW	
			DEBRIS AVALANCHE	
				EARTHFLOW
WET		SAND OR SILT FLOW	DEBRIS FLOW	MUDFLOW
COMPLEX	COMBINATIONS OF MATERIALS OR TYPE OF MOVEMENT			

CLASSIFICATION OF LANDSLIDES

Figure 12–2. Classification of landslides. (From D. J. Varnes, "Landslide Types and Processes," *in* Highway Research Board, Special Publication 29, National Academy of Sciences, 1958 and D. J. Varnes, Proceedings First Inter. Sump. on Landslide Control, Oct. 1972, Kyoto and Tokyo, Japan Soc. of Landslide, Tokyo, 1974.)

Figure 12–3. Soil creep over a weathered schist in North Carolina.

Figure 12–4. A talus slope formed at the base of a cliff, Glacier National Park. (U.S. Geological Survey.)

the base of cliffs may also creep (Figure 12–4). This material has accumulated as pieces of rock fall from the cliff and its mass builds through time. Some of the rocks fall and roll to the end of the talus pile. Others add to the mass until it begins to readjust itself by creeping.

The accumulation of an unsorted mixture of rock, soil, and weathered material on steep slopes and in steep valleys in mountainous areas is called *colluvium*. It gradually moves downslope under its own weight and the influence of gravity (Figure 12–5). This material is generally quite porous and consequently may be a better means of transporting near-surface water than nearby soils. Cutting the toe of a mass of colluvium may result in more rapid types of movement.

Solifluction. Certain types of mass movement occur in arctic or subarctic regions that do not occur elsewhere. *Solifluction* is a mass movement process related to the annual thawing of the upper surface of 3–6 ft (1–2 m) of the permafrost that exists in these areas. As this zone thaws, the thawed portion may move downslope as a lobate mass to produce a corrugated surface that resembles freshly plowed ground (Figure 12–6).

Figure 12–5. Colluvium resting upon a moderate slope, North Carolina.

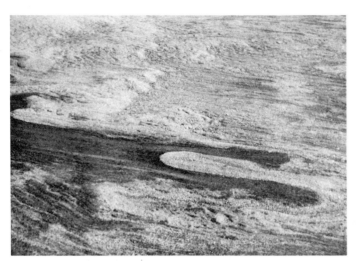

Figure 12–6. Solifluction lobes on eastern Grinnel Peninsula, Devon Island, Northwest Territories, Canada. (J. W. Kerr, Geological Survey of Canada.)

The water–ice interface becomes a lubricated detachment surface over which the lobe travels. Movement ceases during the winter only to be resumed the following summer, until the lobe reaches a low point at which it is in a state of equilibrium with its surroundings.

Rock glaciers. Water may percolate downward into an accumulation of loose rock resting on a steep slope in a high mountain range. If the temperature within this mass is below the freezing point of water, the downward-percolating water will freeze, coating the rock fragments and filling the voids between. As the addition of water adds weight to the mass, so will other fragments of rock that are loosened by freeze–thaw from the upper slopes and tumble onto the existing pile. This added weight, along with the lubricating effects of the ice's melting, refreezing, and also flowing under pressure, may set the mass into a state of slow flow. The pile of frozen rubble has become a *rock glacier* (Figure 12–7). Some rock glaciers, however, are residual true cirque glaciers covered by rubble (see Chapter 15).

Slow earthflows. Mixtures of soil, other weathering products, and loose rock may move slowly downslope over a period of several years. Movement may be detected in several ways. Trees may be tilted, perhaps uprooted, and the growth of saplings interrupted. Roads that had been built on the site of the flow before movement began would be cracked and distorted. Fences would be offset and posts tilted. The toe of the flow would show the greatest evidence of movement and distortion, because it would be overrunning the adjacent area.

A slow earthflow developed in the Gros Ventre River valley in northwestern Wyoming in 1908. It moved down a moderate slope and ceased moving by 1911. Movement was most rapid during the wetter periods, and, instead of moving as a single unit, it moved in sections. This might be expected in a mass that is relatively large in area compared with its thickness and composed of a chaotic mixture of fragments of many sizes. Different segments might move at different rates, some overtaking and partially overriding others, with faster-moving blocks leaving gaps as slower segments lag behind. In the flow area the dip of beds is parallel to the surface slope, and an impermeable shale underlies a porous sandstone at the surface. These factors added to the moderately steep slope and the heavy precipitation during 1908 and 1909, resulted in this event of mass movement.

Rapid types of mass movement

Slumps. A mass of unconsolidated material that moves downward and outward initially as a single unit is called a *slump* (Figure 12–1). Slumps may occur on any scale from features 3–6 ft (1–2 m) across, which bring the slopes of highway cuts and fills into greater stability, to single or composite features that move masses of material 1 mi (1 km) or more in size (Figure 12–8). Slumps involve a rotational type of motion and may

177

Figure 12–7. A rock glacier on Sourdough Peak, Wrangel Mountains, Alaska. (Courtesy of Bradford Washburn.)

change their characteristics at their toe to those of a rapid earthflow or other type of movement (Figure 12–9). They have an arcuate configuration at their head, and movement is concentrated along a single zone or fracture. Slumps may be triggered by earthquakes, heavy rains, oversteepened slopes, slopes undercut by waves, and by other means.

The surface of movement in slumps tends to follow some zone of weakness within the material or a soil–rock interface. The fractures formed in association with slumps and the movement patterns exhibited by them are the same as those of normal faults. Large down-to-basin faults along the Gulf Coast of Louisiana and Texas are essentially large slump structures (see Chapter 18).

Rockfalls. The mass movement process involving rapid downslope movement beginning with a single mass of bedrock and moving for an appreciable distance is called a *rockfall*. Rockfalls may occasionally involve free vertical fall for a portion of their movement. Many rockfalls do not remain intact for very long and become broken into a mass of smaller pieces, becoming a rockslide. Rapid downslope movement generally continues, and the mass may travel 1 mi (1 km) or more from its source.

Slides. The breakage of material along definite shear planes, followed by rapid downslope movement, produces slides of different types. Slumps are actually a type of slide. *Debris slides* involve movement of

Figure 12–8. Large scale slump occurring in poorly consolidated material along the Pacific Coast of California. (Los Angeles Times.)

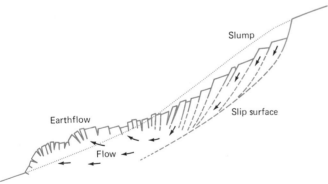

Figure 12–9. An illustration of slumping showing movement of slump blocks near the top but may become an earth-flow along the toe. (From C. F. S. Sharpe, *Landslides and Related Phenomena*, Columbia University Press, 1938.)

loose soil and other unconsolidated material. *Rockslides* are composed of bedrock material. Many slides consist of a mixture of both soil and bedrock. The Gros Ventre,

Wyoming, slide in 1925 involved rapid downslope movement of soil and bedrock.

Flows. Material moving rapidly downslope either in a wet or dry state, but behaving like a liquid in its movement properties, is called a *flow*. Actual surfaces of movement within the mass are not present, as in slides, and the general size of transported particles is small compared with entire mass and the amount of movement taking place.

Rapid earthflows involve mostly soil, whereas a *debris avalanche* and a *debris flow* involve weathered rock and soil. A debris flow has a much higher water content than a debris avalanche. Both represent humid climate equivalents of a mudflow.

A *mudflow* involves fine-grained surficial material that is picked up and moved rapidly during heavy rain, generally in areas with poor vegetation cover, such as most arid and semiarid areas. Mountainous topography provides steep slopes. Drainage courses, as small, normally dry streams, may not be able to accommodate the volume of water and sediment. As the water picks up more unconsolidated material, it becomes more viscous but still behaves as a liquid. Many overflowing stream courses coalesce to form a single rapidly moving mudflow. As more loose material on the slopes becomes saturated and begins to move, the viscosity of the flow increases to a point where its forward progress is slowed by internal friction.

Volcanic ash accumulated during an eruption frequently builds up into an unstable state. Water from rain, melting snow, or volcanic steam may saturate this material and cause it to move downslope as a type of mudflow. Volcanic mudflows are known as *lahars* and have been very destructive in some places. Several well-documented cases of prehistoric lahars can be seen on Mount Rainier in Washington.

Underwater flows have been noted in many areas. The underwater flow at the Lake of Zug in Switzerland began onshore by the caving of a retaining wall built to stabilize the shore. Buildup of water and saturation of the area behind the wall resulted in its collapse, along with portions of houses and streets 200 ft (75 m) inland from the shore,

and the initiation of the flow. The flow actually eroded and transported a much larger volume of material from the lake floor as it traveled along the bottom. Some of the debris was transported more than .5 mi (1 km) over a gentle 3° slope.

Air-cushion-lubricated slides. It was recognized many years ago that slides and avalanches attained velocities that seemed unusually high. Velocities on the order of 200 mi/hr (330 km/hr) were estimated for some slides. How could masses of rock and soil move that fast? An intriguing clue was found when it was discovered that ground surfaces over which these fast slides move are in places undamaged or bear marks (corrugations) that suggest some low-friction nondeforming lubricating layer much more efficient than water or mud. What could that be? There is only one answer: air. Observation of many landslides and snow avalanches reveals that a fluid motion accompanies upward and peripheral air blasts as these slides move rapidly downslope (Figure 12–10). R. L. Shreve first suggested in 1959 that these mass movement phenomena rode on a cushion of air.

The air-cushion effect is most likely to occur in large rockfalls and snow avalanches, particularly where falling material encounters some small flattened portion of an otherwise steep slope. This flattened zone causes the material to spread laterally into a sheet, trapping air underneath, and to rush downslope at high velocity. It is not difficult to visualize the devastating effect of this form of mass movement.

Documented examples of large mass movements

A number of large mass movement occurrences have been either witnessed directly by geologists or studied intensively by geologists shortly after they occurred. The examples described below provide a detailed account of conditions that favored occur-

Figure 12–10. Snow avalanches in the Loveland Pass area, Colorado. The boiling and seething character of the avalanche is due to air escaping upward through the mass. (Courtesy of Jerry Cleveland, Boulder, Colorado.)

rence of the event; factors that initiated it; and size, distance traveled and impact of the area affected (see also Chapter 19). Table 12–1 summarizes characteristics of some large mass movements.

Vaiont Reservoir, Italy, 1963. The reservoir on the Vaiont Canyon in northern Italy was created by the second highest (875 ft or 291 m) dam in the world. It is built in a Pleistocene glacial valley in mountains that were formed during the Tertiary Period (refer to the Geologic Time Scale, Chapter 4). These mountains still are releasing some of the stresses that were applied during their formation. A prehistoric slide had dammed the river in the reservoir basin, and another occurred in historic times in the same area but on the opposite side of the river.

Vaiont Canyon actually lies along a structural syncline (see Chapter 18) so that the rocks dip toward the valley from both sides. During the release of mountain-building stresses, fractures developed that were parallel to the rock beds. These fractures made the valley even more slide prone. Bedrock types are limestones alternating with soft clays and marl. These rocks tend either to fracture

across the beds easily or to prevent the flow of water through them and are otherwise inherently weak. These factors, too, enhanced the slide-prone character of the valley.

Evidences of the earlier slides became known during construction of the dam. In 1960, several years after construction, a small slide of 700,000 yd³ (644,000 m³) slid into the reservoir (Figure 12–11). Exploratory tunneling and drilling in 1961 near the head of the 1960 slide revealed clay beds and slide planes, while some of the drill holes were slowly closed off by creep. The higher groundwater level produced by the impoundment increased the rate of natural creep along the slopes. It also increased the rate of stress release.

The rate of creep was monitored from 1960 until the large slide of 1963. By April 1963 the water level in the reservoir had been raised by 60 ft (20 m). The rate averaged 0.4 in. (1 cm) per week until mid-September 1963, when it increased to 1 cm per day, a sevenfold increase. By early October it was 4 in. (10 cm) per day or more. The rate of creep was estimated at 8–32 in. (20–80 cm) per day on October 9, the day of the big slide. This

Table 12–1. Dimensions of some large mass movement features

Name or locality	Date	Type of movement	Volume (millions of m³)	Velocity (m/sec)	
Elm, Switzerland	1881	Rockfall	10	86	(maximum)
				43	(average)
Frank, Alberta	1903	Rockfall, slide	36	48	
Vaiont, Italy	1963	Rockslide	240	8	(minimum)
West Yellowstone area, Montana	1959	Rockslide	34	46.4	
Turnagain Heights, Anchorage, Alaska	1964	Block glide, translatory slide	11.5	?	
Gros Ventre, Wyoming	1908–1910	Slow earthflow	35	—	

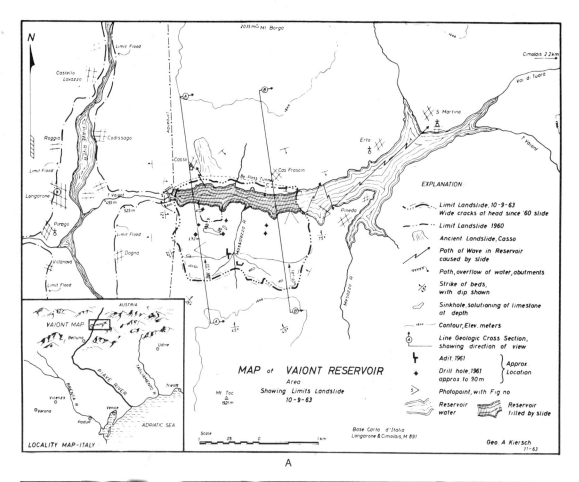

A

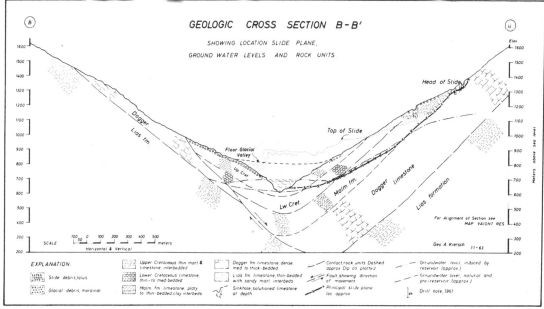

B

This cross-section shows conditions 1.5 km upstream from the dam.

Figure 12–11. *A*. Map of the Vaiont, Italy, reservoir and slide area. *B*. Cross-section through the reservoir and principal slide area showing the bedrock structure and slide plane. (From G. A. Kiersch, *Civil Engineering,* v. 34, 1964.)

slide took place in about 60 sec. During that time more than 260 million yd³ (240 million m³) of rock slid into the reservoir. This slide sent rocks and water more than 843 ft (260 m) up the opposite side of the reservoir. Waves rushed over the top of the dam, rising as high as 300 ft (100 m) above the abutment. The wall of water swept down the valley with such velocity that an air-pressure wave developed in front of it. Several small towns along the path of the flood were destroyed, all in the span of 7 min. About 2000 persons lost their lives.

Remarkably the Vaiont Dam did not collapse. The site was ideally chosen for a dam of this type of construction (thin-wall concrete). Its interior and all tunnels were flooded immediately (including the powerhouse) after the slide only to be opened by the tensional decompression of air currents.

Despite the survival of the dam structure, the reservoir basin should have been more completely explored and the details of the geologic history of the area studied before a decision on whether or not to build a dam on this site was reached. In this instance, the ignoring of the warnings of nature produced disastrous results when the system of rocks and water within the reservoir chose to stabilize itself after the interference of human activity.

Anchorage, Alaska, 1964. The Good Friday earthquake of March 27 affecting the southeast coast of Alaska produced many secondary effects. Not the least damaging of these effects were many slides of several types. They involved rotational slides (Figure 12–12), translational slides (those with the slide plane parallel to the ground sur-

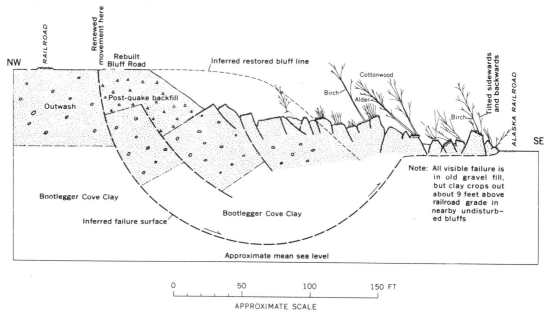

Figure 12–12. Diagram of one of the rotational slides occurring in the Anchorage, Alaska, area. (U.S. Geological Survey.)

face), block glides, and complex slides with more than one type of movement. A submarine slide occurred at the port of Valdez east of Anchorage and removed a portion of the dock facilities there and other types of vertical movement destroyed most of the town. Subsequently the entire village was relocated on more stable ground.

Central in importance to all the slides in the Anchorage area is the Bootlegger Cove Clay, an unconsolidated sedimentary unit of Pleistocene age. This unit contains a zone that is easily sheared, has a high water content, and is highly sensitive so that the earthquake shock waves can readily set it in motion. Added to this, in every locality where a major slide was produced by the earthquake, steep slopes were involved.

The largest and most complex slide was the Turnagain Heights slide. A translatory slide (Figure 12–13), it may have begun as a series of block glides with the surface of movement existing within the Bootlegger Cove Clay. However, the initially intact blocks were disrupted quickly into many directions. Roads, homes, and other structures on the slide were destroyed by tilting, pulling apart, warping, and twisting, as the blocks separated, rotated, and moved up and down relative to one another (Figures 12–14 and 12–15). Pressure ridges were formed by compression along the toe of the slide. This slide destroyed 75 homes in the Turnagain Heights residential area and broke up an area

of about 130 acres, lowering the ground surface an average of 34 ft (11 m). About 12.5 million yd^3 (9.5 million m^3) of material were moved seaward by the slide. Its leading edge glided over the tidal mudflat into Knik Arm. Maximum movement by material within the slide was 2000 ft (615 m).

Damage from the slide was not limited to the area of complete devastation and movement. The ground south of the area surrounding the slide was highly fractured. Many houses suffered structural damage, and underground utilities and other services were disrupted. U.S. Geological Survey geologists found that a much larger slide would have occurred if shaking had not ceased.

Several smaller slides in the Anchorage area were also quite damaging. Some of these involved slumps; others were composite slides (Figure 12–16), beginning as slumps or block glides and then becoming earthflows or debris avalanches. As in the Turnagain Heights slide, the Bootlegger Cove Clay was involved as the zone localizing movement.

Human activity had little to do with the slides at Anchorage. Men and women just happened to be in the vicinity of a major earthquake (Richter magnitude 8.5; see Chapter 19) and suffered the consequences. More than 100 people were killed, and damages ran into the hundreds of millions of dollars. A great deal was learned by hindsight as a result of this disaster. Much of the

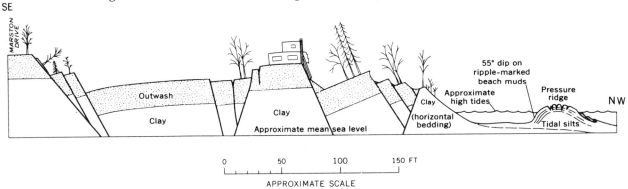

Figure 12–13. Cross-section through the Turnagain Heights slide. (U.S. Geological Survey.)

Figure 12–14. Structural damage due to the Turnagain Heights slide. (W. R. Hansen, U.S. Geological Survey.)

Figure 12–15. Damage to a school in Anchorage. (M. G. Bonilla, U.S. Geological Survey.)

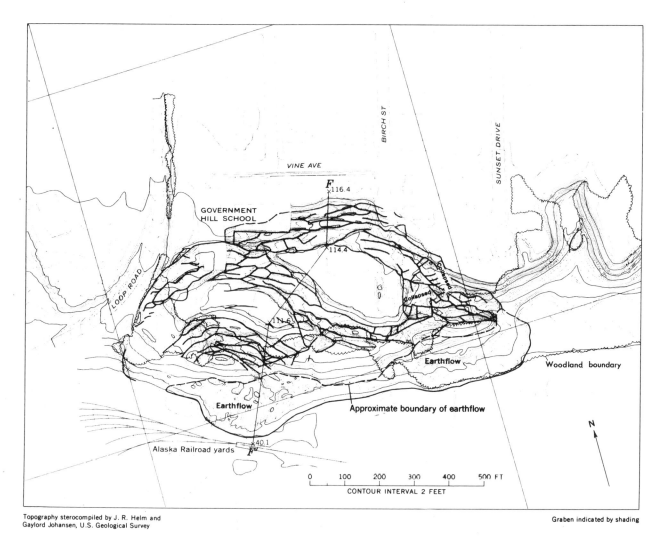

Graben indicated by shading

Figure 12–16. The Government Hill slide in Anchorage, Alaska. (U.S. Geological Survey.)

data gathered are being used for future planning in the Anchorage area and can also be applied to other earthquake- and landslide-prone areas around the world.

Frank, Alberta, 1903. The rock fall and rock slide that took place at Frank, Alberta, in the Canadian Rocky Mountains are thought to have been caused by a combination of natural and human-related circumstances. The slide originated on Turtle Mountain in limestone whose bedding is inclined away from the valley (Figure 12–17). However, fractures in the limestone dip toward the valley; these had been opened up by years of percolating water and frost action. An earthquake in 1901 probably added to the unstable condition of these rocks.

Frank was a coal-mining town. The coal is contained in a series of relatively soft Mesozoic shale and sandstone units over which the limestone had been thrust. The ac-

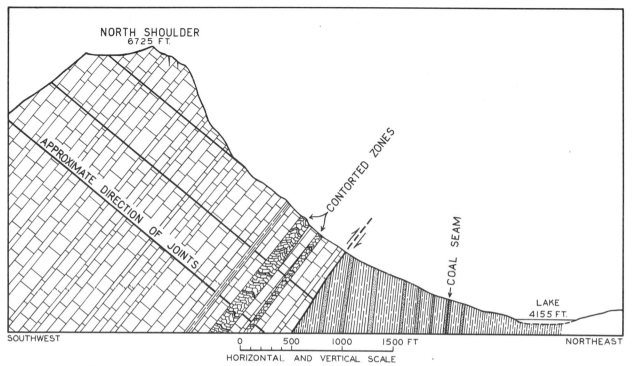

NORTH SHOULDER
6725 FT.

APPROXIMATE DIRECTION OF JOINTS

CONTORTED ZONES

COAL SEAM

LAKE
4155 FT.

SOUTHWEST

0   500   1000   1500 FT

HORIZONTAL AND VERTICAL SCALE

NORTHEAST

Figure 12–17. Cross-section showing the bedrock geology and surface conditions leading up to the Frank, Alberta landslide. (From C. F. S. Sharpe, *Landslides and Related Phenomena,* Columbia University Press, 1938.)

tivities associated with mining are thought to have had some effect in initiating the slide.

The slide occurred as movement in the limestone along joint surfaces that were inclined into the valley. The mass of rock fell and slid down steep talus slopes some 3100 ft (950 m) to the valley floor and then across the valley and 400 ft (125 m) up the opposite side (Figure 12–18). The volume of the slide was estimated at 30 million yd³ (23 million m³). The slide, which took place in the very

Figure 12–18. The Frank, Alberta slide of 1903. (From C. F. S. Sharpe, *Landslides and Related Phenomena,* Columbia University Press, 1938.)

187

early morning, killed 70 people in the town.

Perhaps the Frank slide would have occurred regardless of whether coal was being mined there. This slide came off the main peak of Turtle Mountain. Another potential slide area exists on an adjacent peak of the same mountain. With the same conditions existing there, the town of Frank can only await the moment when it will course down the mountainside. Hopefully it will miss the town.

DESIGN AND PLANNING TO AVOID MASS MOVEMENT HAZARDS

A great deal of information exists concerning causes and circumstances of mass movement processes. We have put some of this knowledge to work for us to prevent slides or to help control the advance effects of these processes. For example, after the Hebgen Lake, Montana, earthquake in 1959 caused a massive slide to dam the Madison River, a channel was dug through the slide and then lined with large boulders before the water level rose over the crest of the dam. This ensured that the water thus impounded would not overflow and then rapidly cut through the dam to flood the area downstream. This measure was modeled on the experience at Gros Ventre, Wyoming, after the 1925 slide blocked the Gros Ventre River, and then was cut through by the river, causing a major flood. Most people believe that a major flood on the Madison River was avoided in 1959 by quickly cutting an outlet channel to control the flow.

Yet every year we read about other areas devastated by landslides. It is likely that other disasters and losses of property could have been avoided by a more detailed knowledge of bedrock and surficial geology during planning stages for industrial and residential developments, highways, and other engineering works. It has long been standard practice to obtain very detailed knowledge of the geology of dam sites. But considerably less effort is spent acquiring knowledge of the mass movement potential that will result in reservoir basins. Seldom does this potential enter into the decision on whether or not to construct a dam on a particular site. Yet this information can be obtained at a small fraction of the cost of the project. Hopefully in the future those who make the decisions on projects involving land use will realize the importance of, and make greater use of, geologic knowledge in their planning.

SUMMARY

Mass movement refers to processes involving the downslope movement of material under the influence of gravity. Steep slopes, water, vegetation, bedrock structure, rock type, and degree of weathering may influence mass movement. Earthquakes, changes in slopes water, changes in properties of materials, and human intervention at the surface of the earth can serve to set off mass movement processes.

Mass movement processes may be divided into two groups: slow and rapid types of movement. Slow types of mass movement include creep, solifluction, rock glaciers, and slow earthflows. Rapid types include slump, rockfalls, slides, and flows. One explanation for very rapid movement of certain types of materials is that they trap a cushion of air beneath them and are lubricated by the trapped air.

A great deal has been learned from large mass movement events in the past. By detailed study of these well-documented events and of the geologic and other situations that caused them to occur, perhaps we will be able to control mass movement better in the future in order to reduce their hazardous aspects.

Questions for thought and review

1. Why must human activity be considered a geologic agent of erosion and deposition today?

2. Why would the existence of one or more factors favoring mass movement not necessarily mean that it will occur?

3. How does water saturation of unconsolidated clays and silts frequently result in mass movement?

4. How can a rockslide take place in an area where bedrock dips into the slope and is not parallel to it?

5. How can humans initiate landslides?

6. Why do solifluction lobes not occur in Kansas?

7. How does a rock glacier differ from a mass of colluvium?

8. How would you prove or disprove that the soil on a steep slope near your home is creeping?

9. Explain the process of slumping. Why do many slumps become earthflows at their toes?

10. Why do mudflows not often occur in the mountains of the eastern United States?

11. What evidence leads to the conclusion that many landslides and avalanches ride on air cushions?

12. What measures could be taken to prevent landsliding in the Anchorage, Alaska, area as a result of future earthquakes?

Selected readings

Ken, P. F., 1963, Quick Clay, *Scientific American*, vol. 209, no. 5, pp. 132–142.

Kiersch, G. A., 1965, Vaiont Reservoir Disaster, *Geotimes*, vol. 9, pp. 9–14.

LaChapelle, E. R., 1966, The Control of Snow Avalanches, *Scientific American*, vol. 214, no. 2, pp. 92–101.

Sharpe, F. S., 1938, *Landslides and Related Phenomena*, Columbia University Press, New York.

U.S. Geological Survey, *Professional Paper 542*. Several volumes present different aspects of the impact of the Alaska earthquake on the area affected.

U.S. Geological Survey, *Professional Paper 435*. Several investigators examine different aspects of the Hebgen Lake, Montana earthquake of August 17, 1959.

13 Streams: the greatest geologic agent

STUDENT OBJECTIVES

On completing this chapter, the student
should:

1. recognize the role of the hydrologic
   cycle in the biosphere

2. be able to identify sources of water
   supply for surface streams

3. discover the relationship between
   stream flow and elements of the
   hydrologic cycle

4. recognize the forces in stream flow that
   determine the work of erosion,
   transportation, and deposition

5. recognize surface features associated
   with erosion and deposition

6. understand the need for surveillance of
   water usage and water quality

All things by immortal power
Near or far
Hiddenly
To each other linked are,
That thou canst stir a flower
Without troubling a star.

Francis Thompson

THE BIOSPHERE

The uniqueness of our planet is attributable in large measure to the presence in our environment of the two fluids—air and water. Without them the biosphere could not exist (Figure 13–1). We could not exist. The idea of the biosphere was introduced by Austrian geologist Eduard Suess in a small book, *The Origin of the Alps*, published in 1875. It appears in the closing chapter as he discusses the various envelopes of the earth. The concept received little attention in scientific circles until the publication of two books by Russian mineralogist Vladimir Vernadsky, the first appearing in Russian in 1926 (*Biosfera*) and the second in French in 1929 (*La Biosphere*).

What is so special about the biosphere? The biosphere, the earth's thin film in which life can exist, is described first as a region in which water exists in large amounts; second, it receives an abundant supply of energy directly or indirectly from the sun; and third, within this region are found interfaces between solid, liquid, and gaseous states of matter. A delicate balance exists among these components, and the maintenance of this balance is essential to our existence. The biosphere is sustained by a series of energy cycles, each depending on harmonious equilibrium.

If life is to continue within the biosphere, materials within it that are basic to existence must undergo cyclic changes. After their

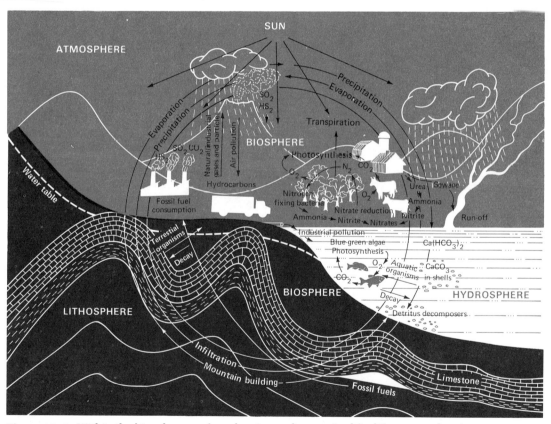

Figure 13–1. Within the biosphere are found major cycles required for life on our planet. The sun supplies the energy required to maintain its operation. The exchanges are carefully balanced. Man's interference with this natural balance could prove disastrous.

193
STREAMS:
THE
GREATEST
GEOLOGIC
AGENT

utilization solar energy must convert them into substances that can be reused.

The natural rate of time required for recycling varies from that of hours, as in the case of the oxygen–carbon dioxide cycle; to days, as in the hydrologic cycle; and even to years, as in the circulation of matter in terrestrial organisms. Recycling solid matter in the earth's crust takes an almost unbelievably long time. For example, calcium is carried from continental rocks by streams and rivers as calcium bicarbonate to the oceans. When conditions become favorable, calcium bicarbonate precipitates as calcium carbonate. Calcium carbonate, found largely in shells of large and small marine animals, forms massive deposits on ocean floors. Replacement of the calcium to the biosphere is believed by many geologists to take place with the movement of the ocean floors toward coastal mountain-building regions. This replacement cycle is measured in hundreds of millions of years.

Artificial injection of matter into the atmosphere and hydrosphere through human activity is occurring at a much faster rate than it did before the industrialization of society. Emissions seen in modern industrial skylines and effluence of industrial wastes into waterways attest to our interference in the natural cycling of matter. These injections, together with newly created cycles of nuclear radiation and toxic elements and compounds (such as methyl mercury, lead, and DDT) are cause for great concern. Only by understanding the geologic implications of energy cycles can we really feel that we are preparing ourselves for an active role in solving the challenging problems that confront us. Knowledge of natural cycles and prudent use of natural resources are now, more than ever before, a prerequisite for survival. (This problem will be reviewed more thoroughly in Chapter 24.)

HYDROLOGY

The most abundant single substance in the biosphere is *water*. In the biosphere water exists as a liquid (fresh and salt), as a solid (ice), and as a gas (vapor). The best estimate of the total volume of the world's water supply is 326 million mi³ (1.5 billion km³). A distribution of this total is shown in Figure 13–2 and in Table 13–1.

What we recognize as water is not simply water; it is water and something else. This combination makes water a unique substance. Water picks up more of the materials

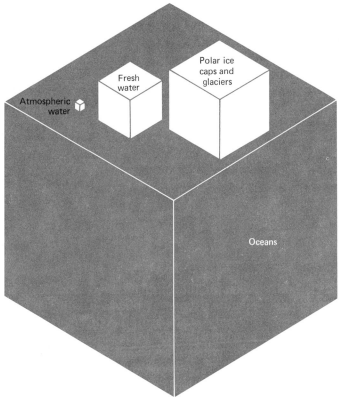

Figure 13–2. The world's estimated supply of water consists mainly of salt water in oceans (97% of the hydrosphere). Of continental waters (3% of the total water supply), 78% is locked in the form of polar ice caps and glaciers. Surface water other than frozen water constitutes 0.5%. Groundwater accounts for the bulk of the remainder. Although atmospheric water, constitutes only $\frac{1}{1000}$ of 1%, it is vital as an energizer of the weather systems that sustain the hydrologic cycle.

Table 13–1. Distribution of the world's supply of water.

Location	Estimated volume		Percentage of total water
	mi³	km³	
Polar icecaps, glaciers	7,300,000	30,660	2.24
Freshwater lakes	30,000	126,000	.009
Salt lakes, inland seas	25,000	105,000	.008
Average in streams	280	1,176	.0001
Groundwater	2,000,000	8,400,000	.612
Soil moisture (root zone)	6,000	25,200	.0018
Oceans	317,000,000	1,331,400,000	97.1
Atmospheric moisture	31,000	13,020	.001

SOURCE: R. L. Nace, 1960, Water management, agriculture and groundwater supplies, U.S. Geological Survey Circular 415, pages 1–11.

that it contacts than any other substance known. Because of this property, water is called the *universal solvent*. Rainwater washes some carbon dioxide from the atmosphere and becomes a weak acid. In passing through the air, other materials such as sulfur dioxide and ammonia are collected by it. This modified water falls on the earth's surface and reacts with materials in rock and soil. Metals that dissolve in weak acids are picked up by fallen water. Sodium, potassium, calcium, magnesium, iron, and zinc dissolve in water and reduce its acidity but in turn increase its concentration of elements. Rainwater is further modified by residues of human activity, such as DDT, oil, chlorides, phosphates, phenol, sewage, and sediment from cultivated fields. These affect water adversely and are among the causes of water pollution.

Water undergoes a series of ceaseless physical and chemical changes. The physical changes involved in evaporation, condensation, and precipitation constitute the hydrologic cycle described in Chapter 4. At the same time, change takes place as water leaches out and carries away minerals in rock and soil.

The long-term operation of the hydrologic cycle has had profound effects. Vast amounts of the most active metals — potassium, sodium, calcium, magnesium — are

now in the oceans. The salt of the seas began to be carried from the continents many millions of years ago, and this process continues today. It is inherent in the hydrologic cycle that most of the metals, salts, and, recently, human residues — substances that are not affected by evaporation — remain in the seas. The buildup of pollutants in the oceans constitutes a danger to future generations.

THE ORIGIN OF RIVERS

Geologists agree that the most important geologic agent in the development of the earth's landscape is running water. Running water in the form of rivers and their numerous tributaries plays an important role not only in the sculpting of the earth's surface but also in the development of civilization. When one examines any map, it becomes apparent that important cities, transportation routes, political boundaries, and distribution patterns of population are all closely related to stream networks.

Rivers and streams of the world were used long before people understood the origin of these waters. The gushing of streams of water from cracks and fissures on mountains or hillsides probably made a deep impression on early peoples. Where did this water come from? One of the early Greeks, Anaxagoras (500–428 B.C.), recorded natural

195
STREAMS:
THE
GREATEST
GEOLOGIC
AGENT

phenomena. He concluded that these streams must have originated from underground lakes or water-filled caverns. References to a great body of subterranean water appear in the books of "Genesis" and "Exodus." In the story of the deluge it is said, "The fountains of the Great Deep were broken up . . . ." Water from under ground welled up to cover the entire earth.

A subterranean source of water to feed rivers did not satisfy all ancient philosophers. Aristotle (384–323 B.C.) objected to this view, for, no matter how large such caverns and lakes might be, eventually their water would be drained by the continuous flow of rivers. He believed that heat (the sun) would evaporate water into the atmosphere and that on cooling it would fall as drizzle or rain. He did not associate precipitation, however, with stream development on the surface of the earth. Instead he applied this principle to the formation of water beneath the earth's surface. He proposed that, as one goes deeper into the ground, coolness of the soil condensed the water vapor found in the soil and that the water thus formed collected into streams and flowed out of the earth as springs.

Perhaps the first writer to bring the origin of rivers openly to the attention of scholars was the celebrated French potter, Bernard Palissy (c. 1514–1589). Palissy had a low regard for the speculative approach used by many medieval philosophers. He relied completely on those facts of nature that he could observe. He states:

> Having made the question of the origin of springs the subject of close and long continued study, I have learned definitely that these take their origin in and are fed by rain and by rain alone.
>
> And these rain waters rushing down the mountain sides pass over earthly slopes and fissured rocks and sinking into the earth's crust, follow a downward course till they reach some solid and impervious rock surface over which they flow until they meet some

opening to the surface of the earth, and through this they issue as springs, brooks, or rivers according to the size of the exit, and the volume of the water to be discharged.

Pierre Perrault, in a book entitled *De l'Origine des Fontaines*, published in Paris in 1674, describes the results of his first attempt to measure rainfall in order to establish a relationship of the amount of precipitation and the amount of water carried off by the rivers. Perrault placed rain gauges throughout the Seine River basin to determine the average precipitation falling during the years 1668–1670. He discovered the amount of water that was carried from this area in the Seine River by measuring the water that passed through the Seine canal at Aigney-le-Duc during this time period. His calculations show that precipitation on the watershed exceeded runoff by six times. Here was evidence that rainfall in the basin was more than enough to supply all the streams in the drainage net.

Recent studies support the findings of Perrault. Leopold (1962) summarized the findings:

> The excess of precipitation over evapotranspiration losses to the atmosphere is a surprisingly small percentage of the average precipitation. The average amount of water that falls as precipitation over the United States annually is 30 inches. Of this total fall, 21 inches are returned to the atmosphere in the form of water vapor through processes of evaporation and transpiration from plants. The balance of 9 inches represents the excess which contributes to the flow of rivers.

THE HYDROLOGIC CYCLE AND STREAM FLOW

The average amount of water from rainfall and other forms of precipitation over the world is approximately 40 in. (1 m). It is estimated that 20% of this precipitation becomes runoff. In the United States, based on measurements made by the U.S. Geological

Survey, one-third of precipitation becomes runoff. The percent of runoff varies greatly in different regions of the United States. Variables such as regional soil and rock composition, amount of previous soil saturation, and intensity of rainfall will determine the percent of surface runoff in a given locality. A short torrential shower will produce more runoff than a long drizzle.

Discharge

Records of stream flow are kept by the U.S. Geological Survey at more than 7300 stream gauging stations. Stream flow is measured by establishing a *stage-discharge rating curve* (Figure 13–3). The flow is measured by recording the height in feet (stage) of the surface of the water above an arbitrary datum plane. This is plotted along the vertical axis. Discharge, measured in cubic feet per second (cfs), is plotted along the horizon-

tal axis of the graph. The equation to find discharge is:

$$Q = AV$$

where Q is discharge, A is cross-sectional area of the river where measurements are taken, and V is mean (average) velocity of water at the measured cross-section. A graph of the plotted points (stage and discharge) can be used to estimate the discharge for any measured gauge height. The total quantity of water flowing past a gauging station in a given time period (year, month, or day) is determined by summing the product of discharge and the time unit. The total amount of water flowing past a station is expressed in acre-feet. A flow of 1 cfs for one day is equivalent to about 2 acre-feet.

Infiltration

Infiltration is closely related to runoff. Infiltration capacity of soil controls runoff in a watershed. Precipitation falling on land surfaces is disposed of in several ways. If soil is not saturated, water will infiltrate into the ground. The rate at which it infiltrates will be determined by such variables as soil structure, texture, vegetative cover, biological structures present (worm borings, animal burrows), and amount of moisture retained in soil from previous rainfall.

Infiltration capacity is the maximum sustained rate at which water will be absorbed by a soil. When infiltration capacity is exceeded and precipitation continues, the excess is rejected by the soil and flows off as surface runoff. Precipitation that falls in amounts less than the infiltration capacity of the soil will produce no runoff.

Initially the infiltration of water into the soil is rapid. As precipitation continues, however, it decreases quickly and approaches a steady value—the infiltration capacity of the soil. The infiltration capacities of two types of soil, sandy loam and clay, are shown in Figure 13–4.

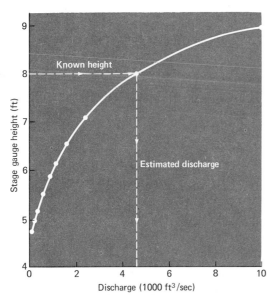

Figure 13–3. Stage-discharge rating curve shows the relationship between height of water at gauge and amount of discharge. At any known river stage the corresponding discharge may be found by reading directly from the plotted curve.

197
STREAMS:
THE
GREATEST
GEOLOGIC
AGENT

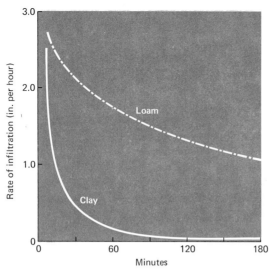

Figure 13–4. Infiltration capacity of soils varies. Infiltration capacity of clay is about 0.05 in. per hour; that of sandy loam is approximately 1.0 in. per hour. The difference may be attributed to the concentration of clay-sized particle. Loam contains 6% and clay 63% of particles finer than 0.002 mm.

THE WORK OF STREAMS

Running water is the most powerful geologic agent. Although it is most often thought of as an erosional force, streams do much more than wear down rocks and soil and transport sediments. Streams also build landforms. The work done by streams can be classified in three categories: (1) erosion, (2) transportation, and (3) deposition.

Erosional work of streams

The energy of running water tears away (Figure 13–5) or dissolves tremendous amounts of the material it touches and transports this debris and dissolved matter along its course until hydraulic factors change so it can no longer hold the load it has been moving. The erosive powers of streams were not recognized generally until the nineteenth century. Prior to that time, De Saussure, between 1779 and 1796, gave detailed accounts

Figure 13–5. A rampaging stream has gained the respect of all who live near it. Floodwaters of great magnitude are rare events, spanning decades or a century. When they do occur, however, their violence leaves a heavy imprint on the environment. (U.S. Geological Survey.)

of his studies of the Alps. He wrote that valleys were products of streams that ran through them. At that time popular belief held that streams flowed in valleys because valleys were already there, formed through a cataclysmic event.

Near the end of the eighteenth century Hutton and Playfair attacked the cataclysmic theory, stating that streams erode their own valleys. In their argument they cited evidence that a catastrophic event would produce valleys that would be straight and have no branching tributary valleys. The irregularity of drainage patterns was proof that valleys were cut by running water.

Geologists exploring the western United States during the second half of the nineteenth century, notably J. W. Newbury, G. K. Gilbert, and J. W. Powell, gathered considerable data supporting the erosional role of rivers. In 1899 W. M. Davis systematized the principles set forth by these scientists into the concept known as the *erosion cycle*. Davis ascribed to streams and valleys developmental periods which he called youth, maturity, and old age, based on the amount of energy present in the streams.

Stream erosion cycle. According to Davis, a stream is young if it has a strong current flowing in a V-shaped, steep-walled valley (Figure 13–6). It carries a load that is small in amount, but the particles making up the load are large in size. The young stream has an excess of energy over the work that it puts into the carrying of its load.

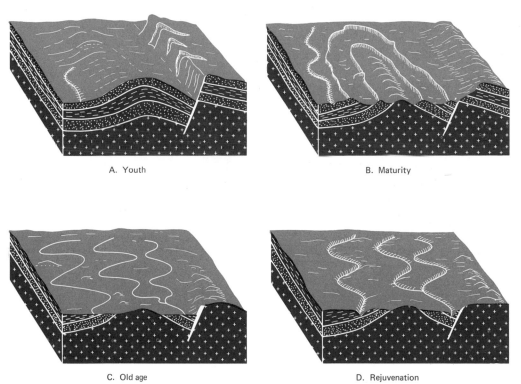

A. Youth

B. Maturity

C. Old age

D. Rejuvenation

Figure 13–6. *A, B,* and *C* show the three stages of Davis's ideal fluvial cycle of denudation. Streams alter the landscape, reducing the angularity of youthful landscape slowly to the flattened landscape of old age. The process can begin again if the region experiences uplift. The entrenched meanders in *D* are characteristic of rejuvenation.

199
STREAMS:
THE
GREATEST
GEOLOGIC
AGENT

Figure 13–7. A river in old age develops many meanders as it follows a crooked course across a wide floodplain. Occasionally meanders become so curved that the river, in its haste, cuts across the narrow neck of land at both ends of the loop and isolates the meander from the river. The isolated meander is an oxbow lake. (U.S. Geological Survey.)

This extra energy is used to down cut the stream bed.

As the gradient of the stream decreases because of downcutting, the amount of energy is also decreased to a point at which it is barely sufficient to transport its load. Here the stream reaches *maturity*. At this stage of development the rate of downcutting slows, and the stream uses its energy to widen the valley. Floodplains appear, and the stream begins to meander systematically.

A stream approaches *old age* when it is incapable of carrying anything but fine debris. Valley widening is dominant over downcutting. The floodplain is wider than the meander belt. Natural levees, oxbow lakes, and meander scars are common (Figure 13–7).

Davis has been criticized for proposing a rather simplistic explanation of stream development. In the 1920s and 1930s Walther Penck and others attacked the Davisian cycle, disagreeing strongly with Davis's assumption that crustal adjustments of the earth were relatively rapid in the initial uplift of the land and were followed by a long, static period that permitted a cycle to run its course. Penck maintained that initial uplift of land is extremely slow and is then followed by an accelerated uplift rate, thus preventing the stages of development described by Davis. A number of geologists also pointed out that rivers may show all stages of development along their course, because streams lengthen by progressive erosion in the headwaters area.

Entropy and dynamic equilibrium.
Erosion performed by running water is extremely complex. Involved in the process of erosion are a number of engineering and physical science concepts. The unifying principle is believed to be the second law of thermodynamics, which describes the concept of *entropy*. In general an increase in entropy is a measure of a decrease in availability of energy in a system to do mechanical work.

Applied to streams, we note that more mechanical (or erosive) work is accomplished in the upper reaches of a stream than in the lower reaches of a stream. This is because the slope of a stream is almost always steeper in the upper reaches, so water and its sediment load have greater available energy to carry out mechanical work on the stream bed and valley. In other words, there is decreased entropy in the upper reaches of a stream system. With distance downstream, slope tends to become more gentle, and energy available for erosive work decreases. Stated simply, entropy increases downstream.

Along the length of a stream system a balance is always present between work done and the load a stream will carry, with adjustments constantly taking place in order to maintain this balance. This is called *dynamic equilibrium*. There is a parallelism between level of entropy and state of dynamic equilibrium.

In an open river system, the rate of increase in entropy equals the rate of dissipation of energy as heat. This may be expressed as the principle of least work in nature. A stable system is one of least work. This does not imply mathematical balance but rather a quasi-state of equilibrium in which adjustments in the variables are continuous.

The distribution of energy available for erosional work in a stream system is reflected in the stream's *longitudinal profile*, a graphic expression of the steepness of slope along the stream's course from headwaters to mouth. Initial downcutting proceeds at an accelerated rate to produce steep grades. As the stream approaches sea level, much of its energy is dissipated; its grade will decrease. The result is a longitudinal profile that is concave to the sky (Figure 13–8).

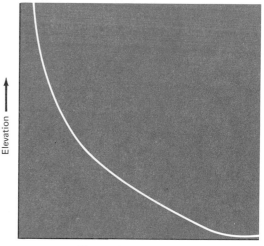

Distance downstream ⟶

Figure 13–8. Longitudinal profile of a stream flowing in a region of normal precipitation assumes a cross-sectional silhouette that is described as concave to the sky.

201

STREAMS:
THE
GREATEST
GEOLOGIC
AGENT

A QUANTITATIVE STUDY OF STREAMS

Stream flow is as changeable as weather, because it is directly related to it. Stream flow may at times be described as a smooth, steady flow. But if weather conditions change, the same stream may become a rapidly flowing, turbulent, murky body of water sweeping away everything in its path. Even under constant weather conditions a stream may flow smoothly in its reaches, tumble crazily in rapids, plunge downward in falls, or pause quietly in deep pools. Although stream flow varies at points along its route, for engineering purposes it is assumed that the stream has a steady and uniform flow. This assumption implies that the depth of the stream is constant over the entire course and that depth does not vary from day to day or from month to month. Obviously these assumptions cannot be accepted as representing true conditions; but, if so viewed, they simplify solutions to engineering problems related to flow of rivers.

If we assume that river flow is steady and uniform, then the rate at which water passes through successive cross sections is constant (Figure 13–9). We can express this mathematically as:

$$A_1V_1 = A_2V_2 = A_3V_3 = Q$$

where A is cross-sectional area at points 1, 2, and 3; V is average velocity at those points; and Q is amount of water discharge.

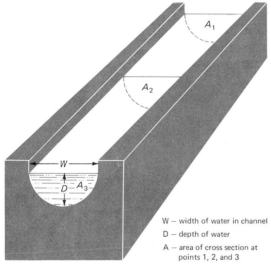

W — width of water in channel

D — depth of water

A — area of cross section at points 1, 2, and 3

Figure 13–9. Stream flow engineering concepts are based on the assumption that, except during flood periods, a stream has a steady and uniform flow. A steady flow further assumes that the depth of the stream is constant over the length of the stream. Obviously these two assumptions are not exact, but, in accepting them as a base for calculating stream flow, hydraulic engineers can quickly find satisfactory solutions for river engineering problems. Assuming a steady, uniform flow, the rate at which water passes through successive cross sections of the stream is constant ($A^1V^1 = A^2V^2 = A^3V^3 = Q$).

Discharge

Discharge, introduced earlier in the chapter in the discussion of runoff, is the volume of water flowing through a particular cross section of the river or stream per unit of time. Discharge depends on the mean (average) velocity of the river at the cross section where the velocity measurement is made. The relationship is expressed as follows:

discharge = average velocity
$\times$ cross-sectional area

or

$$Q = AV$$

An examination of the equation shows that, for a given discharge (Q), a decrease in cross-sectional area (A) will result in an increase in velocity (V). If a stream channel is constricted by resistant rockwalls in a section of its course, the velocity of the stream will be increased (Figure 13–10).

Increased discharge usually means an increase in cross-sectional area and an increase in velocity. Discharge of a river increases downstream as tributary streams empty into it. As a result of the addition of water to the main stream, the river not only widens and

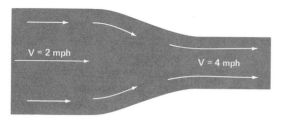

Figure 13–10. Stream velocity increases when the channel narrows.

deepens its channel, it also slightly increases its velocity as it approaches the mouth of the river, even though the gradient (slope) of the river decreases noticeably as it nears the base level. Travelers to New Orleans are amazed at the velocity of the water in what they assumed to be the sluggish Mississippi River.

Laminar flow. A smooth flow of water at relatively low velocity in which fluid elements follow paths that are straight and are parallel to the channel walls of the stream is *laminar flow* (Figure 13–11). Laminar flow in streams is found under conditions where water depth is slight and the velocity is less than .06 mi/hr (96 m/hr). Successive layers of water glide over one another like cards in a sheared deck of playing cards. Each layer remains fairly distinct from the others over

Figure 13–11. Theoretical vertical profile of two types of stream flow. *Left:* In laminar flow the stream lines (or surfaces) remain distinct from one another. It is a smooth flow at relatively slow velocity; the fluid elements follow paths that are both straight and parallel to the channel walls. *Right:* In turbulent flow stream lines are chaotic, with a heterogeneous mixing of fluid elements.

the entire length of the stream. However, laminar flow in nature is not common when compared with turbulent flow. Little or no sediment is carried in laminar flow; thus it has only a minor role in landscape shaping. It is found primarily in groundwater movement.

Turbulent flow. *Turbulent flow* results from resistance or drag caused by roughness of the channel and by an increase in the velocity of the water. Water surfaces in contact with the channel walls and stream bed are stopped by friction, while the rest of the water moves along as before. This disturbance together with numerous irregularities in the channel produces a series of eddies. Turbulent flow is characterized by rapid changes in direction and speed of water particles (Figure 13–11). After a heavy rainfall turbulence in a stream can result in significant changes along the stream valley. Turbulent flow is highly effective both in eroding the stream's channel and in transporting eroded materials.

Velocity, gradient, and discharge. Streams, propelled by the force of gravity, flow downhill. The *velocity* of a stream is measured in terms of the distance its waters travel in a unit of time. It is usually measured in feet per second. The velocity of a stream of 2 ft/sec (2 km/hr) is relatively slow. A velocity of 22 ft/sec (24 km/hr) is relatively high.

Several factors influence the velocity of a stream. These include *gradient* or slope of the stream bed, *roughness* or *smoothness* of the stream banks and bed, and *discharge* or amount of water passing a given point per unit of time.

The velocity of water in a stream flowing down its channel varies throughout its cross section. The velocity of the water touching the stream bed is theoretically zero. Water flowing near the bed is usually a thin layer of laminar flow. The highest velocity is near

203
STREAMS:
THE
GREATEST
GEOLOGIC
AGENT

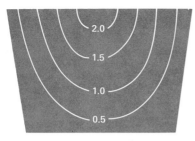

Figure 13–12. Cross section of a stream channel showing velocity differences in various parts. Velocity is indicated in miles per hour. The lines are isovels (lines of equal velocity).

the center of the stream and immediately below the surface of the water (Figure 13–12).

In a cross section of a stream flowing in its course, a typical velocity distribution would approximate the vectors as shown in

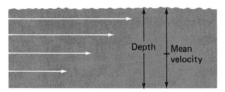

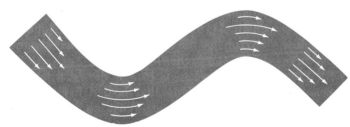

Figure 13–13. *Top:* Vectors show differences in water velocity in a vertical cross section of a stream. The shortest vector (representing lowest velocity) is near the bed of the stream; the longest vector (representing the greatest velocity) is near the surface. Average velocity of the stream is at a point approximately 0.6 of the depth below the surface of the stream. *Bottom:* Vectors illustrate how water velocity of a stream varies as it meanders across the landscape.

Figure 13–13. The average (*mean*) velocity is at a point approximately 0.6 of the depth below the surface of the water. The *shear force* (rate of change of velocity with change in depth of water) is greatest near the bottom.

In reaches (straight flow) the current is most rapid in the center and slowest near the banks (Figure 13–13 bottom). Velocity in bends is greatest on the outside of the bend and slowest on the inside. Notice, however, that frictional resistance at both outside and inside banks slows the prevailing velocity.

LONGITUDINAL PROFILE OF STREAMS

As mentioned before in the discussion of entropy, most streams have a longitudinal profile that is concave to the sky (Figure 13–14). An exception may be found in some ephemeral streams in arid or semiarid regions. Here the volume of water decreases downstream because of ground seepage or rapid evaporation. Sediments carried by the stream are deposited when the volume of water cannot transport the load. Thus streams in dry regions may have a convex longitudinal profile (Figure 13–14).

We might assume from the steep gradient near the headwaters and the slight gradient near the mouth of the river that the velocity of water would be greatest where the slope is greatest. But, although gradients are higher upstream, we find that stream velocity increases downstream and is greatest near its mouth. As tributaries add to discharge, a stream must not only deepen and widen its channel, but it must also increase its velocity.

Deepening and widening of the stream channel reduce the slope of the stream. Were it not for this gradual flattening of the slope, the increase in discharge produced by tributary streams would result in river velocities higher than any observed in nature. In its attempt to find a balance between discharge and other variables (channel, velocity, and

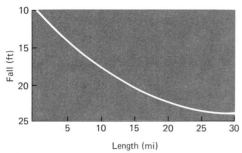

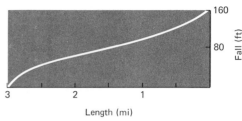

Figure 13–14. *Left:* Longitudinal profile of a stream in a region of normal precipitation is concave to the sky. *Right:* Longitudinal profile of a stream in an arid or semiarid region is convex to the sky. As the volume of water decreases downstream because of ground seepage and rapid evaporation, velocity decreases also. Deposition occurs when water can no longer transport the sediment load (thus giving it a convex profile).

gradient) the river will reduce its gradient and increase its velocity, depth, and width as it flows downstream. The concave longitudinal profile is evidence that the river system tends toward mutual adjustment among these variables in order to result in equilibrium.

Base level of a stream

The base level of a stream is defined as the level below which a land surface cannot be reduced by running water. Anything that checks a stream from lowering its channel may be considered a base level. Layers of resistant rock may serve as temporary base levels. As a stream empties into a lake, its velocity is checked; consequently it loses its ability to erode and cannot cut its bed below the level of the lake. Principal streams serve as local or temporary base levels for their tributaries. Resistant rock strata, lakes, and principal streams are classified as *temporary base levels,* because they may be destroyed ultimately by geologic or biologic processes.

The erosive power of all streams is directly or indirectly influenced by oceans. *Sea level* is considered the *principal base level.* This does not imply that sea level is a permanent base level. At times in the earth's past, worldwide sea levels have changed. If, for example, sea level is lowered, then streams will renew their work of downcutting to the new base level.

Stream adjustment to changing base level

A change in the base level produced either by nature or by human activity will alter the dynamics of a stream. When the base level of a stream is raised by building a dam, the artificial lake produced by this obstruction in the stream's course will be reflected by changes in the upstream profile. The artificial lake slows the velocity of the stream by causing it to deposit its load (Figure 13–15 top). The gradient of the stream above the dam becomes less steep, further reducing stream velocity. Eventually a new stream channel is formed near the dam; its slope approximates the slope of the original channel but is at a higher level.

The reverse is true if the base level is lowered. Suppose that the dam in Figure 13–15 (top) collapsed. What changes in the stream's profile can be predicted? If the lake is destroyed, the stream's velocity will increase, and it will cut down through the sediments it deposited previously (Figure 13–15, bottom). Eventually the stream profile will adjust to its former state.

We can summarize the effects of changing base levels by stating that generally a stream adjusts to a rise in base level by building up

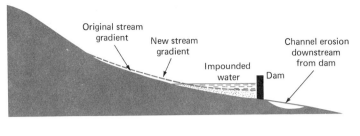

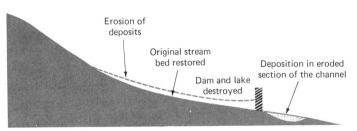

Figure 13–15. *Top:* An artificial lake produced by erection of a dam slows stream velocity, causing deposition of stream load. A new stream channel forms at a higher elevation. *Bottom:* If the dam collapses, velocity of the stream increases. Downcutting of the stream bed resumes through sediments previously deposited, and the original stream bed is restored.

Figure 13–16. The resistant bed of dolomite that forms the lip of Niagara Falls is underlain chiefly by shale. As the less resistant shale deposits erode, the undermined dolomite breaks off in large blocks, causing the lip of the falls to retreat toward Lake Erie.

its channel through deposition of sediment load and adjusts to a fall in base level by eroding its channel downward.

Waterfalls, knickpoints, and rapids

Waterfalls are perhaps the most spectacular of landform features. They first make their appearance in the stream profile as *knickpoints* (points where the stream bed takes a sudden drop). The sudden drop is the result of a number of possible conditions. It may be produced by differences in rock resistance to the erosive power of water on rock strata underlying the stream. It could also result from vertical slippage of rock strata or from faulting (see Chapter 18). Another possibility is topographic changes brought about by an earthquake (see Chapter 19).

Consider the example of Niagara Falls (Figure 13–16). When the North American ice sheet retreated to form the Great Lakes, water flowed from Lake Erie to Lake Ontario across the Niagara escarpment (a steep, almost vertical slope) to produce Niagara Falls. When first formed, the falls lay about 7 mi (11 km) downstream from their present location. Measurements taken during the past century indicate that the Canadian section (the larger of the two falls at Niagara) has been retreating at a rate of 4–5 ft (1–2 m) per year, while the American Falls are retreating at about 3 ft (1 m) per year. The bedrock underlying the falls consists of sedimentary rocks that dip slightly away from the falls and toward the south. The resistant Lockport Dolomite forms the lip of the falls, and the less resistant Rochester Shale underlies it. The more rapid erosion of the Rochester Shale has undermined the Lockport Dolomite, causing large blocks of the dolomite to break off, and has hastened the retreat of the falls. As the falls retreat upstream, the southerly dip of the resistant rock formation will lower the height of the falls. The falls will change eventually into rapids before Lake Erie is reached. It is estimated that the retreat of the falls to Lake Erie will take 25,000–30,000 years.

All knickpoints in the stream profile are transitory. In the process of lowering, they eventually disappear and retreat. Falls thus become rapids. Rapids in turn destroy themselves by impact, eddies, and *cavitation* (sudden collapse of vapor bubbles in a stream that produces extremely strong impact).

Transportational work of streams

Streams are important geologically because they transport sediments. The transportation of eroded sediments to various sites produces many landforms. Streams move eroded materials in a variety of ways: (1) as debris load in suspension; (2) along the bed of the stream channel by rolling, sliding, or saltation (*bed load*); and (3) dissolved and carried in *solution*.

The sum total of material that a stream carries at any time is its *load*. The ability of water to transport material, measured by the quantity it can carry past a given point in a unit of time, is its *capacity*. It would appear that weathering and erosion do not proceed rapidly enough to load a stream to its capacity.

Competency of a stream is measured in terms of the particle of the largest diameter that the stream can move at a given velocity. A small, swiftly flowing stream can move a relatively large particle; thus its competence is great, but the amount of material it transports is small. Conversely, a large, slowly moving stream may carry a great quantity of small particles in suspension; its competence is small, but its capacity is great.

The size of transported particles decreases downstream (even though factors such as discharge, velocity, and capacity increase) because of impact and abrasion. Deposition of transported sediments of a particular size occurs when the velocity of the current falls below the minimum value required for transport (Figure 13–17).

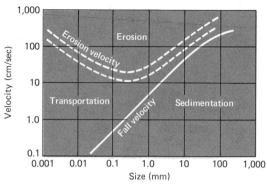

Figure 13–17. Curves of erosion and deposition for uniform material. A wide band is used to define competent velocity, because the value of erosion velocity varies depending on characteristics of the water and the sediment being moved. (Redrawn from F. Hjulstrom, 1935, Studies of the Morphological Activity of Rivers as illustrated by the River Fyris, *Upsala Geological Institute Bulletin* #25, pp. 221–527.)

Critical erosion velocity is the lowest velocity at which loose sediments of a given size will move. Oddly enough, much lower erosion velocities are required to move sand than to move silt. For grains larger than 0.5mm in diameter erosion velocity must increase as the size of particle increases. In Figure 13–17 a wide area rather than a line is used to designate the zone of erosion velocity, because the value varies, depending on such factors as density and depth of water, slope of the stream bed, and density of the particles to be moved.

Some conclusions regarding the transport of sediments can be drawn from Figure 13–17: (1) sand is easily eroded, whereas silts and clays are more resistant to movement; (2) gravel is difficult to entrain (move) because of particle size and weight; and (3) once silts and clays are entrained, they can be transported at much lower velocities.

Settling velocity

The hydraulic principle that may be considered the link between erosional and depo-

207

STREAMS:
THE
GREATEST
GEOLOGIC
AGENT

sitional work of streams is known as *Stokes' law*. Although it applies primarily to small- and medium-sized grains, an analysis of the forces opposing each other (buoyant force versus force of gravity) as they appear in the formula will be helpful in explaining why sediments carried by a stream are deposited along its course. In its simplest form Stokes' law is expressed as:

$$V_s = Cr^2$$

where V_s is settling velocity in centimeters per second, r is radius of the particles in centimeters, and C is a constant relating relative densities of fluid and particle, acceleration caused by gravity, and viscosity of the fluid.

According to this concept the rate of settling of sediment depends primarily on particle size and its density, because other factors of gravity, viscosity, and density of water are constant at a given point in a stream for a given time. We can further simplify this with the following notation:

$$V_s \propto r^2$$

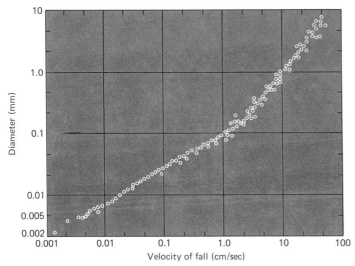

Figure 13–18. A plot of some experimental data on measured fall velocity of quartz sand grains of varying size. (Redrawn from E. W. Lane, 1938, Notes on formation of sand, *Am. Geophys. Union, Trans.*, vol. 18, pp. 505–508.)

or, settling velocity of particles carried in suspension is proportional to the square of the particle radius.

Figure 13–18 plots some experimental findings on measured fall velocity of quartz sand grains of varying size. The data shown indicate that fall increases regularly with grain sizes up to 0.1 mm. Above 0.1 mm, fall velocity increases less rapidly with grain diameter. We can infer from such experimental studies that, when upward components of velocity drop below settling velocity for a given grain size, the particle will drop from suspension and will be deposited.

Moving the sediments

We discussed earlier in this chapter the three principal ways that a stream transports matter: (1) in suspension, (2) through movement of its bed load, and (3) in solution. A further discussion of these modes of transport may be helpful in our understanding of the geologic work of streams.

Suspension. The suspended load is composed of finer sediments that have been previously eroded. It includes those particles supported by the stream and carried above the layer of laminar flow. Once such particles are entrained, very little energy is required to transport them. The suspended load tends to decrease the inner turbulence of the stream and reduce frictional losses of energy. Thus it makes the stream more efficient in its work.

Concentration of the suspended load varies with the depth of the stream below its surface. The concentration is highest near the bed, and it decreases rapidly toward the surface of the stream. The actual point of highest concentration is directly dependent on the area of greatest turbulence. The concentration at any depth depends on the settling velocity of particles and on a measure of turbulent mixing. Suspended sediment is more concentrated when velocity is high and water depth is low.

There is little change in the concentration of very fine particles with depth. In a study of suspended sediments in the Mississippi River, Lane and Kalinske reported that silt particles less than 0.005 mm in diameter remain fairly constant throughout the depth of water. They also reported a very narrow range in concentration from 47.2 ppm (parts per million) at the bottom to 46.1 ppm near the surface.

Movement of bed load. Bed load is composed of sediments moving along the channel bottom in the lower layers of laminar flow. The sediments move by sliding, rolling, or saltation. Do not be confused by the word "saltation"; it is derived from the Latin *saltare*, "to jump." Imagine a particle of sediment suddenly lifted from the stream bed by a turbulent eddy, flung upward, and then dropped back to the stream bed (Figure 13–19). This process is repeated over and over and the particle of sediment travels great distances along, on, or near the stream bed.

Bed load moves at a rate slower than the velocity of the stream. Particles may move

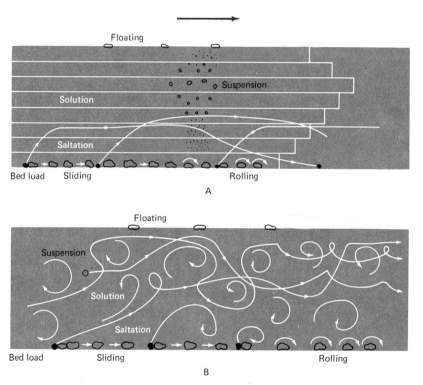

Figure 13–19. Material that a stream picks up from its channel, as well as that which slips into it from its banks, is carried downstream in a variety of ways. Very light substances float on its surface. Fine particles are carried in suspension. Some of the material becomes dissolved in water and is carried in solution. Materials moving along the stream's bottom constitute its bed load. Sediments in the bed load move by rolling, sliding, or saltation. Transport in laminar flow (*left*) varies somewhat from that in turbulent flow (*right*) in streams.

Figure 13-20. Large boulders in a slow-moving stream. These boulders were carried by the stream by floodwaters during a nonrecorded rare event in the distant past and were deposited here when the stream's competency decreased. Another great flood in the future may move these boulders farther down the stream channel.

individually or in groups. Once in motion, large particles move more easily than smaller ones. Rounded particles travel faster than irregularly shaped sediments.

Movement of bed load is related to the competency of a stream. Competency varies along its course as well as seasonally. A stream may move large particles during floods, when volume and velocity of water increase. At such times the competency of the stream increases greatly. We frequently see large, apparently immovable, boulders lying in a stream bed (Figure 13-20). Chances are excellent that these boulders were carried by floodwaters during some nonrecorded rare event in the distant past. It is probable that, if another great flood occurs, the increased competency of the stream will move them farther down the stream channel.

Solution. The dissolved load of a stream is not generally visible, but the corrosiveness of water is an important factor in erosion of land masses. D. A. Livingston in a recent study reported that a total of 4304 million tons (3905 million metric tons) of soluble material is carried annually from continents by streams.

Some streams carry a dissolved load greater than that which they carry as solid particles. The amount of load carried in solution depends to a large degree on the relative contributions of groundwater and surface runoff to the stream discharge. If river flow is stable, resulting principally from groundwater, concentration of dissolved salts is often high. If surface runoff is the primary contributing source of the river's discharge, concentration of dissolved salts is usually low.

Effects of floods on the work of streams

In our discussion of transportation of sediments we mentioned that, the greater the stream velocity, the greater becomes its erosive and transporting power. Hence it follows that, when a stream is at flood stage, the fast-moving water will pick up sediments that may be resting on the stream bed and sweep them downstream. The increased discharge of water will produce an increase in the cross-sectional area of the stream bed. This usually means that the banks of the stream are worn away, but frequently channel depth is also increased by scouring of bedrock. It can be safely stated that erosional work of streams is greatest during flood periods.

The problem of floods is one of probability. Knowledge of the probability of occurrence of floods is becoming increasingly more important as more community developments move into floodplain regions. Floods must be treated as random events. In statistical language the concept is explained thus: Floods taking place during a given time period constitute a sample of an infinitely large population of such events in time. To

Figure 13–21. The intensity of flooding accompanying hurricanes or heavy rainstorms cannot be forecasted precisely. Flood damage may run into millions of dollars in a populated community. (U.S. Geological Survey.)

illustrate, we can say that, if in a time period of 15 years a stream has an extremely large flood, it is probable that within the next 15 years it will have another flood of equal magnitude.

It is likely that very large floods occur only once a year in a given region, so precautions can be made to minimize flood damage. Catastrophic events of rare frequency upset predictions. The devastating floods produced by rainfall accompanying Hurricane Agnes in the eastern United States (Figure 13–21) and by a freak midsummer storm in Rapid City, South Dakota, underscore the complexities involved in predicting floods. Understand that predictions should not be confused with forecasts. Predictions are long-term expectations.

Even though rare occurrences such as the floods initiated by Hurricane Agnes perform only a moderate proportion of the total erosional work of streams, they have a great effect on landforms in the vicinity of the streams. These floods tear away complete tree and shrub root systems, making it possible for the stream to erode its channel at higher rates after the floods subside. Because raging floodwaters offer inhospitable conditions for research, effects of these rare-frequency occurrences are not so well known as are other hydrologic processes.

Data concerning flood frequency have been compiled for many states and for principal river basins in the United States. These data are continually increasing; at present we can estimate the probability of flood discharges from published sources for approximately two-thirds of the United States.

211
STREAMS:
THE
GREATEST
GEOLOGIC
AGENT

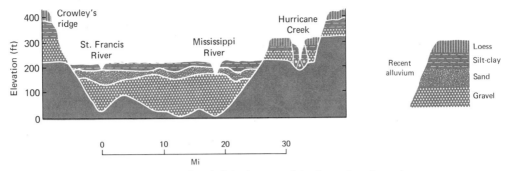

Figure 13–22. The Mississippi River floodplain is one of the largest and most thoroughly researched floodplains in the Western Hemisphere. Its bedrock valley is rugged and is filled with alluvium. Deposits are all young; it is believed that much of the sedimentation occurred during the last glacial epoch. (Redrawn from H. N. Fisk, 1944, *Geological Investigation of the Alluvial Valley of the Lower Mississippi River,* Mississippi River Commission, Vicksburg, Mississippi.)

Depositional work of streams

Sediments picked up and carried by streams give rise to a variety of depositional features along the stream's course. These features include alluvial floodplains, deltas, bars, islands, alluvial fans, natural levees, bahadas, and braided streams (Figures 13–22, 13–23, 13–24).

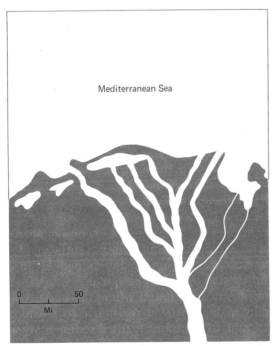

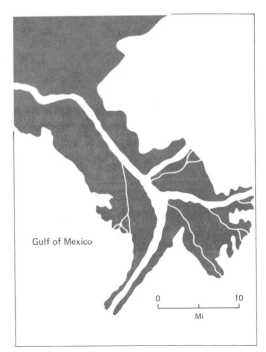

Figure 13–23. The term *delta* is from the Greek letter delta (Δ). Many river deposits take this shape. *Left:* The Nile River delta is an excellent example. *Right:* The Mississippi delta is composed of many branching channels and resembles a bird's foot.

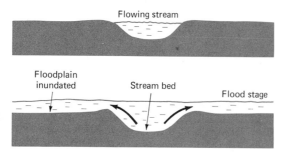

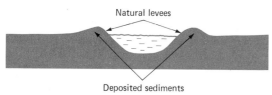

Figure 13-24. A feature associated with stream deposition, common in low-lying lands, is the natural levee. During flood periods, as water rises over the banks and spreads out widely on the floodplain, it loses velocity rapidly. The loss of velocity causes the water to drop much of the sediment it is carrying near the bank. Over a period of time a small ridge, called a natural levee, forms.

The primary cause for deposition is a decrease in stream velocity with a consequent loss of transporting power. The loss of competency (ability to move sediments) may be attributable to:

1. a decrease in gradient (slope) of the channel
2. a decrease in discharge (water is either lost or is spread out over a wider area)
3. damming of the channel (natural or man-made obstructions to flow)
4. increased caliber of the load (when load exceeds stream capacity as found in glacial outwash or rapid denudation)
5. increase of vegetation, produced by high-nutrient pollutants, along banks and in the bed
6. loss of velocity as a stream enters a plain, a lake, an ocean, or standing backwaters of a flooded area

STREAM PATTERNS

The number of different natural stream drainage patterns or nets is surprisingly small considering the wide variability in the size of drainage basins, rock structural patterns, types of bedrock underlying the streams, and climatic conditions found on the earth. The most common drainage pattern is *dendritic*. As the Greek word base implies, the dendritic pattern resembles the branches of a tree.

Other types of drainage patterns are *trellis, rectangular, radial,* and *annular* (Figure 13-25). Trellis and rectangular patterns are characterized by the presence of many right-angle bends in minor streams and linearity in major streams. Radial patterns resemble the spokes of a wheel and develop outward from a central area, such as the peak of a volcanic cone or the top of an intruded dome. As the dome structure erodes away, a less frequent type of pattern appears. The radial drainage pattern of a youthful dome develops into an annular (ringlike) pattern (Figure 13-26).

EFFECTS OF ROCK STRUCTURE ON DRAINAGE

Variations in resistance to erosion by rocks and rock structures of a region will produce surface variations that reflect these differences.

Consequent streams

The course (a consequence of the original slope of ground surface) of a *consequent stream* is determined by the slope of the land. Consequent streams are most likely to be found where the underlying rock is uniform in resistance to erosion and where a uniformly tilted sedimentary unit is present. Consequent drainage systems develop as straight streams flowing downslope or as a radial pattern down the flanks of a volcano (Figure 13-27).

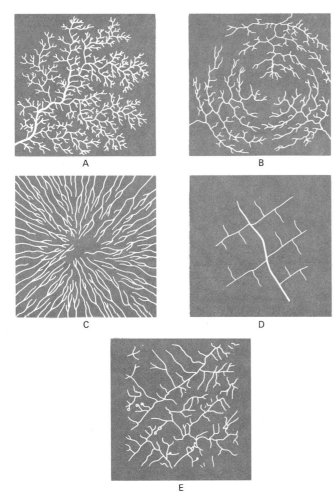

Figure 13-25. Types of drainage patterns: *A.*
Dendritic. *B.* Annular. *C.* Radial. *D.* Trellis. *E.*
Rectangular.

Subsequent streams

Over long periods of time, changes because of uplift or subsidence become evident on the surface of the earth. When a main stream maintains its original course following uplift of the land, it is called an *antecedent stream* (its course anteceded the uplift). But tributaries of an antecedent stream, however, may change their courses subsequent to uplift. The new streams are known as *subsequent streams* (Figure 13-28).

In the Appalachian region many tributaries of major streams flow in valleys floored with less-resistant rocks such as shale and limestone. Resistant parallel ridges of sandstone and quartzite separate these streams to form trellis or rectangular patterns (Figure 13-27). Some tributaries in trellis patterns flow in a direction that is opposite to the dip (see Glossary) of the underlying rock bed. These are known as *obsequent streams.*

On the Colorado Plateau the underlying rock structure, in the form of fractures in sandstone layers, alters direction of stream flow to produce subsequent streams.

Other subsequent drainage patterns include the annular pattern found in domes and basins and in parallel streams flowing on a uniformly dipping rock sequence.

Stream piracy

With the passing of time, streams will shift their course and produce changes in drainage nets. One of the more interesting effects produced by shifting stream direction is stream capture or *stream piracy.* When one of two streams flowing in adjacent valleys is able to deepen its course more quickly than the other, it may then extend its valley through headward erosion. The more rapidly eroding stream erodes backward into the less aggressive stream and steals a portion of its course. The capturing stream is designated as the *pirate stream*, and the stream that has lost its upper section is called the *beheaded stream.*

WATER QUALITY AND DEMANDS ON THE WATER SUPPLY

How much water is available for use by the public and industry? If we multiply the total area of the earth by average rainfall in feet, the resulting product is 140 million billion gallons of water. This amounts to 40 million gallons of water per year for every

213

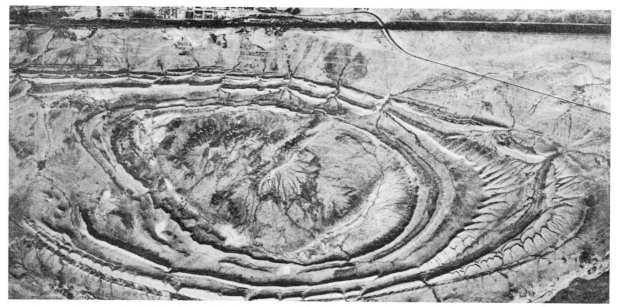

Figure 13–26. The Grenville Dome near Sinclair, Wyoming, is in the late maturity stage of landscape development. Radial drainage is no longer present, and the Grenville Dome has developed an annular drainage pattern. (U.S. Geological Survey.)

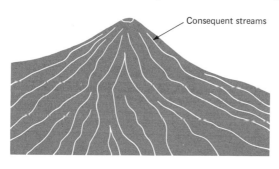

Consequent streams

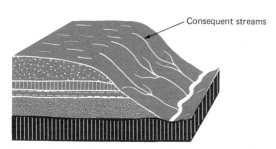

Consequent streams

Figure 13–27. *Top:* A stream flowing in a radial drainage pattern down the slopes of a volcano is a consequent stream. *Bottom:* Consequent streams can be found in drainage systems other than radial.

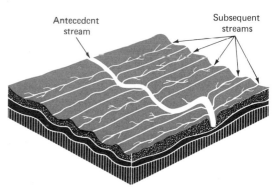

Antecedent stream

Subsequent streams

Figure 13–28. Some rivers can maintain their courses across resistant rocks after regional uplift has taken place. Because its course anteceded uplift, this stream is called an antecedent stream. The tributaries of an antecedent stream, lacking the eroding capacity of the main stream, find it easier to erode channels in the weaker beds. Because the tributaries alter their courses subsequent (after) to the uplift, the streams in the altered courses are called subsequent streams.

215
STREAMS:
THE
GREATEST
GEOLOGIC
AGENT

person on earth. In the United States one-third of the flow of all streams is used at least once. At the present rate of population growth we will be using almost the entire stream flow by the year 2000.

Individuals who are aware of the rapid rise in population (in the United States the population is doubling with a 62-year interval) may not be too surprised to learn that demands on water usage are doubling every 10 years. Industries alone use 17 trillion gallons of water annually; less than one-third of this water is treated before it is discharged back into our source of freshwater. It can be seen that demands on our water supply will increase at an exponential rate and that, unless we use our freshwater sup-ply prudently, we will be faced with a critical shortage of potable water in the near future.

SUMMARY

Running water, in geologic terms, is water moving under the influence of gravity over the earth's surface. It possesses a tremendous amount of kinetic energy. This energy, working on surface landscape, carries various-sized particles of soil and rock fragments from one place to another. The initial thrust for the energy of running water can be traced to the hydrologic cycle that is set into motion by the sun.

The study of work performed by running

Figure 13–29. A region having hills and valleys may, in time, be completely covered by sediments. A stream flowing across the newly deposited sediments may possess sufficient force to continue eroding downward through these ancient hills and valleys even though it may encounter resistant rocks. Because the course of the stream is superimposed across the old landscape it is called a *superimposed* stream. Erosion later excavates the ancient valleys exposing the resistant ridges. Superimposed streams flow through the ridge in a narrow gorge called a *water gap*. (Buffalo Society of Natural Sciences.)

water is divided into three categories: (1) erosion, (2) transportation, and (3) deposition. Erosional work of running water implies the cutting and transporting of surface materials. Depositional features appear when a stream loses competency (ability to carry foreign matter) and sediments are deposited along the channel, in the floodplain, or at its mouth.

The ability of a stream to erode depends on its velocity and capacity to carry rock material: (1) in suspension; (2) by rolling, sliding, and saltation; and (3) in solution. Velocity of a stream is controlled by its volume of water and by its slope (gradient). Downcutting continues with decreasing vigor, primarily because of the increase in entropy (energy available for useful work) as the stream approaches base level. As its slope flattens, a stream will begin sideways (lateral) cutting. This change in activity produces the typical longitudinal stream profile best described as concave to the sky.

The Davisian erosion cycle divides stream development into three stages: (1) youth, (2) maturity, and (3) old age. Young streams exert most of their energy by cutting down. One outstanding characteristic of a *youthful stream* is its narrow V-shaped valley. When streams reach *maturity* the region attains its maximum relief. Many tributaries are present, and drainage in the region is adequate. Streams are classified as *old* when their gradient becomes low. Valleys become broad to a point at which interstream relief is flat or gently rolling. Meanders are conspicuous, and oxbows are common. Large floodplains will have many lakes and swamps.

Streams have complex histories and the Davisian concept of youth, maturity, and old age must be considered only as an idealized model rather than reality.

A number of natural stream drainage patterns can develop. Any pattern developed is controlled by rock structure, types of rocks present, and climatic conditions in the region. The most common drainage pattern is dendritic. Other patterns found are trellis, rectangular, radial, and annular.

The human impact on streams has been environmentally negative. Much concern for future source of fresh surface water has been voiced in recent publications. It is feared by some that drastic reductions in the availability of freshwater will appear by the turn of the century. These indicators should be examined carefully and remedial steps taken to maintain the present level of freshwater availability.

Questions for thought and review

1. Describe the role of water in the biosphere.
2. What stream characteristics are changed most by an increase in discharge?
3. Describe the changes that may occur both upstream and downstream when a stream is dammed.
4. Discuss strong and weak points in the erosional cycle proposed by Davis.
5. Under what conditions may the longitudinal stream profile be convex upward?
6. Explain, using the concept of entropy, why most streams in regions of normal rainfall have a longitudinal stream profile that is concave upward.
7. Describe various ways that a stream can transport sediments.
8. Explain the difference between the terms "prediction" and "forecast" when used in the study of floods.
9. List the most common depositional features of streams.
10. The loss of competency in a stream may be caused by changes occurring in or near the stream. Discuss some of the changes that will effect the stream's ability to move sediments.
11. Describe four commonly found stream drainage patterns and explain the principal causes for each pattern of formation.

217
STREAMS:
THE
GREATEST
GEOLOGIC
AGENT

12. How do consequent streams differ from subsequent streams?
13. What is stream piracy? How does it develop?
14. Discuss the future demands on stream flow in the United States. What changes do you recommend to conserve our freshwater supply?

Selected readings

Adams, F. D., 1954, *The Birth and Development of the Geological Sciences*, Dover Publications, New York.

Davis, W. M., 1902, *Geographical Essays*, Ginn, Boston. (Reprinted 1954, Dover Publications, New York.)

Dury, J. T., 1959, *The Face of the Earth*, Penguin Books, Baltimore.

Flemal, R. C., 1971, The Attack on the Davisian System of Geomorphology: A Synopsis, *Journal of Geological Education*, vol. 19, no. 1.

Hack, J. T., 1960, Interpretation of Erosional Topography in Humid Temperate Regions, *American Journal of Science*, vol. 258-A, pp. 80–97.

Hutchison, G. Evelyn, 1970, The Biosphere, *Scientific American*, vol. 223, no.3 pp. 45–53.

Leopold, L. B., Wolman, M. G., and Miller, J. P., 1964, *Fluvial Processes in Geomorphology*, Freeman, San Francisco.

Morisawa, Marie, 1968, *Streams, Their Dynamics and Morphology*, McGraw-Hill, New York.

Penck, W., 1953, *Morphological Analysis of Landforms*, Macmillan, London. (Translation of 1927 ed.)

Schumm, S. A., 1966, The Development and Evolution of Hillslopes. *Journal of Geological Education*, vol. 14, no. 3, pp. 98–104.

14 Underground water

STUDENT OBJECTIVES

On completing this chapter, the student
should:

1. recognize the importance of
   groundwater in the formation of geologic
   features and as a source of fresh water

2. be able to formulate hypotheses related
   to the sources of groundwater

3. identify the variables that determine
   groundwater storage

4. distinguish between confined and
   unconfined groundwater and related
   flow patterns

5. identify geologic features produced by
   groundwater

6. recognize environmental problems
   related to groundwater pollution

*There is always mystery attached to the desert
oasis, a spot made green by fresh water in a land
where little rain falls and few streams survive . . .
Whence comes this living water?*

Agar, Flint, and Longwell

WATER BENEATH THE GROUND

Running water on the surface of the earth is admired for its beauty in streams, brooks, and spectacular waterfalls plunging over a precipice. But there is also something magical about a stream mysteriously disappearing into the ground or water suddenly emerging from beneath the earth's surface.

Several million tourists every year visit great caverns that have been carved by underground water. These subsurface labyrinths exert a strange fascination on the eye and mind of the visitor. In the deep silence of the caverns, one can see such oddities as *stalactites*, iciclelike features hanging from cavern ceilings (Figure 14–1).

Caves are not devoid of life. More than

Figure 14–1. The most magnificent and spectacular underground chambers known can be found in Carlsbad Cavern National Park in southeastern New Mexico. Stalactites and stalagmites cover almost every inch of ceiling and floor space. Cavern features such as flowstone, draperies, cave grapes, and helictites are abundant. These subterranean features emerging from darkness create eerie scenes of great natural beauty. (Courtesy of New Mexico Department of Development.)

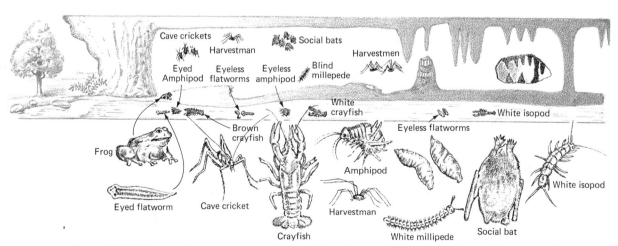

Figure 14–2. Cave environments vary inward from the cave opening. Differing living conditions, as one penetrates deeper into the cave, produce a system of life zones, with each zone supporting its own distinctive community of plants and animals. It is common to find 100 or more species of subterranean creatures living in a cavern system.

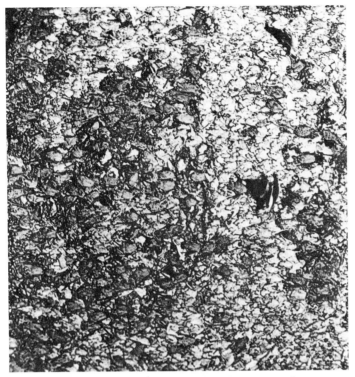

100 species of subterranean creatures inhabit cave environments (Figure 14-2). Translucent and eyeless fish navigate through cavern waters, using a sonarlike system so acute that it can detect nearly invisible microorganisms in the water on which these fish feed. Other creatures include crickets, crayfish, flatworms, and salamanders. In some caves, hanging head down from ceilings, are thousands of bats (Figure 14-3). Summer visitors at Carlsbad Caverns, New Mexico, wait until dusk to watch a half-million bats rush from the main entrance and fly toward the Black River in search of food.

The behavior of groundwater is often puzzling, as illustrated by the case of the disappearing water in Cephalonia, an island in the Ionian Sea off Greece. Seawater pours into openings on this mountainous island and gurgles between cracks into the ground below. One such hole sucked the seawater so fast that an enterprising resident erected a grain mill with a water wheel to harness this energy. Two Austrian geologists solved the mystery in 1963. Using tracer dyes, they learned that dyed water appeared two weeks later on the other side of the island after traveling underground for a distance of 10 mi (16 km). See Figure 14-4.

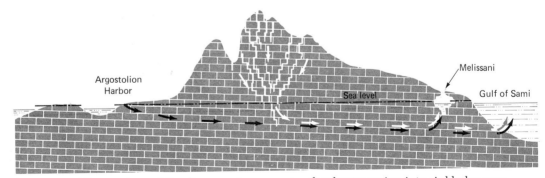

Figure 14-4. The mystery of the disappearing seawater that kept pouring into sinkholes near Argostolion Harbor on the Ionian island of Cephalonia was solved by tracer dye movement and disclosed an elaborate underground water system. The cavern system that serves as a conduit for the sea water was carved by fresh water in the Ice Age, when sea level was much lower than it is at present. Midway through its course, seawater is joined by fresh water filtering downward in the mountainous region. Faster-flowing fresh water increases the speed of seawater by producing a suction effect, thus drawing seawater in at Argostolion.

Many people believe that spring water possesses medicinal and therapeutic values. High mineral content of some spring waters encourages its use as a natural elixir capable of curing many afflictions. It is not uncommon to find bottles of "mineral spring water" in some drug stores.

Underground water is as necessary a natural resource as petroleum, coal, iron ore, or natural gas. Although much more water is found underground than on the surface, a continuous withdrawal of underground water may cause less water to be available for stream replenishment and also may result in subsidence or other effects on land.

Mexico City is an example of subsidence. Land in the old section of the city has subsided a minimum of 13 ft (4 m) and in some places as much as 25 ft (7 m) since 1891 because of excessive withdrawal of underground water. Sewer lines that were designed to drain waste water out of the city now slope into the city. Visitors walk downstairs rather than upstairs when entering some cathedrals (Figure 14–5). The undulat-

ing school building on the "Street of the Good Death" in Mexico City is another example of the effects of underground water withdrawal (Figure 14–6).

Underground water withdrawal on Long Island, New York, has produced another modern problem associated with population growth. Excessive well drilling, reduction of rainwater infiltration because of surface construction, and other factors lowered groundwater pressures, upsetting the equilibrium between subsurface freshwater and adjoining salt water. The result was the intrusion of salt water into the freshwater supply.

Urbanization decreases the quality of both surface and subsurface water. In the Long Island episode, population spread, and some of its effects on water infiltration of the soil are notable. The large areas of highways, roofs, and shopping center and school parking lots, neatly covered with tar or asphalt, that are found in urban and suburban areas greatly increase the percentage of land surface that is impervious to water. This pre-

Figure 14–5. Subsidence may be massive or local. The Basilica of Our Lady of Guadalupe in Mexico City is a good example of subsidence affecting a small area. The western bell towers, built on bedrock in 1709, stand as firm as they did then. The eastern towers have fallen victim of the sinking soils so prevalent in Mexico City.

Figure 14–6. Constant removal of underground water from subsurface volcanic ash has caused some areas of Mexico City to sink as much as 25 ft. (8 m) This 400-year-old building, now a school, once housed a hospital. The terminally ill and infirm were cared for here by the friars of San Camilo until death would overtake them. Thus the street became known as the "Street of the Good Death." Undulating roof lines along the street show the variance in subsidence that can be found in a small region.

vents infiltration of water into the ground. Runoff into streams increases to produce floods in regions that seldom saw flood waters. High flood peaks and low rate of infiltration reduce the recharging of groundwater reserves. This eventually is reflected as a decrease in the amount of water flowing in surface streams.

Natural underground water reservoirs are threatened by urbanization in ways other than depletion. Urbanization introduces into groundwater industrial wastes, pathogenic bacteria, and increases in water temperature.

To understand problems related to underground water that face any industrial society, it is necessary that we become familiar with the geologic work of subsurface water, the basic physical laws operant, and various changes produced by this interaction.

THE SOURCE OF UNDERGROUND WATER

Zone of aeration

As the force of gravity pulls water down from the surface, some of it is caught by rocks and soil. Here the spaces between particles are filled partially with air and partially with water. This sector of the earth's crust is known as the *zone of aeration* (Figure 14–7). Water so suspended is prevented from moving deeper into the ground by two physical forces: (1) cohesion of water particles to other water particles and (2) adhesion (molecular attraction) between earth materials and water. This action is possible because water is a wetting liquid. The mixture of air and water in the zone of aeration is responsible for decay of organic matter in the soil and for weathering of rock fragments. These conditions make farming of the land possible.

Zone of saturation

Not all precipitation is captured by the zone of aeration. Much of the water con-

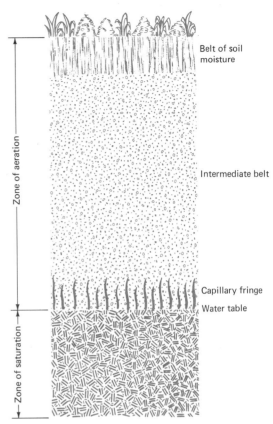

Figure 14–7. Underground water reacting to physical laws of nature separates into two major zones: (1) zone of aeration and (2) zone of saturation. The zone of aeration is further subdivided into three regions. The uppermost is the belt of soil moisture, the source of most water for surface plants. Beneath it lies the intermediate belt. In this region water is held securely by molecular attraction. Here there is little movement of water except during periods of rain or during snow melting. Bordering the zone of saturation is the capillary fringe. In this region water can rise from the water table to various heights depending on the size of the interstices (see Glossary).

tinues downward because of gravity and the weight of the water above it. This process continues until all of the spaces between particles become filled with water. This is the *zone of saturation*, and the water per-

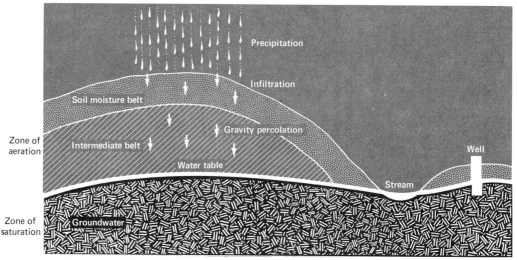

Figure 14-8. The water table is the upper surface of the zone of saturation. It has an irregular surface with a shape and slope that conforms generally with the ground surface above it. It is highest beneath hills and lowest beneath valleys. It is exposed in wells and permanent streams.

meating this zone is *groundwater*. The surface between the zone of aeration and the zone of saturation is the *groundwater table* (Figure 14-8).

Meteoric water

Most groundwater in the upper part of the earth's crust comes from precipitation of one form or another. This source is called *meteoric water*. As the most important source of groundwater, this water has moved through the zones of aeration and saturation.

Connate water

In addition to meteoric water is another minor source known as *connate water*. It is fresh or marine water trapped in sediments when they were deposited at the bottom of lakes or seas. Much connate water is salty, because most sediments were deposited in marine waters. This source of water may be found in deeper sedimentary rocks of the earth's crust. In many cases connate water is associated with oil accumulations. Oil floats on it; when an oil pool is tapped, connate water rises upward together with the oil.

Juvenile water

Another minor source of groundwater is the water associated with volcanic activity. Such water is called *juvenile* water, because it is reaching the surface of the earth for the first time and has not yet participated in the hydrologic cycle. Not all of the water issued by volcanoes is juvenile. Much of it, perhaps most, comes from groundwater in the vicinity of the volcano. It can be concluded, nevertheless, that much of the water present during primeval stages of the earth's history was trapped within the earth and is slowly being released by magmatic action.

THE STORAGE OF UNDERGROUND WATER

Although a few underground streams do exist, most water is stored in a variety of intergranular pore spaces, or other openings.

Porosity

The volume of pore space in a rock is called the *porosity* of the rock. Porosity is measured by the amount of space occupied by these voids or cavities compared with the total volume of the rock. The more porous a rock is, the greater is the percentage of open spaces within it. Groundwater works its way through these open spaces. The amount of pore space will depend on size, shape, sorting, and type of packing of rock components (Figure 14–9).

In general a large assortment of sediment sizes in a rock results in lower porosity than if the sizes were confined to a narrower size range. Packing arrangements, even when sediments are of the same size and shape, can produce a wide range in porosity.

Should each grain be placed directly above the center of another, maximum porosity can be obtained. If the centers are offset so that the grains are alternately spaced, minimum porosity occurs (Figure 14–10). Various clastic sediments offer various porosity ranges: soils, 50–60%; clay, 45–55%; silt, 40–50%; sand, 30–40%; gravel, 20–40%; sandstone, 10–20%; shale, 1–10%; limestone, 1–10%.

Permeability

A rock can be porous and still not permit water or other fluids to pass through it. One of the most important properties of a sediment is its permeability. *Permeability* is a measure of the ability of rock to transmit fluids. It is the relative ease with which fluids can pass through rock pores. Gravel has large openings to provide easy passage for fluids. As grains become smaller, pore spaces also become smaller, so a greater force or a longer time span is required to move a volume of fluid through the sediment. In the petroleum industry the porosity of the rock determines how much oil is available, but the permeability ultimately determines how easily the oil is recovered.

In the case of underground water, the rate at which a rock transmits water will depend on its porosity and whether or not the pore spaces are interconnected. To illustrate the necessity for both factors to be operational, we can compare the porosity of clay and sand. Clay has higher porosity than sand. However, clay's particles are minute flakes, and the spaces between these flakes are very small. Here molecular attraction on water is strong. Because pore spaces in sand grains are interconnected, water can move freely through sand. Sands also are generally well sorted. Most well-sorted granular rocks have high porosity and high permeability.

Figure 14–9. The percentage of space occupied by voids or cavities compared with the total volume of the rock determines its porosity. The amount of pore space depends on the shape, size, and packing potential of clastic particles.

MOVEMENT OF GROUNDWATER

The flow of surface water is measured in terms of feet per second (Chapter 13). But

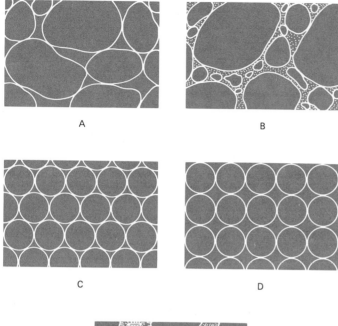

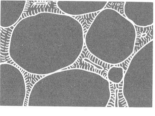

Figure 14–10. Packing arrangements of sediments will determine porosity. A. Well sorted, high porosity. B. Poorly sorted, low porosity. C. Close packed, 26% porosity. D. Well sorted, grain centers above each other, high porosity. E. Cemented sediments (cement of clay, calcite, or limonite), porosity near zero.

underground water, because it must travel through small passageways, moves at a very slow rate. The unit "feet per second" is appropriate only in cavern flow. Most underground water movement is measured in units of feet (meters) per day. In some instances feet (meters) per year would be appropriate.

When rock structure and lithology are such that a continuously connected hydraulic system is created through which water can freely move, underground water is said to be *unconfined*. When permeability is such that water can move into rock strata from which it cannot freely escape, the water becomes *confined*. Water pressure is built up in confined water.

Unconfined water

A region that is underlain by a thick permeable and porous rock composed of granular material provides ideal conditions for unconfined groundwater. The lower boundary of the unconfined water is underlying impervious rock strata. The upper contact of this saturated zone of water is the water table, a somewhat flattened replica of the ground surface. The upper surface coincides with the surfaces of lakes, streams, and marshes in the area. Only rarely is the water table truly flat (Figure 14–11).

Confined water

Water becomes confined when it is trapped within a porous and permeable rock unit that has an impervious rock formation below and above it (Figure 14–12). Water enters such a rock unit at the surface or at a point where it is connected with water-bearing units that permit free movement of water. The porous and permeable rock unit is called an *aquifer*, which literally means "a bearer of water."

As water moves down through an aquifer, the water-bearing layer is filled, and pressure begins to build up. When the aquifer is tapped by a well, water under pressure will flow from it. Groundwater that is under sufficient pressure to rise above the zone of saturation is *artesian water*. The term derives its name from the region of Artois, France, the location of some of the first artesian wells.

Because artesian water travels a long distance and is protected against surface contamination by impervious rock layers above it, artesian well water is generally safe to drink without further treatment. Artesian

Figure 14–11. The water table is a somewhat flattened replica of the land surface above it. It is also closely associated with surface features such as streams, springs, and lakes all of which draw on its supply of water.

water, however may be hard because of large amounts of dissolved minerals. This presents problems in laundering clothing and other processes requiring soap or detergents for sudsing. The water must be softened in many cases. But artesian wells tend to be more reliable water sources than simple wells; they are not dependent on local rainfall.

Laminar flow

Because the flow of underground water is usually very slow, its movement is described as being a laminar flow (see Chapter 13).

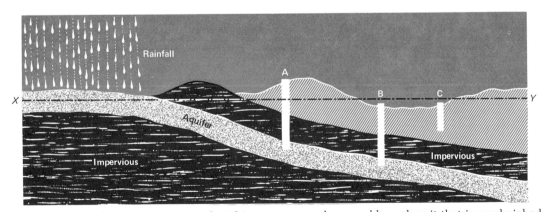

Figure 14–12. When water is trapped within a porous and permeable rock unit that is sandwiched between two nonporous rock units, it is called confined water. This porous rock unit is an aquifer. The slope of the aquifer produces pressure. When the aquifer is tapped, pressurized water will flow to the surface. It cannot rise, however, above a point on the ground surface that is above the highest point in the aquifer. In the figure, the artesian well at A will not flow. The well at B will flow under pressure. The well at C is not an artesian well, because it does not enter the aquifer. It draws its water supply from local groundwater; water must be pumped to the surface.

(Movement of surface water is generally turbulent.) The one exception to this is the flow of water in underground caverns: Here the flow may be turbulent.

In laminar flow the water near the outer walls of rock strata is presumed to be motionless because of molecular attraction to the walls. Water particles farther from the walls move more freely in smooth, threadlike patterns. Resistance to the movement of water decreases toward the center, with the most rapid flow reached at the exact center.

Darcy's law

Just as surface water needs a slope for its movement, so must groundwater. The force that moves water to flow along a slope is that of gravity. Gravity moves water downward to the water table and then propels it through the ground until it is discharged into a spring, a stream, a well, or a lake.

The slope required for this movement is the *hydraulic gradient*. It is calculated by dividing the length of the flow (point of intake to point of discharge) into the vertical distance between these two points. The vertical distance is called the *head* (Figure 14–13). If the head is 10 ft (3 m) and the length of

flow is 1000 ft (300 m), the hydraulic gradient is 0.01 or 1%.

Experiments related to the rate of flow of groundwater through rock were performed by Henry Darcy in 1856. His work showed that the velocity of movement of water between two points on top of the water table is equal to the product of rock permeability and the hydraulic gradient. This relationship is known as *Darcy's law*:

$$V = P\frac{h}{l}$$

where V is velocity, h is head, l is length of flow, and P is coefficient of permeability.

Rock permeability is measured in a unit known as a *darcy*. Permeability is 1 darcy when 1 cm^2 of rock surface releases 1 cm^3 of fluid per second under a pressure differential of 1 atmosphere. The permeability of a consolidated sand $\frac{1}{2}$ mm (25.4 mm equal 1 in.) in average diameter is approximately 1 darcy.

Although the rate of movement of groundwater is generally slow, it varies greatly from one place to another depending on the permeability and structure of rocks. Movements of 3–30 ft (1–10 m) per year are found frequently. However, rates of several hundred feet (meters) per day do occur in some alluvial fills. This is comparable with the velocity of some sluggish surface streams. At the other extreme, when porosity and permeability of rock strata are low, water may be almost static.

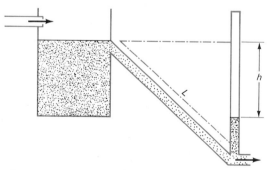

Figure 14–13. Darcy's law states that the velocity, *V*, of the flow of water is proportional to the permeability of the rock, *p*, and the hydraulic head, *h*, and inversely proportional to the length of the path of flow, *L*. Darcy's law enables us to predict the amount of water that can be pumped from an aquifer.

SPRINGS

Springs usually form wherever the water table intersects the land surface. A variety of subsurface conditions, however, can produce spring flow. Thus we have fracture springs, solution-openings springs, depression springs, contact springs, and artesian springs (Figure 14–14).

Silver Springs, in central Florida, is the largest spring in the United States and is one

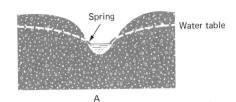

A

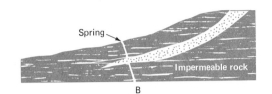

B

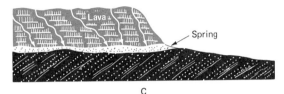

C

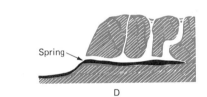

D

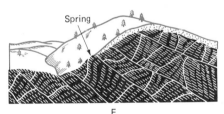

E

Figure 14–14. A variety of subsurface conditions can produce spring flow. More common types include- (A) depression spring, (B) artesian spring, (C) contact spring, (D) solution spring, and (E) fracture spring.

of a number that penetrate the Ocala Limestone. This formation is extremely permeable because of solution cavities. The Ocala is overlain by impervious rock units and is therefore under artesian pressure. The main pool at Silver Springs is about 65 ft (20 m) wide and 12 ft (4 m) deep. Silver Springs water is noted for its clarity. The main opening from the Ocala Limestone is clearly visible from the surface. It discharges approximately 500 million gallons (1,900,000 m³) each day (Figure 14–15).

The largest effluence of spring water in the United States is credited to a 10-mi (16-km) stretch of springs feeding the Fall River in California. More than 900 million gallons (3,420,000 m³) of water enter the Fall River daily.

Figure 14–15. The largest single spring in the United States is Silver Springs, Florida. Ocala limestone of the Florida penisula, with its large solution cavities, permits underground water to move freely and vast quantities of water flow to the surface under artesian pressure. Environmentalists are waging a battle with oil interests to preserve the beauty and purity of this spring. Oil companies believe that rock strata in this region are likely sites to search for new oil reservoirs. Vertical relief is greatly exaggerated.

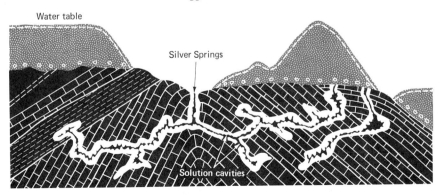

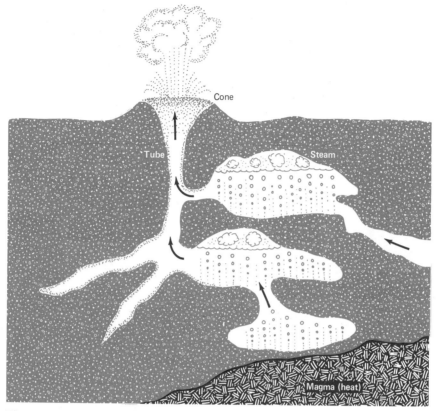

Figure 14–16. A geyser has an intricate underground plumbing system. Geysers are found in regions in which faults or fractures have caused sections of rock to move up, down, or sidewise in relation to neighboring rock. Faults and fractures provide conduits for escaping hot water and steam. In addition to these passageways other conditions must be present to produce a geyser. Volcanic magma (liquid rock) must lie beneath surface rocks. Cavelike cavities, strong enough to act as pressure boilers, must adjoin fractures in the rock. Finally there must be an underground source of water that flows so that magma will heat it and force it upward.

HOT SPRINGS AND GEYSERS

When groundwater is heated by subsurface conditions and then finds its way to the surface before cooling, a thermal spring is produced. These are commonly called hot springs or warm springs. A spring is classified as a thermal spring if its water is at least 8C° higher than the average temperature of the air. Of the thousands of thermal springs in the United States, most are found in the western mountains.

Western thermal springs derive their heat primarily from hot magma (and associated rocks) that has pushed its way upward to relatively shallow depths. *Geysers* are a special type of hot spring in their ability to intermittently throw steam and water into the air. The underground structure of a geyser, it is believed, consists of a crooked tubelike opening leading from the interior to the surface. Several caverns may be connected to the tube (Figure 14–16).

Groundwater partially fills the tube and some of the connecting caverns. The heated

Figure 14–17. Strokkur geyser. This eruption is 60 feet (20m) in height. (Courtesy of U.S. Geological Survey.)

water is trapped under pressure in the crooked tube. Continued heating produces a water temperature above the boiling point. Steam produces sufficient pressure to eject a small amount of water to the surface. This expulsion of water in the initial upsurge reduces pressure on the superheated water in the tube. The reduction in pressure causes water to boil explosively, driving a column of hot water and steam into the air (Figure 14–17). The eruption continues until water and steam are driven out of the tube and storage caverns. The process is repeated as long as there is a supply of groundwater, with a highly constant time interval between ejections.

Geysers are found typically in *geyser basins.* These are grabens (see Glossary) that are formed by parallel faults (Figure 14–18). Almost all of the geysers known are found in Yellowstone National Park, Iceland, and New Zealand. Structural condi-

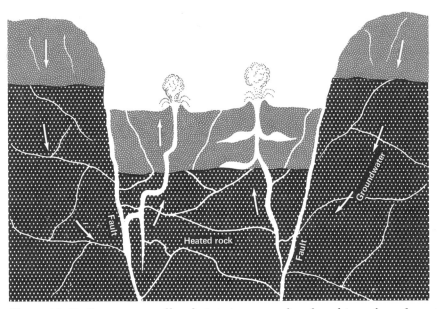

Figure 14–18. Geysers, as well as hot springs, are often found in grabens formed by parallel faults. Heat from buried volcanic magma rises to the surface, heating rock and groundwater above it. Faulting and jointing in the rock produces tubelike escape routes for superheated water and steam trapped in the many cavities found beneath the surface.

Figure 14–19. Geyserite forms around geysers and thermal springs. Most of these deposits are siliceous. (Courtesy of Ward's Natural Science Establishment, Inc.)

Figure 14–20. Hot waters emerging from thermal springs are often saturated with calcium carbonate ($CaCO_3$). When cooled, the calcareous material in the water precipitates to form travertine. Small pools near springs produce interesting terrace formations.

tions mentioned above prevail in these regions.

The warm water of geysers and hot springs can dissolve some minerals readily, and, in regions where thermal springs are common, certain characteristic rock deposits can be seen. A large amount of silica (quartz) dissolved in underground water is brought to the surface in solution. The silica drops from solution (precipitates) at the surface as the water cools. At first a colloidal gel forms. The gel consolidates into an amorphous form of quartz known as *geyserite* (Figure 14–19).

The calcite that dissolves in hot water on reaching the surface forms *travertine*, also known as dripstone (Figure 14–20). Extremely porous or cellular varieties are known as calcareous tufa or calcareous sinter.

GEOLOGIC FEATURES PRODUCED BY UNDERGROUND WATER

If water were a pure substance, it would not readily dissolve rocks and minerals except for a few such as halite, gypsum, and anhydrite. We learned from Chapter 13 that water is not water alone but is water and something else. As precipitation passes through the atmosphere, carbon dioxide dissolves in it, and thus some rainwater becomes carbonic acid. Rain passing through sulfur dioxide fumes dissolves this gas and eventually forms sulfuric acid. Before any form of precipitation reaches the ground, it is altered chemically a number of times. When it begins to infiltrate the ground, alterations in both the physical and chemical compositions of water continue. So a variety of chemical processes, such as carbonation, oxidation, and hydration, operate freely as groundwater circulates through rocks and soils.

Rocks composed chiefly of the mineral calcite (such as limestone, marble, and dolo-

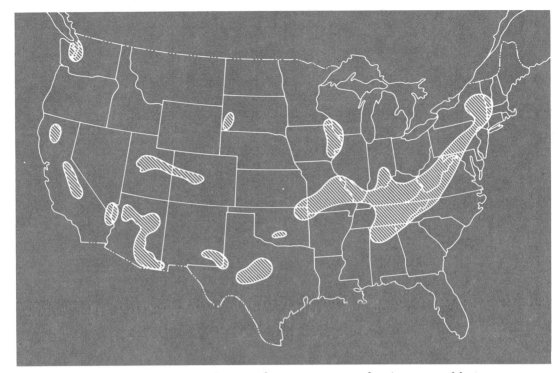

Figure 14–21. Scattered throughout the United States are caves of various types. Most common (above) are limestone caves; other types include sea caves, sandstone caves, lava caves, and talus caves. Regardless of where you live, you are within a one-day drive of a cave. Many of the best caves are found in national parks or state parks. Some, however, are privately owned and are run as commercial enterprises. *

mite) are especially subject to carbonation, the great cave maker.

Caves

Caves are naturally formed underground voids or cavities. They are abundant in limestone rock formations that were formed in the Paleozoic era in the region of the Appalachian Mountains and Plateau. Here we find the spectacular Mammoth Cave in Kentucky and the Luray Caverns in Virginia. Other limestone caverns are found in central Florida, in the Ozarks, and on the Carlsbad Plateau in New Mexico (Figure 14–21).

Geologists generally agree that caves are formed by the solution of rock by groundwater. However, geologists do not agree on the conditions that existed at the time of cave formation. Debate still continues whether caves were formed above the water table, at the water table, or below the water table. Although consensus is lacking, most geologists believe that caves are usually formed below the water table and are then exposed to air when the water table is lowered by the downcutting of surface streams or lack of groundwater recharge.

Proponents of the hypothesis that caves were formed above or near the water table present this evidence for support:

1. The geometric outline of some caves is similar to the alignment of fractures in bedrock. Fractures provide passageways

for movement of water through the rock.

2. Blocks of fractured rock that have collapsed into caverns are found.

3. Underground streams found in many caves can be traced to the ground surface.

Streams are diverted from the surface into caves. These may later resurface at a lower elevation in the terrain.

4. Streams in caves carry sediments that can produce erosional features such as pot-

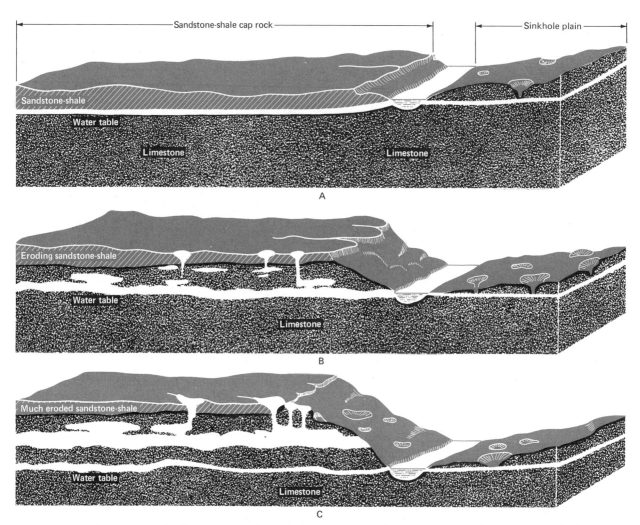

Figure 14–22. In the two-cycle hypothesis of cave development (A and B) are included in the first phase of the process. In the early stages of cave development (A) the water table lies close to the surface. Acidic water finds its way through fractures in the rock, dissolves limestone in that region forming unconnected cavities in the saturated zone below the water table. Later (B), periodic floods enlarge these cavities and many are joined together forming a continuous channel. The second phase (C) follows many centuries later, after the water table has dropped because of downcutting of streams in the region. The once water-filled caves are left high and dry, while another cave system begins to form at the new water table. Features such as stalactites and stalagmites form while the caverns are above the water table.

holes and meanders. Dissolved limestone could not produce this result.

In 1930 W. M. Davis proposed a two-cycle hypothesis for cave formation. According to him, first stages of cave development appear below the water table when acidic water penetrating down from the surface, dissolves pockets of limestone. The second cycle begins after the groundwater level drops, because downcutting of surface streams erodes stream beds to a new level. After this phase many typical cave features such as stalactites and stalagmites appear (Figure 14–22).

Evidence supporting the two-cycle hypothesis has been found by its proponents. The two-cycle hypothesis is strengthened by the following observations:

1. Caves beneath the water table and filled with water are present in the Ocala Limestone in central Florida. Water-filled caves are also found in the Appalachians.
2. Crystals lining walls of caverns apparently grew under water.
3. Longitudinal profiles resembling those of surface streams are rare in caves.
4. Caverns have a three-dimensional pattern with blind pockets and dead ends rather than a stream pattern.
5. Solution cavities are frequently found in massive, unfractured rocks.

A third hypothesis for the formation of caves must be considered. This hypothesis proposes that caves form very close to the water table because of the lateral movement of groundwater. Evidence supporting this hypothesis can be found among those listed for the first two hypotheses. Caverns form when the water table is high. Features such as stalagmites form when groundwater level drops. This up-and-down movement of the water table can produce a wide variety of cavern formations.

The debate over the geologic history of caves provides an excellent backdrop for an exercise in inferential techniques in scientific inquiry. Based on information supplied here, how would you explain the development of cave systems found in so many areas of the United States?

Not all caves are formed in limestone. Sandstone caves, such as those found in Arizona, form near the base of sandstone cliffs. When sand grains are weakly cemented, rainwater and wind gradually erode cavities in the rock. Many shallow caves were used by prehistoric Indians as shelters.

Still another type of cave feature is Subway Cave in Lassen National Forest, California. This is a remnant of past volcanic activity. Caves similar to Subway Cave are formed when a flowing stream of lava hard-

Figure 14–23. This cave passageway in Carlsbad Caverns National Park, New Mexico, with its stalactites, stalagmites, drapery, and grape features, is one of many awe-inspiring experiences awaiting a cave explorer.

ens on the outside, while the inner lava flow continues in its movement. Eventually the inner lava passes through the solidified lava, leaving behind an empty tunnel. Several hundred tunnels of this type can be found in Lava Beds National Monument, California.

Strong coastal wave and surf action also produces caves. Relentless pounding against cliffs wears away weak spots in the rock to develop unusual cave features. Pirate stories abound with secret caves used by buccaneers in their efforts to hide their loot. Less romantic, but perhaps more aesthetic, are the sea caves in Acadia National Park, Maine.

Characteristic cave deposits

Among features that attract tourists to many caves in the United States are the weird and oddly shaped stone deposits found there. The imagination is highly stimulated by bizarre ceilings, floors, and walls (Figure 14–23). Cave deposits are the products of millions of years of slow deposition of minerals found dissolved in groundwater. Many, but not all, cave deposits were formed while caves were above the water table.

Groundwater seeps into the cave through rock and slowly drips from the walls and ceilings to form pools or streams on floors. Constant dripping of groundwater through openings in the rock leads to sites for the deposit of calcite. Drippings from the ceiling produce calcitic deposits that are initially strawlike and are known as *soda straws* (Figure 14–24). These thin-walled hollow tubes are about ¼ in. (1 cm) in diameter. As mineral-laden water runs through the centers, they grow to a length of 4 or 5 ft (2 m).

Soda straws eventually develop into stalactites. Calcitic deposits form from groundwater flowing over the outside of soda straws after their centers have become plugged (Figure 14–25).

Water that drips from the end of a stalactite, falling to the floor of the cave and evaporating, leaves behind a deposit of calcite.

Figure 14–24. Dripping calcitic water from cave ceilings forms strawlike deposits called soda straws. They are thin-walled hollow tubes approximately .25 in. (0.6 cm) in diameter. The hollow center eventually clogs causing the flow of water to switch from the center to the outer surface. After this change in drip pattern the soda straw grows into a stalactite.

This deposit slowly grows larger to form a *stalagmite* (Figure 14–26). The tops of stalagmites are stumplike and usually have a small saucer-shaped top. Should the stalactites and stalagmites continue to grow, they will merge to create a *column* (Figure 14–26). Groundwater flowing down the sides of the column enlarge it by adding layers of *flowstone* on its surface. A variety of other deposits may develop (Figures 14–27 and 14–28).

Karst topography

Subsurface movement of water frequently affects surface landscapes. If enough ground-

Figure 14–25. The stalactite is a cave ceiling deposit. Minerals in solution in groundwater slowly build the stalactite drop by drop. Usually the stalactite forms when the center of the soda straw becomes plugged. Stalactites look like giant carrots hanging down from the cave ceiling. (Courtesy of Ward's Natural Science Establishment, Inc.)

Figure 14–26. Mineral-laden water dripping from the ceiling or from a hanging stalactite will slowly form a stump like structure on the cave floor called a stalagmite. Continued growth of a stalactite and a stalagmite directly below it will eventually form the column, a pillar like speleothem. (Courtesy of Ward's Natural Science Establishment, Inc.)

water is present in a region to produce cave features, surface drainage downward into these caverns will produce mass wasting and give rise to a number of geologic features that are classified as karst topography. (The name is derived from the Karst region of northern Italy and Yugoslavia, where this type of topography is well developed.)

Karst topography is most likely to develop in regions in which limestones or other relatively soluble rocks are exposed or near the surface. A karst region will form

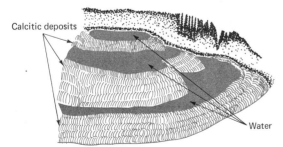

Figure 14–27. Groundwater loses its carbon dioxide as it flows through the cave. As a result of this change, a calcitic deposit resembling a small dam will form along the floor of the cave. This is known as a rimstone dam. Usually the slow flow pattern of groundwater in the cave will continue over a long time period; consequently rimstone dams are usually seen as a series of steplike terraces.

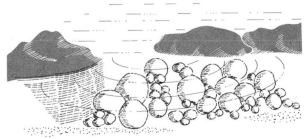

Figure 14–28. Cave pearls are calcite deposits that form around sand grains and other particles in pools along the cave floor. They form only when agitation in the pool is sufficiently great to prevent the pearls from becoming attached to the floor. They are generally 1 in (2.54 cm) in diameter and grow slowly in size as layer upon layer of calcite attaches itself to the slowly rolling pearl.

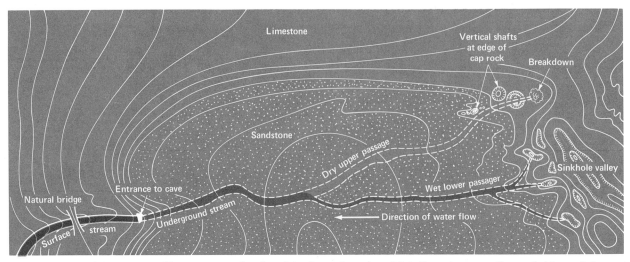

Figure 14-29. Karst topography develops in regions in which limestone or other soluble rock is exposed to water in sufficient amounts to produce caves. A map of Mammoth Cave, Kentucky, is superimposed on a geologic map of the region. The map shows that a layer of sandstone caps the ridge above the cave. Because little water can penetrate the sandstone roof of the cave, the upper passage is dry. Moving beyond the sandstone cap, limestone found in this region reacts with water, especially during spring floods, to produce the wet lower passage. The flow of water is generally from the sinkhole region in the east toward the cave entrance.

Figure 14-30. Collapsing cave ceilings produce a karst feature called a sinkhole or sink. Numerous sinkholes in a region such as the Mammoth Cave area of Kentucky provide striking evidence of a subterranean cave system. (Courtesy of Ward's Natural Science Establishment, Inc.)

whenever water is present in sufficient amounts to produce caverns from these rock types.

Karst topography is visible in the Mammoth Cave region of Kentucky (Figure 14-29). Here a resistant sandstone overlies limestone. Partial erosion of the sandstone has exposed the limestone sufficiently for it to have been attacked by groundwater. Many of the caverns that formed in the limestone have collapsed. This collapse is reflected at the surface as a circular to irregular depression known as a *sinkhole* or *sink* (Figure 14-30). Surface drainage into sinkholes may continue down through the cavern system. Should these outlets be closed by the breakdown debris, sinkholes may fill with water to form a small lake.

As more and more sinkholes form in a karst region, a larger amount of surface drainage goes underground. Small tributaries gradually become less prominent,

238

Figure 14–31. Collapse features in karst country produce uniquely pretty landscapes. A karst window is formed when the ground above an underground stream sinks into the stream channel, exposing it to surface view. The karst window gradually expands to form a karst tunnel. The stream becomes exposed for long distances with only an occasional natural bridge, such as this one, to remind us that it once flowed beneath the surface of the earth. (Courtesy of Ward's Natural Science Establishment, Inc.)

leaving only their valleys as proof of earlier occupancy. Water carried originally by tributaries reaches a main stream through underground channels. Stream valleys may end abruptly when a stream is lost to a subsurface channel. These are known as *disappearing streams* and *blind valleys*.

When the surface over an underground stream collapses to expose the stream channel, a feature known as a *karst window* forms. This depression becomes wider and longer with time to produce a *karst tunnel*.

When tunnels are only partially collapsed, *natural bridges*, such as the one at Natural Bridge, Virginia (Figure 14–31), form.

ENVIRONMENTAL PROBLEMS RELATED TO GROUNDWATER

Deep-well disposal

In the petroleum industry it is common practice to dispose of salt water recovered with oil by injecting it back into the subsurface. This is accomplished by using old wells in the field or sometimes by drilling new ones for this purpose. In the case of an oil field this form of waste removal is desirable for a number of reasons. A major reason, of course, is to be rid of the brine, and this is accomplished usually without harm to the natural environment. Another desirable effect is to restore subsurface equilibrium that was disturbed by oil withdrawal and to prevent subsidence of the ground. Also, brines or other wastes are confined in the subsurface to the limits of the rock structures that entrapped the oil initially; thus the wastes cannot migrate to contaminate normal underground waters reserved for human use.

As more and more public and governmental attention was focused on the increasing pollution of surface waters, several industries began to drill deep disposal wells as a means of getting rid of undesirable, even dangerous, by-products of manufacturing processes. Although some industries may have been unconcerned about polluting subsurface waters, others disposed of wastes in this manner out of ignorance.

Deep-well disposal cannot be undertaken indiscriminately by individuals, industries, or municipalities. A few problems are involved. The location of the disposal well should be above a subsurface rock structure (see Chapter 18) that will confine fluids; this is the usual case in the oil field situation. The well should also penetrate rock units with sufficient porosity and permeability to hold wastes. Depending on the situation, only a certain depth range in the subsurface may be suitable for disposal. Also, the chemical nature of both the wastes and the host rocks should be known so that the effect of possible underground chemical reactions can be evaluated. An obvious question that must be resolved before drilling begins is whether or not nearby existing water wells will be affected.

In summary, some deep-well disposal operations can be satisfactory, given intelligent consideration of geologic factors and careful planning under knowledgeable auspices. On the other hand, deep-well disposal as a general and unregulated practice can lead to far more serious consequences than other forms of pollution.

Nuclear aspects and the environment

Demands for nuclear energy have introduced a new element into the problem of managing underground water reserves—danger that underground testing of radioactive materials may contaminate subsurface waters. Long-lived radioactive substances could conceivably move long distances in the subsurface and affect groundwater. At one time the U.S. Army disposed of dangerous wastes by dumping them into shallow pits. Eleven years later the noxious wastes were discovered in subsurface waters and had apparently contaminated several square miles (square kilometers) of groundwaters in Colorado.

It can be seen that critical problems exist for an expanding, energy-hungry society. There are no easy answers to the problem of achieving harmony between our use of resources and disposal of undesirable by-products without harm to the natural environment.

SUMMARY

Groundwater accumulates when rain or meltwater infiltrates into the ground. Water percolates downward until it reaches a point below which the voids and other open spaces in rocks are filled with water. This point marks the upper surface of the groundwater table or water table. The water table generally parallels surface topography—higher under hills and lower beneath valleys.

Although groundwater is chiefly meteoric, it also includes minor quantities of connate and juvenile water. Springs and lakes can result when the surface of the earth intersects the water table.

Artesian wells are drilled through impervious layers of rock to an aquifer. They do not depend on local precipitation for their water. The aquifer is a permeable layer that crops out at a higher elevation and at a great distance from the location of the well. Artesian water is under pressure because of the difference in elevation between its source and its position. This vertical distance is called the head.

Subsurface volcanic action or chemical processes produce geysers and hot springs. Groundwater that has been heated by subsurface conditions and then finds its way to the surface before cooling is called a hot spring or thermal spring. Geysers are a special kind of hot spring. Hot water has a higher mineral content than surface waters, accounting for the large mineral deposits found around geysers and hot springs.

Groundwater does not remain stagnant but moves slowly through permeable rock strata. In its passage through rocks groundwater may carry out geologic work, especially if it encounters limestone or dolomite. These rocks are soluble in the presence of groundwater. The solution of limestone and dolomite initially results in widening of cracks and other openings and eventually leads to the formation of caves.

Karst topography develops in regions where groundwater has dissolved soluble rocks. Features such as sinkholes, disappearing streams, and blind valleys are notable.

Two environmental problems of great concern include deep-well disposal of industrial wastes and nuclear hazards to the groundwater supply.

Questions for thought and review

1. What surface changes take place when excessive groundwater is removed? Give an example.
2. Explain the process of recharging of groundwater.
3. Although clay has much greater porosity than sandstone, the latter might yield more water. Why?
4. Distinguish between porosity and permeability.
5. What happens to the water table during a dry season? During a wet season?
6. How does an artesian well differ from a normal well?
7. Explain the conditions that produced salt water encroachment in the Long Island area.
8. How does the existence of sinkholes in an area hasten the expansion of cave systems?
9. How are stalactites and stalagmites formed?
10. List some possible threats to the groundwater supply.

Selected readings

Adams, F. D., 1954, *The Birth and Development of the Geological Sciences*, Dover, New York.

Evans, David M., 1969, *Disposal Wells — Not the Place for Poisonous Industrial Wastes*, Paper presented at AlChe 66 National Meeting, Portland, Ore.

Ferguson, G. E., Lingham, C. W., Love, S. K., and Vernon, R. O., 1947, *Springs of Florida*, Florida Geological Survey Bulletin 31.

Grosvenor, M. B., 1973, Homeward with Ulysses, *National Geographic*, vol. 144, no. 1, pp. 1–39.

Heath, R. C., Foxworthy, B. L., and Cohen, Philip, 1966, *The Changing Pattern of Groundwater Development on Long Island*, New York Geological Survey Circular 524.

Livesay, A., 1953, *Geology of the Mammoth Cave National Park Area*, Kentucky Geological Survey Special Publication 7.

Mather, Kirtley F., 1939, *A Source Book in Geology*, McGraw-Hill, New York.

Schneer, Cecil J., 1969, *Toward a History of Geology*, M.I.T. Press, Cambridge, Mass.

Sheldrick, Michael G., 1969, Deep-Well Disposal: Are Safeguards Being Ignored? *Chemical Engineering*, April 7, 1969, pp. 74–78.

Todd, D. K., 1967, *Ground-Water Hydrology*, Wiley, New York.

15 Glaciation and the pleistocene

STUDENT OBJECTIVES

On completing this chapter, the student should:

1. understand the nature of the Pleistocene Epoch

2. be able to describe how glaciers form

3. understand how glaciers move

4. recognize erosional and depositional features of glaciers

5. know the difference between stratified and unstratified glacial drift

6. recognize depositional and erosional landforms produced by glaciers

7. be conversant with hypotheses relating to future ice ages

The best prophet of the future is the past.

John Sherman

THE MYSTERY OF THE ICE

To many the most mystifying of all geologic phenomena is the one surrounding the great ice sheets that covered most of northern North America, as well as many other parts of the world. Beginning approximately 2 million years ago, two great masses of ice began developing on this continent. In eastern Canada an ice mass now known as the Laurentide ice sheet formed; in the northwest the Cordilleran glacial complex appeared. The two ice sheets combined at times to cover, except for a few isolated areas, the entire northern portion of North America from the Atlantic to the Pacific (Figure 15–1). Only the driftless area of southwest Wisconsin, parts of Alaska, and the Arctic coast escaped glaciation (Figure 15–2). The thickness of the ice sheet has been estimated at about 10,000 ft (3 km). This corresponds with the thickness of the ice now covering Greenland.

One cannot help but wonder why such vast sheets of ice formed in the past, disappeared, and then returned again in cyclic intervals. It is believed that we have undergone at least four cycles of advance and retreat of glacial ice from the polar regions. What planetary or cosmic changes took place during the Pleistocene? Can these conditions appear again? How and when did civilized peoples first learn of our planet's encounter with the Ice Age? What evidence can we find today that much of northern North America and

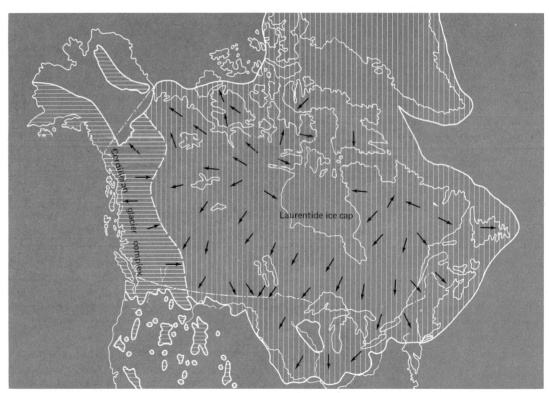

Figure 15–1. Approximately 2 million years ago two great ice masses appeared in North America. In the northwest one formed the Cordilleran glacier complex, and in the northcentral and east North America the Laurentide ice cap. These ice sheets covered most of Canada and about one-third of the northern United States.

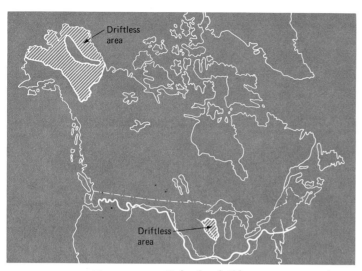

Figure 15-2. Only the driftless area escaped glaciation in northern United States. The driftless region includes the southwest corner of Wisconsin and the land in Minnesota and Iowa bordering this region. A highland area, the Superior Highland, north of this region interfered with the southward movement of the ice sheet from Canada. The lowlands west and east of this highland area now occupied by Lake Superior, Green Bay, and Lake Michigan provided easy routes for the ice to travel around the driftless area. Unglaciated regions in Alaska and the Arctic were also regions where topography prevented easy movement of glacial ice.

Europe were covered by this heavy blanket of ice?

One does not generally consider the present as an age of glaciation, but in reality it must be so considered. About 10% of the land area of the earth is covered by ice. Of all the earth's water, 1.5% is in the form of ice. A sheet of ice about 2 mi (3 km) thick covers most of the continent of Antarctica. Other ice sheets cover much of Iceland, Greenland, and the islands in the Arctic Ocean. Glaciers are found in high mountain valleys in many locations on most continents. Glaciers can even be found near the equator, as exemplified by Kilimanjaro in Africa. What is most important, however, is that recent estimates of the ice-budget (balance between moisture supplied to a glacier or ice sheet and moisture removed from it) indicate that in the southern hemisphere the ice-budget is positive (growing) and in the northern hemisphere the ice-budget is nearly balanced. One can see that glaciers are here to stay for quite a long time.

HISTORICAL BACKGROUND

Our awareness of the presence of these vast ice sheets formerly covering what is now fertile land is recent even if measured in terms of historical time. Speculation leading to modern-day hypotheses regarding the ice ages began in Europe. At first, glacial deposits found so prominently in northern Europe were attributed to the Great Flood. Marine fossils found in bedrock were thought to be debris associated with Noah's trip in the Ark. These may appear to be simple-minded explanations today. But if we were not aware of the existence of great ice sheets in several parts of the earth, it is doubtful that we would believe that morainal deposits (glacial debris load left after it retreats) and many other physical changes produced by glaciers were the handiwork of moving sheets of ice.

Alpine glaciers received early attention from inquisitive naturalists. In 1723, J. J. Scheuchzer projected his hypothesis to explain movement of ice in the alpine valley as the product of a force resulting from the expansion of water as it froze in glacial fissures and cracks. His contemporaries believed that movement was simply that of ice sliding downhill.

People living near alpine glaciers were aware of blocks of highly polished rock, many with scratches on their surface, that were strewn around the valley floor great distances from ice sheets. They also noted ridges of gravel-like debris in front of valley glaciers and where glaciers touched valley walls. Not until 1821 were the bits of the puzzle pieced together by Ignaz Venetz in a paper he presented to the Swiss Society of

Natural History. He argued that valley glaciers in the Swiss Alps formerly occupied positions much farther down the valley than their present boundaries. Several years later he proposed that all of northern Europe had been covered by an ice sheet.

Among the naturalists influenced by Venetz was a young man named Louis Agassiz. Agassiz drove wooden stakes across a valley glacier to test the rate and direction of movement of glacial ice. He identified such glacial deposits as moraines and erratics (glacially transported rocks found deposited on other rock types) in both Europe and North America.

Glacial theory developed rapidly during the balance of the nineteenth century, but not until Arctic and Antarctic regions were explored were the extents of ice coverage on our planet fully appreciated. The study of glaciers is still in its infancy. We can expect to find answers to many of our problems in data now being collected in Antarctica by glacial geologists. A major effort is being conducted in Greenland under the direction of geologist Chester Langway.

MAJOR GLACIAL ADVANCES

At least four periods of glaciation have been established. These occurred in the final interval of earth history, the Quaternary Period. (We are in the Quaternary Period now.) Its 2 million years is divided into two epochs, the Pleistocene, which ended approximately 10,000 years ago, and the Recent (or Holocene). Glaciation was prominent in the Pleistocene, and we now have in the Recent Epoch surface deposits that were left behind by these glaciers.

The cyclic record of advancing, then retreating ice sheets in North America during the Pleistocene is preserved clearly in a number of regions. The stratigraphy of the Upper Mississippi Valley is, perhaps, our best record of glacial movement. The advance of a glacier is recorded by characteristic deposits of *drift* (sediments transported by glaciers) that is spread across land surfaces. When the forward edge of the ice sheet retreated, these deposits were exposed to weathering and erosion. This exposure changed rock fragments to a soil capable of supporting plant life. When the glacier moved across the area a second time, a second layer of glacial drift was deposited. Now there is a record of two glacial drift strata separated by an ancient soil layer. Pleistocene stratigraphy in the Upper Mississippi reveals four glacial till layers, each separated by a soil or a weathered layer. This record is interpreted as evidence of four major glacial advances during the Pleistocene. Each glacial advance was named after the region where its deposits are most prominent. From the oldest to the youngest, these are: Nebraskan, Kansan, Illinoisan, and Wisconsin Stages (Figure 15–3).

From recent studies in eastern Canada and in the Arctic, evidence is mounting that more than four ice advances occurred. These findings call for a reexamination of the present subdivision of the Pleistocene glaciation periods. For example, the Iowan substage of the Wisconsin may be a separate period of glacier advance.

The last major advance of glacial ice saw its maximum strength about 18,000 years ago. About 30% of the earth was covered by ice, and at least 8% of the water was frozen. Deposits and erosional effects of this glacial advance have produced much present topography in many parts of the world. Because these deposits are at the surface today, our knowledge of late Pleistocene geology is more detailed than that of other epochs and periods on the Geologic Time Scale.

HOW DO GLACIERS FORM?

We know that glacial ice covered most of Canada and more than one-third of the United States. Because much of this area is now free of ice, we may wonder how the

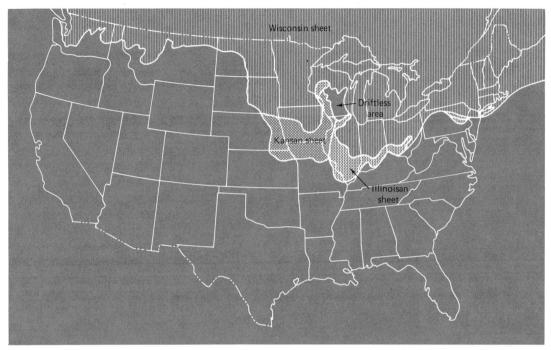

Figure 15–3. The margins of the major ice sheets that moved into the United States during the Pleistocene. The oldest of these, the Nebraskan, failed to move as far south as the younger ice fields. The general extent of Nebraskan ice can only be estimated. Its southern margin was well within the borders of the post-Nebraskan ice fields.

climatic conditions during the ice age periods differed from those of today. We may also ask, "Will the ice age return again?"

It is hypothesized that one condition is necessary for growth and expansion of glacial ice: the amount of snow and ice accumulating in a region must be greater than the amount lost through melting or sublimation (loss to the atmosphere) over an extended period of time. Glaciers will grow readily in size, if increased precipitation in the forms of snow and sleet falls at a temperature that is low enough so that the ice thus formed can be preserved throughout the year. Extremely low temperatures are not necessary for glacial formation and growth. In fact, glaciers are more likely to form when year-round temperatures are only slightly below freezing; long-term ice formation on

bodies of water reduces glacier formation. In addition to large water sources, the atmosphere in this temperature range can hold more water than can extremely cold air.

If the annual mean temperature over the face of the earth were to fall about 10C°, it could bring about the formation of an ice sheet over much of Canada, northern United States, and northern Europe. Increasing amounts of volcanic dust and the present rate of growth of particulates and aerosols introduced into the atmosphere by automobiles and industry can reduce the annual mean temperature to the point required for glacier formation. The ice sheet would form first in high mountains where temperatures at present are close to those required for mountain glaciation. With increased size of mountain glaciers, downslope movement would commence. Eventually glaciers from various valleys in that region would unite and, constantly fed by glacial ice forming at high altitudes, move across the plain.

Movement of glacial ice

Snow is a crystalline form of water (Figure 15–4A). When temperatures remain below freezing after a snowfall, each snow-flake recrystallizes. Extremities of the snow-flake and water present move toward the center of the snowflake. This nucleus recrystallizes into a small grain that is part snow and part granular ice; this substance is called *firn* (Figure 15–4B). As more snow accumulates, pressure of the snow and firn compacts ice grains together. These newly rounded grains tend to rotate and slowly move downhill because of the weight of snow above them. Large crystals of ice will grow as ice grains rotate, until the atomic arrangement within abutting grains has the same orientation. Adjacent grains then fuse to become a single crystalline mass.

When the thickness of the snow and firn exceeds 195 ft (60 m), the weight is so great that the firn compacts to form a solid mass of ice. It is hypothesized that the movement of this ice layer can be accomplished by both internal and external movement of individual ice crystals. Slippage may occur between layers of atoms by their gliding on crystallographic planes within the crystal. After the glide is completed, recrystallization in the new alignment takes place. Glacial ice movement can also occur by shearing along planes in the ice and recrystallization of ice in the new position. The net result of these changes is a flowage similar to that observed in plastics and viscous liquids. This type of behavior is called *pseudoplastic* or *plasticoviscous* flow. The changes thus produced correspond to changes of rocks in metamorphic process.

R. P. Sharp lists several possible mechanisms for glacier movement (Figure 15–5). These may work independently or in combinations. Sharp reports that glacial movement is accomplished by the sliding of ice over bedrock, through intergranular rotation of firn, by movement of water as it recrystallizes into ice, by phase change and slippage within ice crystals, and by shearing through ice.

It must be emphasized that ice and glacial ice behave differently because of the various

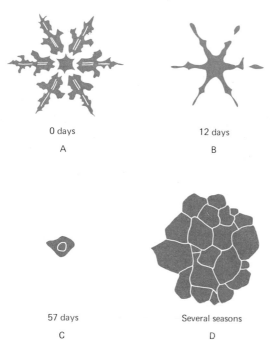

0 days	12 days
A	B
57 days	Several seasons
C	D

Figure 15–4. A snowflake (*A*), while temperatures remain below freezing, will transform (*B*) into granular ice called firn or névé (*C*). In this transformation process the arms of the snowflake melt into drops of water that migrate toward the center of the snowflake. This process is called sublimation. The compact grains of ice particles so formed as water and snow recrystalize and fuse together to form glacial ice. Most geology texts use *glacier ice, glacier movement*, etc. rather than glacial ice, glacial movement, etc., but speak of glacial deposits, glacial periods, glacial features, etc. we can accept either usage. (*D*) The transformation of snowflakes into firn may occur within a single season. The transformation of firn into glacial ice requires many seasons of accumulating snow cover, which ultimately provides sufficient pressure to change the loose granules into a solid mass of glacial ice.

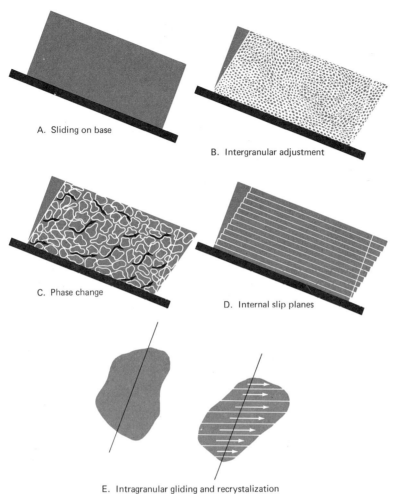

A. Sliding on base

B. Intergranular adjustment

C. Phase change

D. Internal slip planes

E. Intragranular gliding and recrystalization

Figure 15–5. Movement of glacial ice may be explained by such mechanisms as sliding (*A*) and several internal adjustments (*B–E*). (Redrawn from R. P. Sharpe, 1960, *Glaciers*, Eugene, Oregon, University of Oregon Books, .78p.).

structural changes described above. Ice normally behaves like an elastic and brittle solid. When under pressure, or a load equivalent to the weight of 195 ft (60 m) of ice, it will deform continuously and will flow.

The ice that covered Canada during the Pleistocene accumulated to a thickness that is comparable with that found in Antarctic glaciers. This thickness produced enough pressure on the lower layers to cause flowage. As long as the icefield is fed by snow and sleet, it will continue to spread in an irregular path over plains and down stream valleys in mountains. How fast and how far the icefield moves depends on the rate of accumulation of snow in snowfields and on atmospheric temperature. If the rate of precipitation is reduced, or if the temperature rises, ablation (mostly by melting) begins at the terminus of the glacier, and the glacier retreats.

In an icefield the behavior of the ice differs at various depths. The upper crust is brittle and reacts like normal ice. Crevasses

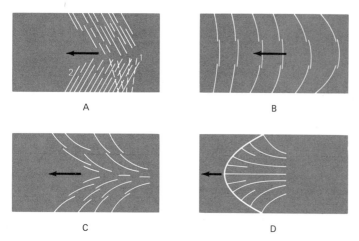

A   B

C   D

Figure 15–6. A number of crevasse patterns may develop as glacial ice moves down a valley. Most crevasses are produced by the extension of the upper surface of the ice. *A* Marginal, (1) old rotated crevasses, (2) newly formed crevasses. *B* Transverse crevasses. *C* Splaying (oblique) crevasses. *D.* Radial splaying. Arrows indicate ice flow direction. (Redrawn from *R. P. Sharpe*, 1960, *Glaciers*, Eugene, Oregon, University of Oregon Books, 78 p.

and fractures form near the top of the glacier (Figure 15–6). Below a depth of 195 ft (60 m), it assumes pseudoplastic characteristics with its associated flow. Crevasses and fractures in the upper zone will open and close as ice in the pseudoplastic zone shears and extends (Figure 15–7).

Movement of debris by glaciers

Most glaciers erode bedrock below them and then carry along this load in their migration. Evidence of great erosional capabilities and capacities of glaciers can be seen in numerous debris deposits found in many formerly glaciated areas. Striated or polished bedrock, plucked blocks, erratics, and gouged-out stream valleys are further evidence of the great erosional force of a moving glacier. Rock and soil debris so removed becomes a part of the moving glacial mass.

In addition to debris picked up by the glacier's plucking, scraping, and pushing

Figure 15–7. *Left:* Roosevelt and Coleman Glaciers are severely crevassed, very active ice tongues descending from a common accumulation area on the northwest slopes of Mt. Baker, Washington (10,778 ft; 3285 m). These glaciers respond rapidly to small changes in climate. *Right:* The quick response to climatic changes results in the development of the crevasse field. (Courtesy of U.S. Geological Survey.)

250

along bedrock and valley walls, frost action accompanying glaciation will increase the rate of *wastage*. Mass wasting is further accelerated when valley walls are undercut by the razorlike edges of the glacier. Slumping and sliding that follow will dump large quantities of *talus* (a heap of rock fragments) onto the top surface of the glacier. Some of this top debris remains in the ice, but much of it works its way down slowly through crevasses and fractures to the bottom of the ice. As time passes, the glacial icefield becomes a great abrasive machine. The abrasive character of the glacier, together with its bulldozerlike properties of scraping and pushing material that lies in the path of its movement, produces many topographic changes in the landscape.

EROSIONAL FEATURES OF CONTINENTAL GLACIERS

Erosional features resulting from glacier movement have been identified and classified with regard to the type of glacier that produced them. Some features are associated with continental glaciation, others with valley (alpine) glaciation; some are common to both. Many small-scale features, such as grooves, striations, and polished surfaces, are common to both types. Large-scale erosional features produced by the continental ice sheet, however, differ from large-scale features produced by valley glaciers.

Erosional features common to continental glaciers are not so easily recognized as those produced by mountain glaciers. Continental ice sheets tend to streamline the topography over which they move, rounding off peaks and ridges. Rounded, streamlined hills called *roches moutonnées* (sheep-shaped rocks) are common in hill areas that have undergone continental glaciation. These hills, often of solid rock, are teardrop in shape. The hill has at one end a smooth, streamlined *stoss* side, and at the other end a steep, more angular side. The streamlined slope

Figure 15–8. Glacial erosion of bedrock produces in many places asymmetric hills that have a gently sloping, striated, and polished surface on one side and a steep, plucked slope on the opposite side. These hills are called *roches moutonnées*. The gentle slope of the *roche moutonnée* faces the direction of the source of the glacier, and the steep slope points in the direction that the glacier advanced. (Courtesy of U.S. Geological Survey.)

points to the direction from which the glacier came. The steep slope points in the direction that the glacier moved (Figure 15–8).

A second continental glacier feature, the *drumlin*, is also a streamlined feature (Figure 15–9). Unlike the *roche moutonnée*, it is not solid rock but is composed of glacial *till* (unsorted, usually poorly stratified glacial debris). Its unconsolidated matter is made up of soil and rock mixtures that are shaped into a whalelike feature as the glacier moves over it. Because the glacial debris must be in place before it can be shaped into a drumlin, the problem of whether this is an erosional or a depositional feature has been the subject of long debate among glacial geologists. The direction of the moving glacier can also be established by the drumlin. Unlike the *roche moutonnée*, the streamlined side of the teardrop (tail side of a whale) faces the direction in which the ice moved, whereas the steeper

Figure 15-9. A drumlin near Lyons, New York, clearly shows the teardrop topographic feature produced by an ice mass moving across previously deposited glacial debris. The long axis of the drumlin parallels the direction of ice flow. Drumlins are not usually found as isolated features but appear as drumlin fields. Four major drumlin belts are located in North America: in New England; in New York and Ontario; in Nova Scotia; and in the Wisconsin, Iowa, Minnesota region. (USDA Soil Conservation Service, Syracuse, N.Y.)

slope of the drumlin faces the direction from which the ice came (Figure 15-10). A classic example of drumlin landscape can be found in the Palmyra region of New York State.

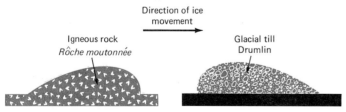

Figure 15-10. Both the *roche moutonnée* and the drumlin can be used to establish the direction of flow of a former ice field. The most famous drumlin in America is Bunker Hill.

EROSIONAL FEATURES OF VALLEY GLACIERS

Erosional features formed by mountain glaciers can be readily recognized because they do not blend as well into the landscape as do the continental features. At the head of a glaciated valley can be found a hollowed-out, amphitheatre-shaped *cirque* (Figure 15-11). When a glacier begins its trip down-slope, it pulls a vertical section of the slope from the mountain top. This action is called glacial *quarrying* or *plucking*. The walls of a cirque may drop almost vertically thousands of feet before they curve out onto the floor of the valley.

Cirques are most likely to form at drainage divides of ancient streams, although they can also form on valley sides where no stream valleys preceded them. Cirques forming on opposite sides of an ancient drainage divide will form a spiny ridge known as an *arête* (Figure 15-12).

The bottom of a cirque is frequently gouged out and forms a depression in the floor of the valley. As the glacier melts, the depression fills with meltwater forming a mountain lake called a *tarn* (Figure 15-13). It is possible that rock formations in a mountain ridge may be such that the longitudinal profile of a glaciated valley will show selective quarrying downslope because of fracture intensity. Thus a series of cirquelike features each with a tarn below it, is produced at intervals. This string of mountain lakes is referred to as *pater noster* lakes, because they resemble the beads of a rosary.

At times two cirques converge, cutting into the same wall. A mountain pass will form as the surface is lowered beneath the level of the remainder of the summit area. The pass is called a *col* (Figure 15-14). If several cirques enlarge and converge, an isolated peak surrounded by the typically vertical cirque wall is produced. This feature is known as a *horn*; the best-known horn is the Matterhorn in Switzerland (Figure 15-15). It has been a challenge to both novice and ex-

Figure 15–11. Boston Glacier, the largest single glacier in the North Cascades, occupies a broad cirque northwest of Buckner Mountain (9111 ft; 2777 m). (Courtesy of U.S. Geological Survey.)

perienced mountain climbers for many years. Horns are also formed during late stages in glaciation as remnants of an arête. In this case several horns are aligned along the former spiny ridge.

A mountain glacier will follow the path of least resistance, so it commonly begins its downslope course following a stream valley. The valley walls must accommodate the mass of ice that is moving down. The ice gouges out walls, smooths out curves, and removes all obstacles in its path (Figure 15–16). When ice is forced around sharp curves,

Figure 15–12. An arête is a sharp ridge produced by glacial action. Cirque walls forming on opposite sides of a mountain ridge may move backward toward each other to form the spiny ridge called an arête.

253

Figure 15–13. As glacial ice melts, the depression below the cirque wall in the valley floor is often filled with water. The lake formed by the meltwater is called a tarn. If a series of cirquelike features are produced at intervals and fill with meltwater, this string of mountain lakes is called pater noster lakes. Lakes photographed are in Snowy Range, Wyoming. (Courtesy of U.S. Geological Survey.)

Col

Figure 15–14. Two cirques cutting into the same wall from opposite sides of a mountain ridge may converge to form a sharp-edged gap in the ridge. This gap (or pass) is called a col.

Figure 15–15. When several cirques enlarge on the flanks of a mountain and converge, a glacial feature known as a horn or matterhorn is formed. The Matterhorn in Switzerland, near the Italian border, is a classic example. (Courtesy of SWISSAIR.)

Figure 15–16. An icefield moving through a valley smooths out curves, gouges out valley walls, and removes all obstacles in its path. A glaciated valley is U-shaped. Often hanging valleys such as the waterfalls at the extreme right are found along the valley wall. (Courtesy of Ward's Natural Science Establishment, Inc.)

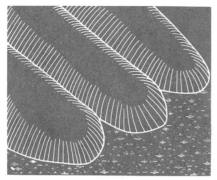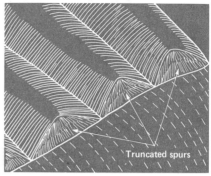

Truncated spurs

Figure 15–17. The massive weight of a valley glacier moving downslope shears off any spurs of land that extend into the valley. Cliffs produced by moving ice are shaped like large flatirons with their points directed towards the sky. These features are called *truncated spurs*. *Left:* Before glaciation. *Right:* Glacial shearing produces truncated spurs.

the mountain spur that extends into the path of the glacier is cut back. This feature is called a *truncated spur* (Figure 15–17). The glaciated valley is U-shaped rather than the V-shaped valley so characteristic of mountain streams.

The valleys of tributaries of a main stream also become glaciated. Because these tributary valley glaciers are smaller than main glaciers, erosion is not as effective, and their valleys are not cut so deeply as is the main valley. When the glacier retreats, valleys of the tributary glaciers open into the main valley above the valley floor. These are called *hanging valleys*. Scenic waterfalls can be found in regions where hanging valleys exist after streams renew their flow (Figure 15–18).

When a valley glacier flows directly into the sea, the glacier ice is able to carve its valley below sea level before the ice breaks away to form icebergs. The breaking off of ice is called *calving*. The submerged valley is called a *fjord* (Figure 15–19). Fjords can be found in both the northern and southern hemispheres, notably in Greenland, Norway, British Columbia, Patagonia, and New Zealand.

Figure 15–18. Bridal Veil Falls in Yosemite National Park is a picturesque waterfall flowing down from a hanging valley. The stream flowing through the hanging valley prior to the period of glaciation was a tributary of the stream that flowed through the main valley. Because the tributary valley glaciers were smaller than the glacier that moved through the main valley, they did not cut down so deeply as did the main glacier. After glaciers receded, valleys of the tributaries opened into the main valley from high above the valley floor. (Courtesy of Ward's Natural Science Establishment, Inc.)

Figure 15-19. Valley glaciers flowing directly into the sea carve the valley floor below sea level. After the ice retreats, the valley is submerged, and a fjord (fiord) forms. (Courtesy Scandinavian National Tourist Offices.)

DEPOSITIONAL FEATURES OF GLACIERS

The debris load of a glacier is a wide range of rocks, soil, sand, gravel, and other unconsolidated material that is found in the path of the moving glacier. This assortment is plucked up and pushed along by the icefield. Deposition begins when the ice begins to melt. The melting process is most noticeable at its terminus. Meltwater from the en-

tire icefield flows downslope to the forward edge of the ice, carrying along with it smaller particles found in the icefield and depositing them out beyond the ice.

The terminus may move slowly forward during the early stages of the melt process, pushed forward by the tremendous inertial energy trapped in the icefield. After inertia has been expended, the terminus may remain static for a period of time. The debris load, however, continues to move toward the forward edge of the icefield as the meltwater seeks lower elevations and by conveyorlike movement of the debris within the ice. When *ablation* (melting and sublimation) exceeds the rate of movement to the end of the glacier, the terminus of the glacier will begin to recede.

Glacial deposits are called *drift*. The name "drift" originated before the former existence of glaciers was known. It was erroneously believed that this material had drifted into the area by water or had been carried in on floating ice.

Glacial drift may be classed as (1) stratified and (2) unstratified.

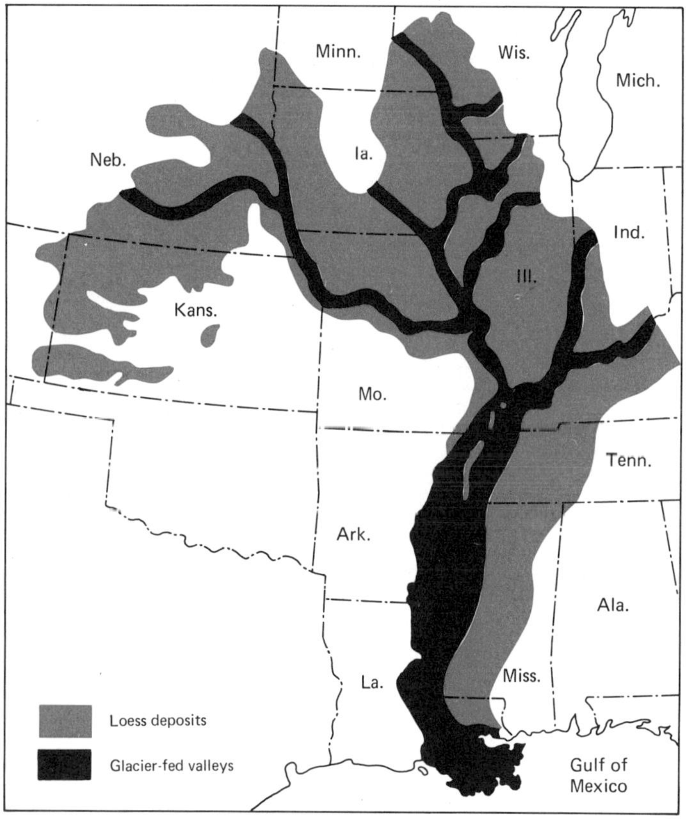

Stratified deposits

Deposits laid down by meltwaters of glaciers are *glaciofluvial deposits* and are found under and immediately beyond the glacier. They consist of particles of rock that have been broken down by disintegration rather than by decomposition. These deposits are sorted and stratified and may have cross-bedding or other sedimentary structures present.

A milklike water usually flows from under both the valley and continental glaciers. The milky appearance is caused by a powdery substance called *rock flour*, which is carried in meltwater. Streams carry rock flour beyond the edge of the glacier and deposit this silt-sized sediment in the outwash plain. It is believed that rock flour is the source for surficial, windblown deposits of *loess* (Figure 15–20) found in the central part of the Mississippi River valley and across extensive portions of China.

Meltwater flowing on the surface of a glacier during summer or a warm spell in winter will work its way down around the margins or seep through the ice or cracks in the ice to form streams. These streams will tunnel through the ice, following sinuous and branching paths. All kinds of rock materials make up the load carried by these streams. After the glacier melts, it will leave behind a ridge marking the ice stream channel. This is an *esker*. Eskers may be short, no longer than a football field, or they may be more than 100 miles in length. Their height varies from that of a tall man to that of a four-story building. Regardless of their height, their width is usually measured in a few feet, perhaps the width of a country road.

Hills of stratified drift that collect in the openings on the tops of glaciers or are deposited as a steep alluvial fan against the edge of

Figure 15–20. The bulk of loess, wind-deposited dust and silt, found in central United States, is believed to be associated with glacial outwash. The composition of deposits is similar to that produced by the grinding up of rocks. This rock flour is then carried in meltwater. Surficial winds distribute rock flour from the outwash plain for great distances and redeposit it in sheets.

an ice sheet are called *kames* (Figure 15–21). A *kame terrace* is produced by stratified sand and gravel that has been deposited between a melting glacier and the valley wall. After the glacier disappears, these deposits stand as terraces along the side of the valley (Figure 15–21). A kame terrace differs from a lateral moraine, because it is *sorted* drift. A kame terrace, a product of fluvial action, is a stratified deposit. A lateral moraine is composed of unstratified rubble that drops down onto the glacier from the valley walls above it.

Glacial melting is a continuous process. A glacier may grow by the addition of ice in its upper regions, but at its lower end it is always melting. When ice is added faster than water is lost, the glacier grows and advances. When water is lost faster than ice is added, the glacier shrinks and retreats. In either case, meltwater will continually flow from the lower end of the glacier. It will carry with it glacial drift and deposit it in a large fan-shaped *outwash plain* (Figure 15–21). The southern part of Long Island, New York, is a good example of this form of glacial deposition. The outwash plain is formed by rock flour, sand, and clay that drops out of the slow flowing meltwater. The streams that criss-cross the outwash plain are braided. Braided streams also occur in regions that are not glaciated.

Outwash plains are often pitted with de-

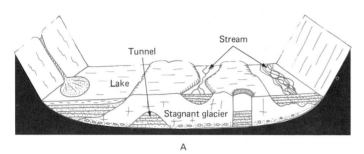

A

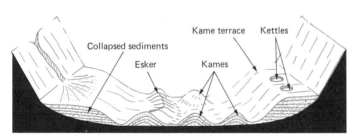

B

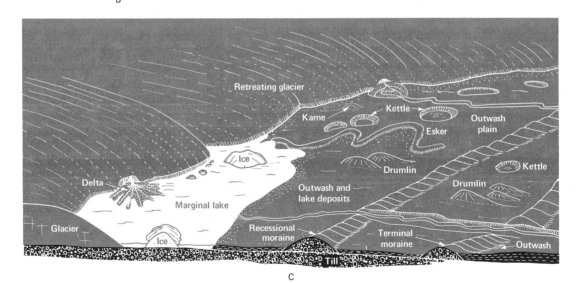

C

Figure 15–21. Relative location of erosional and depositional glacial features formed by valley glaciers (*A* and *B*) and by continental glaciers (*C*).

pressions called *kettles* (Figure 15–21). Frequently, assorted debris in a glacier completely covers a large block of ice that has cracked away from the main ice mass. As the ice melts, during the late stages of the melting, the debris sinks down and produces a basin. Its shape resembles a large kettle.

A long, narrow body of outwash confined within a glaciated valley is called a *valley train*. A valley train reaches its greatest height along its middle and slopes slightly toward the valley walls as well as downvalley. The buildup of outwash in some large valleys filled them to depths of many hundreds of feet. Outwash moved into the mouths of the tributaries, forming fans or deltalike deposits and blocking the discharge of tributary streams into the valley. Stream patterns on valley trains usually take the form of braided channels.

Unstratified deposits

Nonsorted and unstratified, or poorly stratified glacial drift, is called *till*. Two types of till can be found; each is a result of a different depositional process.

Glacial drift that has been transported under the glacier or near its base undergoes considerable pressure. As these rock fragments become freed by melting ice, they become pressed or lodged in the deposits under the glacier. These till deposits are known as *lodgment till*. Elongated, unsorted rocks are aligned with their long axis in the direction of ice movement. Individual sediment particles may be crushed. These deposits will show fissility similar to that found in shale.

Drift deposits formed from debris transported on or within the glacial ice is called *ablation till*. This debris is gradually lowered and then is deposited as ablation, which causes the glacier to shrink. The deposit is loose and does not show the effects of pressure found in lodgment till.

A receding glacier will deposit drift that will form a wide variety of topographic landforms. *Moraine* is a general term applied to large landforms composed chiefly of glacial till.

Terminal moraines (Figure 15–20) are ridges deposited along the outer edge of a glacier at the position of the glacier's farthest advance. Terminal moraines form when the glacier reaches an equilibrium between wastage (melting) and ice supply. The front of the glacier is stable as ice continues to move downward, delivering, as if on a conveyor belt, a continuous supply of glacial drift. Because the forward edge of the glacier is in a zone of wastage, the drift is dropped there; the terminal moraine continues to grow. The size of the ridges so formed will depend on the period of time the glacier remains in equilibrium. These ridges vary from hundreds of feet high to low walls of drift. When the glacier begins to retreat (wastage exceeds ice supply), the meltwater forms a lake between the glacier and the terminal moraine. Eventually the lake overflows the dam produced by the terminal moraine, cutting notches into it. The escaping meltwater deposits low alluvial fans. Streams flowing beyond this point typically flow as braided streams.

If the equilibrium balance is upset in favor of increased ice supply, the forward movement of the glacier will destroy older moraines in the path of its advance. Till from these moraines will be redeposited in new positions. However, if the glacier retreats because wastage exceeds ice supply, periods of temporary pauses will produce ridgelike masses of drift accumulations called *recessional moraines*. A recessional moraine is similar to a terminal moraine except for its position relative to the forward edge of the glacier and its point of maximum advance. There may be a number of recessional moraines present, each one marking a point where the glacier front was temporarily stabilized.

In valley glaciers normal downslope movement of *talus* (fallen materials forming a slope at the foot of a steep valley wall or cliff) forms a ridge on the outer margins of the glacier. Undercutting of the valley wall by glacial movement greatly increases the amount of the debris far beyond that found in unglaciated valleys. When the ice melts, this material is laid down along the edges of the valley in ridges called *lateral moraines* (Figure 15–22).

When two or more valley glaciers join in their downslope movement, the debris in the adjacent sides of the merging glacier will produce a single medial ridge. When the glacier melts, the medial ridge drops to the valley floor. This deposit, midway between valley walls of the enlarged glacier, is a *medial moraine* (Figure 15–23).

As the supply of new ice decreases, the movement of debris toward terminal or recessional moraines slows down and then stops. With continued melting of the main body of the glacier, the debris on its surface, and that within the glacier, is slowly lowered to the base of the valley. Till so depos-

Figure 15–22. As a valley glacier moves downslope, debris from the sides of the valley falling on the edge of the glacier is carried by the glacier. When the glacier retreats, an elongated body of drift is deposited near the lateral margins of the valley. Well-developed lateral moraines are shown on both sides of this valley beyond the junction of the Cowlitz and Ingraham Glaciers, Mt. Rainier, Washington. (Courtesy of U.S. Geological Survey.)

Figure 15–23. When two or more valley glaciers join, the lateral debris in the two adjacent sides will combine to form a single medial ridge. After the glacier retreats, a medial moraine develops midway between the valley walls. What appears to be a huge highway is Yentna Glacier, Alaska. It is a good example of a large valley glacier with many tributaries. (Courtesy of U.S. Geological Survey.)

ited is termed a *ground moraine*. The unsorted till is thin, and, because it has a generous mix of assorted-sized boulders, it is ill suited for farming. Ground moraines are commonly found in New England. They may form gently rolling plains, or they may completely clog preglacial valleys. The topography produced will depend on the glacial debris carried.

Erratics are boulders of various size that differ from the underlying rock type. The term implies something odd or queer, and that is how they appear to a viewer when he encounters a boulder in a most unexpected place (Figure 15–24). These rocks have been moved, in most instances, from the place of origin to their new location by a glacier.

Erratics vary in size from those weighing

Figure 15–24. Glacial till contains rocks that are different from the rock on which the till rests. These different rock fragments are called erratics ever since Charpentier in 1841 described such rocks and boulders as *terrain erratique.* This erratic shown is a boulder of granite. Similar erratics are widely distributed in parts of United States once covered by glaciers.

less than a hundred pounds (45 kg) to those weighing several tons. Near Conway, New Hampshire, is an erratic that weighs approximately 10,000 tons (9000 metric tons). Most erratics were carried relatively short distances from their source. Several, however, have been carried by the glacier for hundreds of miles (kilometers). Erratics containing native copper from Michigan's Upper Peninsula are found in southeastern Iowa and southern Illinois, 500–600 miles (800–1000 km) from the source region.

A series of erratics from the same source region, with common characteristics that can be used to identify the place of origin, is called a *boulder train.* Boulder trains may appear as a line of erratics stretching along a valley, or they may form a fan-shaped pattern with the apex of the fan located near the source area. Mapping of boulder trains, as has been done in New England, provides a good clue to the direction of glacial movement.

CHANGES PRODUCED BY GLACIERS

The importance of the Ice Age is not to be measured in terms of erosional and depositional features produced by glaciers but in the tremendous effect it had on the geology of North America. If we can visualize a field of ice reaching a height of 2 mi (3 km) slowly but steadily advancing from Canada into the northern United States, we can appreciate the changes that must have taken place on the earth's surface in the path of this icefield.

Observe the changes produced in east-central North America. Before the invasion of this region by glacial ice, rivers and streams followed a pattern of flow completely different from that at present. Many of the streams drained to the north. In what is now the Finger Lakes region of central New York were found a series of parallel rivers flowing northward. They emptied into a westward-draining river valley, which later was to become Lake Ontario. The southward advance of the glacier from Canada blocked this stream drainage pattern and gouged out the stream valleys into wide, deep gorges. As glaciers retreated, great deposits of drift dammed the northern sections of these valleys. These dammed rivers are now the Finger Lakes (Figure 15–25).

Across the east-central area that is now the boundary between Canada and the United States, the glacier carved huge basin-like depressions. As the glacier melted, its water occupied these depressions to form a number of lakes; these lakes eventually formed the Great Lakes system.

The Great Lakes now drain through the St. Lawrence valley. Because the St. Lawrence region was among the last to be freed of ice, water from the Great Lakes had to find other outlets. Water from as far east as central New York flowed down the Illinois River to the Mississippi. Later in the recession of the glacial ice front, much of the water drained through the Mohawk valley to the Hudson River. When the St. Lawrence

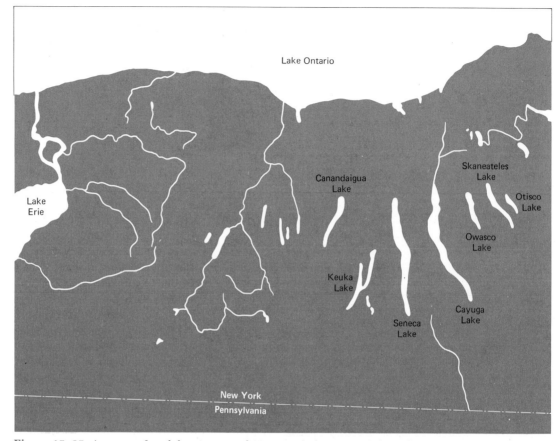

Figure 15–25. A group of 11 lakes in central New York, because of their digitate arrangement, are called the Finger Lakes. The lakes are glacially gouged-out streams that were dammed at the northern end by glacial deposits. (Only names of the larger lakes in this group are given.)

valley became free of ice, the Mohawk outlet was abandoned.

The retreat of the glacier from the Great Lakes region reduced the weight on the land, and it slowly rebounded. Rising land produced changes in the form of the ancestral Great Lakes. It is believed that this process of uplift is still taking place.

One of the more famous remnants of the Ice Age is Niagara Falls. The retreating glacier in this region uncovered a steep, one-sided ridge known as the Niagara Escarpment. The water that was dammed by the ice in the Lake Erie basin spilled over this ridge, after the retreat of the glacier, to form the falls. Niagara Falls has migrated a distance of

7 mi (11 km) from its first site. The spectacular Niagara Gorge traces the movement of the falls from its original position near Lewiston, New York, to its present site by the city of Niagara Falls, New York (Figure 15–26).

Similar changes are visible in other regions of North America that were covered by the ice sheet. Each has an exciting story to tell of the world during the Pleistocene.

WILL THE ICE AGES RETURN?

Climatologists report that the world's weather is turning sharply cooler. Signs of this are evident. Drifting icefields have hindered access to Iceland's ports for the first time in this century. Since 1950 the growing season in England has been shortened by two weeks. Director Reid Bryson of the Institute for Environmental Studies at the University of Wisconsin reports that, if this trend continues, it will affect the whole human populace.

A long-term study of climatic conditions would place the first half of the twentieth century into an exceptionally warm period. The warming trend peaked in 1945, and the temperatures have been dropping since. The drop to date is only 1.5C°, far from the 10C° drop necessary for a new Ice Age. If this trend is not reversed, however, the planet may be caught in an ice-forming cycle similar to that of the Pleistocene.

Long-range weather forecaster Edward M. Brooks believes that the present cooling trend follows a 40-year cyclic pattern. He feels that this trend will continue until 1985. We will not need to wait much longer to see

Figure 15–26. Niagara Falls, in its retreat from the Niagara Escarpment, has carved a magnificent river gorge for a distance of 7 mi (11 km). (Courtesy of American Airlines.)

if the trend will reverse. Both Bryson and Brooks are in agreement, however, that the world is headed into a period of weather unfavorable for agriculture. This is extremely bad news because of the explosion of population in many countries of the world. It appears as if we will be producing less, rather than more, food. As food reserves dwindle, we may move into a period of massive, unimaginable tragedy. Long-range plans to feed an ever-growing population must be made.

SUMMARY

The Ice Age that began approximately 2 million years ago covered the northern part of North America from the Atlantic to the Pacific except for a few isolated areas. The depth of the glacial ice sheet is estimated at about 10,000 ft (3 km). The Ice Age took place in the Pleistocene Epoch. The retreat of the glacial ice, about 10,000 years ago marked the end of the Pleistocene. Geologists have found evidence that the ice sheet underwent at least four cycles of advance and retreat during that time.

Glaciers grow when the accumulation of snow and ice in a region is greater than the amount lost through melting. They retreat toward the earth's poles when this trend is reversed.

The formation of glacial ice is the result of alternate freezing and thawing of snow. The recrystallized substance so formed is called firn. When the thickness of snow and firn exceeds a depth of 195 ft (60 m), the weight becomes so great that the firn compacts to form a solid mass of ice.

Several mechanisms are possible for glaciers to move from one place to another. Glacial movement is accomplished by sliding, by rotation of firn, by movement of water as it recrystallizes, by phase change and slippage within the crystals, and by shearing. Glacial ice continuously deforms and flows.

Two general types of glaciers include valley (Alpine) and continental. Glaciers erode bedrock below them, pick up debris, and move it to another place. The work of a glacier can be classified as erosional and depositional.

Roches moutonnées and drumlins are erosional features of continental glaciers. Cirques, arêtes, cols, horns, and fjords are erosional features of valley glaciation.

Glacial deposits are classified as stratified, and unstratified. Stratified deposits are fluvial in origin. These deposits may have cross-bedding or other sedimentary structures present. They are formed beyond the edge of the glacier in the outwash plain. Other stratified landforms are eskers, kames, kame terraces, and valley trains.

Drift is a general term for material deposited by glacial meltwater. Till is a glacial deposit that has not been stratified or sorted by water action.

Unstratified deposits found after the retreat of glaciers are moraines, drumlins, boulder trains, and erratics. A kettle is a depression remaining after large blocks of ice buried in glacial drift have melted.

The great ice sheet that covered much of northern North America produced many surficial changes. Drainage systems were reversed, the Great Lakes and the Finger Lakes were formed, and Niagara Falls began its movement from Lewiston, New York, to its present site in Niagara Falls, New York.

It is difficult to forecast the outcome of the present cooling trend. Climatologists differ regarding whether a new Ice Age lies ahead. There is agreement, however, in predictions of shorter growing seasons and lower crop yields for the next 10 years.

Questions for thought and review

1. Compare the capacities of glaciers and streams to erode, transport, and deposit sediments.
2. What effects would large ice sheets have on climates near the edges of the ice?
3. What are some differences between gla-

cial drift and sediments deposited by streams?

4. How is glacial ice formed?
5. What is the difference between till and stratified drift?
6. Describe the formation of an outwash plain.
7. How is a kettle formed?
8. Describe the formation of hanging valleys.
9. Compare the topography of land that has been overrun by a continental glacier with that subjected to mountain and valley glaciation.
10. Would a ground moraine make good farmland? Why? ·
11. Why are moraines used in selecting cemetery sites?
12. Describe the differences between a *roche moutonnée* and a drumlin.
13. Explain the arrangement of sediments in a valley train.
14. Are we in another Ice Age? Why or why not?

Selected readings

Davis, J. L., 1969, *Landforms of Cold Climates*, M.I.T. Press, Cambridge, Mass.

Embleton, Clifford, and King, C. A. M., 1968, *Glacial and Periglacial Geomorphology*, St. Martins, New York.

Fairchild, Herman L., *New York Moraines*, Geological Society of America Bulletin 43, pp. 627–662.

Flint, R. F., 1957, *Glacial and Pleistocene Geology*, Wiley, New York.

Leverett, Frank, 1902, *Glacial Formations and Drainage Features of Erie and Ohio Basins*. United States Geological Survey Monograph 41, pp. 672–709.

Sharp, R. P., 1960, *Glaciers*, University of Oregon Books, Eugene, Ore.

16 Exploration of the oceans

STUDENT OBJECTIVES

After studying this chapter, the student should:

1. be familiar with modern developments affecting our knowledge of the oceans

2. be aware of factual data, such as salinity and depths, concerning oceans

3. know something about major topographic features of the ocean floors

4. be familiar with the distribution and origin of ocean floor sediments

5. be cognizant of different types of oceanic currents and their interrelationships with ocean basins

6. understand how waves are generated in the oceans and how they affect coastal areas

7. understand why the oceans are a great equilibrium system and be aware of the potential for upsetting this system

The sea never changes, and its works,
for all the talk of men, are wrapped
in mystery.

Joseph Conrad

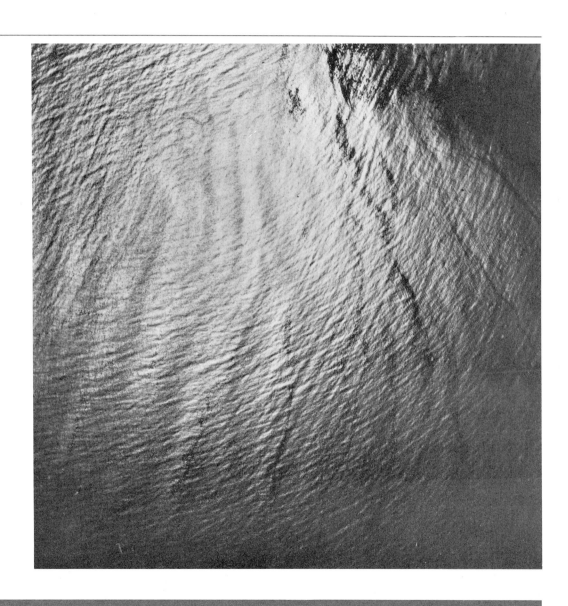

Most of us have visited a beach. If we made our first trip when we were very young, the ocean appeared as an immense, mysterious, even fearsome body of water from which waves rose and crashed on the shore (Figure 16–1). Whenever we became sufficiently brave to enter the surf, the first breaker to slap us in the face told us by its taste and the burning sensation in our eyes that seawater is salty. A thin salty powder remained on our skin as the salt water dried. We watched the waves break, then wash bubbling and foaming onto the beach, spread out in a thinning sheet, then retreat to be overrun by another advancing sheet. We noticed the ripples formed by waves in the sand and the coarse shell material or mica that accumulated in the troughs of the waves. Such is generally the nature of a first experience with the ocean.

People have been fascinated and awed by the seas for centuries. Oceans cover more than 70% of the earth's surface. Yet sea floors remain the least known parts of the earth.

In the last 30 years we have learned more about the oceans than in all previous time. Still a great deal remains to be discovered.

Oceans are of considerable importance to humankind. For many centuries people have harvested fish from the oceans for food. We have increased this industry to the point where many former great fishing areas, such as the Grand Banks of Newfoundland, have been overused. The seas may provide other types of food for us in the future. Oceans also serve to control the climate of much of the land by moderating air temperatures along coasts and by serving as a source of moisture in the hydrologic cycle (see Chapter 13). Oceans are also responsible for maintaining a large part of the oxygen content of the atmosphere.

Oceans will become more important to us in the future; they will serve not only as sources of food but also as sources of valuable metals and fuels. For years the shallow sea areas near the edges of the continents have produced large amounts of petroleum.

Figure 16–1. Breakers along part of the Atlantic Coast of the United States.

The likelihood is excellent that many more oil fields lie undiscovered beneath sea floors along continental margins. Yet we must be very careful about the manner in which these resources are exploited, for the sea is in a delicate state of balance. Life began in the seas; to destroy this balance may endanger the survival of the life that began there as simple forms billions of years ago.

SOME STATISTICS

Oceans contain a considerable variety of ups and downs (Figure 16–2). The greatest depth is considerably greater than the highest mountain on any continent (Table 16–1). If Mount Everest could be placed within the Marianas Trench, more than a mile (kilometer) of water would separate its summit from the surface of the sea. The average depth of the oceans is likewise considerably greater than the average elevation of the continents.

Ocean water is salty, containing various elements that have accumulated in the seas. Chlorine and sodium are the most abundant ions in seawater, followed by magnesium, sulfur, and others (Table 16–2). All the gases found in the atmosphere are found dissolved in seawater.

We refer to the total concentration of salts

Table 16–2. Average concentrations of major components in sea water (in parts per thousand)

Chloride	19.354
Sodium	10.76
Sulfate	2.712
Magnesium	1.294
Calcium	0.413
Potassium	0.387
Bicarbonate	0.142
Bromide	0.067
Strontium	0.008
Fluoride	0.001
	35.137

From K. K. Turekian, *Oceans*. Prentice-Hall, 1968 by permission.

in sea water as *salinity*. The average salinity of sea water is 35 ppt (parts per thousand). This value is remarkably constant within major oceans. It decreases at the mouths of rivers and increases in areas of high evaporation, such as in the Mediterranean Sea (Table 16–3).

In addition to returning water that was previously evaporated from oceans, rivers contribute large amounts of sediment and dissolved material each year to oceans. The total sediment contribution by all major rivers is 30×10^{15} grams per year (3.3×10^{10} tons).

Table 16–1. Comparison of depths in the oceans and elevations of continents.

	Average depths of oceans	Greatest depths of oceans (in meters)
Atlantic	3,926	9,200
Pacific	4,282	11,022
Indian	3,963	6,090
All oceans	4,117	

	Average elevations of continents (in meters)[1]	Highest elevations of continents (in meters)
Africa	757	5,895
Asia	969	8,848
Europe	343	4,810
North America	726	6,194
South America	343	6,960
All continents	769	840

[1] From L. E. King, 1967, *Morphology of the Earth*, 2nd ed., Hafner, New York. (By permission)

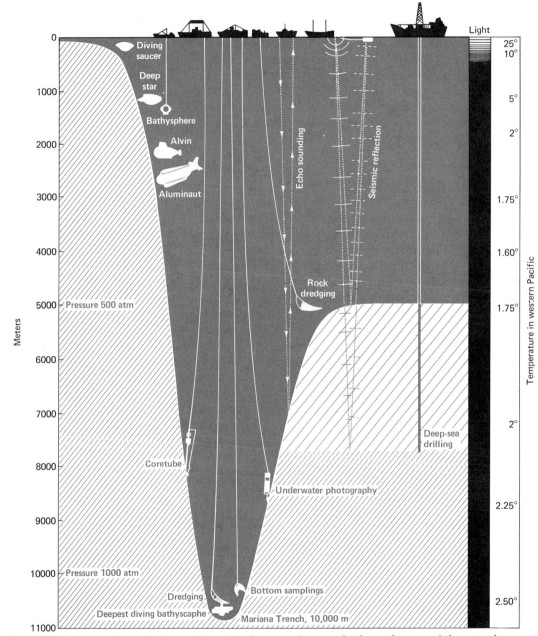

Figure 16–2. Comparison of the sea floor depths, sampling methods, and some of the vessels being used for exploration of different parts of the oceans. (From *The Face of the Deep* by Bruce C. Heezen and Charles D. Hollister. © 1971 by Oxford University Press, Inc. Reprinted by permission.)

Table 16–3. Salinities of the seas and oceans (in parts per thousand)

Arctic Ocean	32
Atlantic Ocean	35.8
Caspian Sea	37
Gulf of Mexico	36–37
Indian Ocean	35
Mediterranean Sea	37
Pacific Ocean	35.0

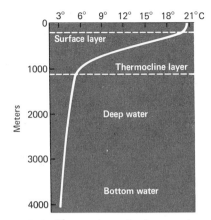

Figure 16–3. Change in water temperature with depth in the oceans. (From *An Introduction to Physical Oceanography* by William S. von Arx, Addison-Wesley Publ. Co. Used with permission.)

The temperature of ocean water varies considerably from place to place and with depth (Figure 16–3). The surface waters of the Arctic and Antarctic oceans are very cold, whereas those of tropical oceans are quite warm (Table 16–4). Warm and cold currents cause further variations in temperature. Temperatures at depth are generally near the freezing point.

Table 16–4. Average surface temperatures in the oceans.[1]

Arctic Ocean	0–5°
Atlantic Ocean	15.7°
Indian Ocean	25–27°
Mediterranean Sea	15–22°
Pacific Ocean	16.6°

[1] °C, summer temperature.

OCEANOGRAPHY

The science of oceanography has experienced a great awakening over the past three decades. During and after World War II a systematic exploration of ocean basins was begun by scientists of several countries, notably the Soviet Union, France, and the United States. Initially the principal reason was military: We were producing naval vessels capable of operation under water at greater depths and for longer periods of time than ever before, so it became necessary to know what was under there. Later the potential of the deep ocean as a source of economic wealth has prompted continued study. In addition, the tremendous food-producing ability of the oceans has created still another reason to study the seas in the light of expanding population.

Oceanography is the study of the oceans. As a discipline it may be subdivided into several subdisciplines, including biological, chemical, geological, and physical oceanography. Our present concern, geological oceanography, overlaps all the others to some degree.

Early attempts to understand ocean basins were limited to soundings by dropping a weighted line over the side of a ship and determining the depth when it touched bottom, along with some crude sampling of sediments and life forms at shallow to moderate depths.

Today we have the assistance of some very sophisticated equipment to help us learn more about the oceans. Perhaps the most important instrument yet developed has been the *continuous echo sounding profiler*. The sound wave that it sends to the sea floor is reflected and recorded by the ship. The amount of time needed for a signal to bounce off the sea floor and to return is a measure of the water depth. Continuous measurements of depth allow a topographic profile to be constructed along the course of the ship. Instruments to measure heat flow, earth magnetism, and gravity and to obtain various types of water, sediment, bedrock, and biological samples have also contributed important information.

The Deep Sea Drilling Project was begun in 1968 to sample sediments and crust of the sea floor by drilling holes from a ship-mounted platform and recovering cores of the material penetrated. This project has been financed by the United States, but scientists from about 75 countries have participated in the voyages of the *Glomar Challenger* (Figure 16–4). The sea floors of all but the Arctic Ocean have been cored. Many preliminary questions concerning ages of different parts of the sea floor and sediments that cover it; rates of sedimentation; and potentially economic deposits of manganese, copper, and cobalt have been answered. Still, these cores represent a tremendous reservoir of information that will be gleaned from future detailed studies.

The development of equipment to carry us to different depths has greatly enhanced our knowledge of the sea floor. This equip-

Figure 16–4. *Glomar Challenger* (Deep Sea Drilling Project/SIO).

ment ranges from the familiar scuba gear to the bathyscape *Trieste*, capable of withstanding the tremendous pressures of the deepest parts of the oceans. The *Trieste* has carried its occupants to a depth of greater than 35,000 ft (10,000 m) in the Mariana Trench. Vessels such as *Alvin*, *Deep Star*, and others are capable of probing the sea at moderate depths of less than 10,000 ft (3,000 m). Many oceanographic vessels from the United States and other countries increase our knowledge of the ocean each year. Today oceanography enjoys a status of considerable importance. We have already begun to reap many practical dividends of oceanographic research; many more will likewise be forthcoming in the future.

OCEAN BASIN TOPOGRAPHY

The early concept of sea floor topography was simple: a flat, featureless plain from the edge of one continent to another with an occasional volcano rising above the waves. However, with the invention and implementation of seismic and sonar equipment for use in determining water depths and later the continuous echo sounding profiler, this concept was quickly abandoned. The margins of continents are submerged by seas to depths that increase seaward. Typical changes that occur with depth vary from continent to continent and with position along the continental edge. Along most of the continental margins is a *continental shelf*, which varies in width from a few miles (kilometers) to more than 100 miles (160 km) (Figure 16–5). In this flooded portion of the continent, depths reach 600 ft (200 m) on the shelf edge, average 400 ft (130 m), and may be as shallow as 270 ft (90 m).

Beyond the continental shelf is the *continental slope*, which slopes more steeply than the shelf and extends to a depth of 6,000–9,000 ft (2,000–3,000 m). The continental slope is relatively narrow, and, although it slopes seaward more sharply than

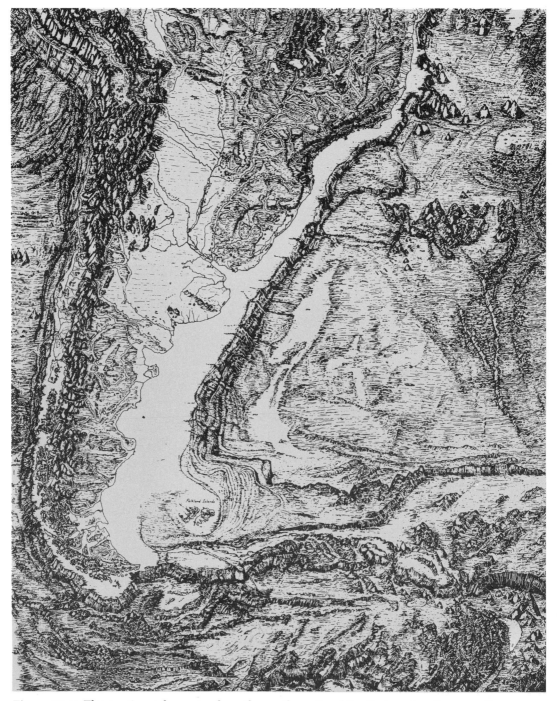

Figure 16–5. The continental margin along the southern tip of South America. Compare the continental shelf on the east coast to that on the west coast of the continent. (From Bruce C. Heezen and Marie Tharp, *Physiographic Diagram of the South Atlantic Ocean*, Geological Society of America).

the relatively flat continental shelf, in many places it would be difficult for an observer standing on it to detect more than a gently inclined surface. In others the slope is definitely steep.

The continental slope along many continental margins gives way to the *continental rise*, which slopes more gently to the *abyssal plain* at depths of 12,000–15,000 ft (4,000–5,000 m). This is the case in eastern North and South America. In other areas, as along the west coast of South America, the continental slope continues downward into an oceanic *trench* (Figure 16–6). Trenches are long, narrow features whose depths exceed those of abyssal plains. The greatest depths in trenches are the greatest depths in ocean basins. Trenches may appear along the edges of continents and also on the ocean sides of island arcs, such as Japan or the Philippines.

Rising from the floors of ocean basins are elongate oceanic *ridges*, which are interconnected between one ocean basin and another (Figures 16–7 and 20–5). These features are made up of a series of smaller ridges paralleling the trend of the main ridge, fractures that offset the ridge and a central rift along the axis of the ridge. The Mid-Atlantic Ridge bisects the Atlantic Ocean; the East Pacific Ridge is closer to Central and South America

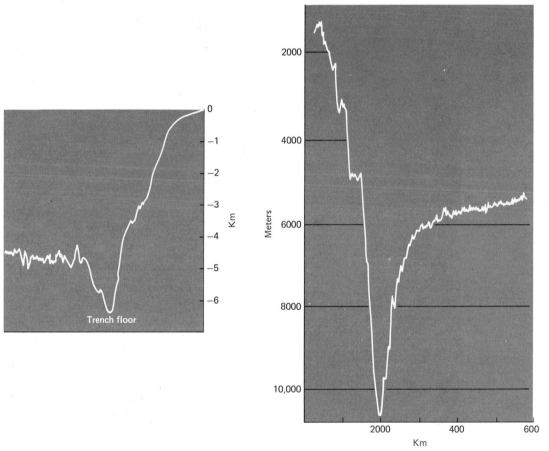

Figure 16–6. Profiles across the Peru-Chile and Tonga Trenches. (From *The Face of the Deep* by Bruce C. Heezen and Charles D. Hollister. © 1971 by Oxford University Press, Inc. Reprinted by permission.)

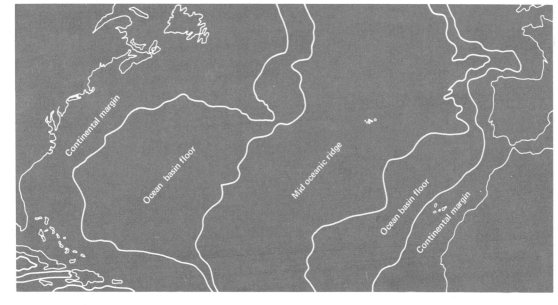

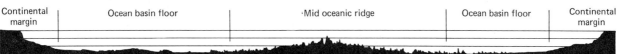

| Continental margin | Ocean basin floor | Mid oceanic ridge | Ocean basin floor | Continental margin |

Figure 16–7. Profiles across the Atlantic Ocean showing the Mid-Atlantic Ridge and the continental margins. (From Bruce C. Heezen et al., 1959, Geological Society of America. Special Paper 65.) By permission.

than it is to Asia and Australia. The East Pacific Ridge actually is traceable closer and closer to the coast of Central America; it then extends into the Gulf of California and finally disappears beneath North America. It reappears off the coast of Oregon and Washington and then disappears beneath North America again.

Ridges, then, do not have to reside far out into an ocean basin. Although one end is generally traceable into another ridge system in the Southern Ocean, then into one in another ocean basin, the other end may disappear under a continent or may simply terminate. The volcanic activity that is very common along ridges builds some parts of ridges above sea level. (The significance and origin of ridges are discussed in Chapter 20.)

Scattered throughout the ocean basins, as either isolated features or in clusters, are conical mountains called *seamounts*. These volcanoes have been built on the sea floors by repeated eruptions of lava in the same way as volcanoes are built on land. Some of these reach the surface and become volcanic islands. After volcanic activity ceases, these islands tend to sink partially into the sea floor because of their weight and the inability of the ocean crust to support them. Some of these features become flat-topped because of their original shape, wave and subaerial erosion, or a combination of these processes. A number of flat-topped seamounts, called *guyots*, have been found in the oceans. Many of their crests are submerged 3000 ft (1000 m) or more. Yet shallow-water fossils have been dredged from them.

Charles Darwin in 1842 proposed an evolutionary scheme for coral reefs that develop around and on these volcanic islands (Figure

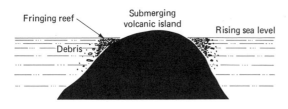

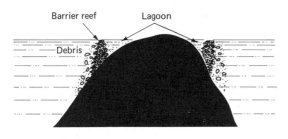

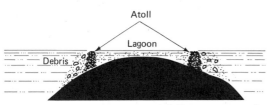

Figure 16–8. Development of different types of coral reefs on a volcanic island according to Charles Darwin.

16–8). A *fringing reef* first develops against the land mass. As sea level rises, or as the island sinks, the reef grows upward and effectively becomes separated from the island by a lagoon, although it is still attached to it at depth. The fringing reef has evolved into a *barrier reef*. With further sinking of the island and upward growth of the reef, the land mass disappears, and a circular reef and central lagoon remain to produce an *atoll*. Although this scheme has been questioned since it was originally proposed, most of the available data on the histories of volcanic islands now favor Darwin's idea.

Along the continental margin frequently V-shaped canyons cut into the continental slope (Figure 16–9). These are *submarine canyons*, and their origin has been the subject of considerable debate since they were discovered during the latter part of the nineteenth century. These canyons may or may not be near the present mouths of major rivers. Several theories have been advanced, including: subaerial erosion during extensive lowering of sea level during the Pleistocene, opening of fractures in the earth (faulting), tidal scouring, erosion by submarine springs, turbidity currents (see page 286), submarine mudflows, and landslides. Of these, erosion by intermittent turbidity currents and submarine mudflows is presently thought to be the best explanation for these features.

Closely associated with many submarine canyons are *deep-sea fans*, which extend from the mouths of submarine canyons out onto continental rises and even onto abyssal plains. These fans resemble alluvial fans or deltas in their overall shapes and vary in size from a few tens or hundreds of miles (kilometers) wide to the Ganges Cone (Figure 16–10), which spreads over the Indian Ocean floor for a distance greater than the length of the Indian subcontinent. On one leg of the Deep Sea Drilling Project a thickness of more than 58,000 ft (18,000 m) of sediment was determined by seismic measurements in one part of this feature. Nowhere on earth is a greater thickness of undeformed sediment known. In areas where the continental rise is missing and the continental slope continues into a trench, deep-sea fans are likewise absent. Sediment brought down the slopes in these areas accumulates in the trenches; thus no fan develops except perhaps as a small feature at the end of a nearly filled trench.

OCEANIC SEDIMENTS

Continental margins and abyssal plains are cloaked in a mantle of sediment, except for bare rock exposed in the walls of the submarine canyons that cut continental slopes. Oceanic sediment has several sources and is composed of many different materials. As expected, it accumulates very slowly far

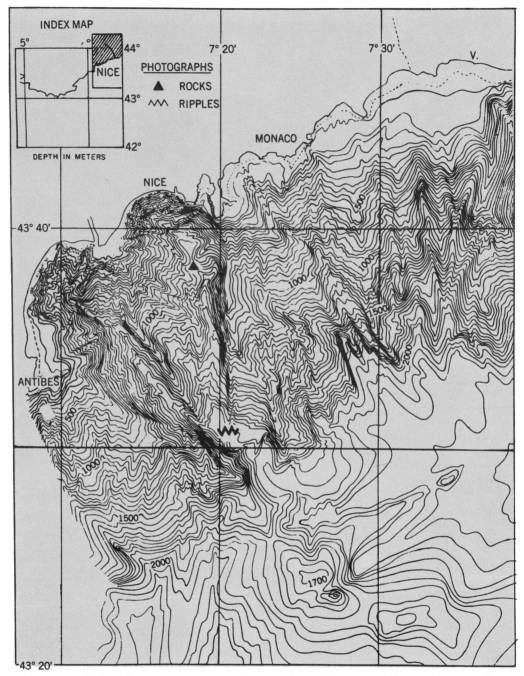

Figure 16–9. Submarine canyon topography along the Mediterranean Coast of southern France. (From *The Face of the Deep* by Bruce C. Heezen and Charles D. Hollister. © 1971 by Oxford University Press. Reprinted with permission.)

Figure 16–10. The Great Ganges Cone in the Indian Ocean. Note that its length exceeds that of the sucontinent of India. (From Bruce C. Heezen and Marie Tharp, *Physiographic Diagram of the Indian Ocean*, Geological Society of America.)

from any land source and more rapidly near a landmass.

Sediments of the continental margins

Most of the sediments that accumulate on continental shelves and slopes are land-derived clastic sediments: muds, sands, and occasionally pebbles. Yet in some areas this is not the case. For example, lime mud is accumulating in the Bahama Banks and off northern Australia.

It might be expected that coarse sediments would be found closer to shore and finer sediment farther offshore. To accept this would be an oversimplification. A fluctuation in sea level—of as much as 450 ft (125 m) below its present level to 200 ft (70 m) above it—is a result of the taking up of water in ice and subsequent melting of that ice several times during the Pleistocene. Also, in many places parts of continental shelves either have been covered by glaciers or have had sediment bulldozed out onto them by glaciers. Anomalously large boulders and cobbles have been dredged from the floor of the North Atlantic too far from shore to have been carried by anything other than icebergs, which dropped them as they melted. Polished and grooved surfaces of many of the boulders tell us of their glacial encounters.

Sediments of continental margins are generally clastics, but their size distributions have been made less predictable by fluctuations of sea level, by glaciation of continental shelves, by presence of islands on continental shelves, and by submarine transport processes, such as turbidity currents. Organisms may also have an important bearing on the kinds of sediment that accumulate in an area. When certain types of organisms are abundant, their skeletons may constitute a major part of the sediment in an area. These may include fragments of shells, calcareous algae, coral, and echinoids. Skeletons of microscopic plants and animals may also add significantly to the sediment in some areas.

Pelagic sediments

Sediments accumulating in deep-ocean basins, whether on abyssal plains, continental rises, deep-sea fans, or channels, are collectively known as *pelagic sediments*. They may be brought to the oceans by streams and rivers, by volcanic and glacial processes, and by winds; or they may form within the oceans through chemical or biological processes. Sediments may be divided into three groups: inorganic sediments, glacial sediments, and biologically derived sediments.

Inorganic sediments. Several types of inorganic sediments accumulate on sea floors. Clays derived from land are probably the most common. They are made up of clay minerals, along with particles of quartz and other minerals so small that they occur in the clay-size range. Some of these minerals in sea floor sediments precipitate directly from seawater. Most clays in the Atlantic Ocean basin are land derived. Kaolinite is a common constituent of deep-sea sediments.

Fine detrital quartz is also common in sea floor sediments. Most of it is in the silt- and clay-size range, but some is fine sand. Detrital feldspars and other minerals may also occur in these sediments.

Volcanic materials and their decomposition products are common constituents of sea floor sediments in areas of high volcanic activity, such as the Central and South Pacific. Volcanic ash settles into the sea after it is blown into the air by volcanoes. Much of this material is cooled very rapidly and forms glass. Volcanic glass in a hydrated form is relatively unstable and breaks down through reaction with seawater to form several minerals. One decomposition product may be a zeolite mineral; another may be the clay mineral montmorillonite. Here are examples of minerals that form within ocean basins.

Minerals that form through inorganic processes at the site of deposition are known as authigenic minerals. Manganese nodules are another example of authigenic material that forms within ocean basins. They may prove to be of considerable value in the future as sources of manganese, copper, cobalt, nickel, and, to a lesser extent, iron. The nodules are formed by precipitation of impure manganese and iron oxide and hydroxide around a nucleus of any material that is chemically different from their surroundings. Volcanic rocks, sharks teeth, bones, and even discarded man-made objects, such as artillery shells, have served as nuclei for precipitation.

Glacial sediments. Sediments of at least partial glacial origin are common in areas of some oceans. This is particularly true in the Southern Ocean and the Arctic, where icebergs transport sediment seaward. As icebergs melt, sediment is dropped and deposited on the sea floor, frequently many miles (kilometers) into the ocean basins. Large boulders of continental rocks are frequently found in areas of the abyssal plains where only fine sediments should occur.

Biologically derived sediments. Most creatures live within 300 ft (100 m) of the surface of the oceans. Yet sediments in the deepest parts of ocean basins may contain appreciable quantities of skeletons of microscopic organisms that have accumulated on the sea floor after the organisms died. These microscopic animal and plant creatures are collectively known as *planktonic* (floating) organisms.

Some of these creatures secrete siliceous skeletons; others have skeletons of calcium carbonate. A few have phosphatic skeletons. Microscopic animals called *radiolarians* and microscopic plants called *diatoms* secrete skeletons of silica (Figure 16–11). *Foraminiferans* (animals) and *coccoliths* (plants) secrete calcareous skeletons (Figure 16–12). All these are very important in sediment accumulations, but calcareous material tends to redissolve at depths greater than 13,500 ft

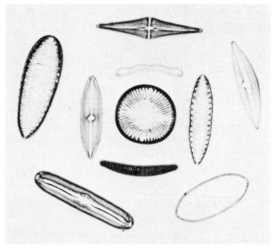

Figure 16–11. Electron photomicrograph sketch of several diatoms. (Courtesy of Ward's Natural Science Establishment.)

(4,500 m) because of undersaturation of $CaCO_3$ in seawater at these depths.

Phosphatic material from bones and other skeletal materials of larger organisms rarely make up an appreciable percentage of sea floor sediments. This material is apparently consumed by other organisms or redissolved into seawater, so it does not have the opportunity to accumulate. Some bones, teeth, and other phosphatic materials do occur in sediments, but they are unimportant in terms of total volume of sediment.

Organic matter may accumulate when bottom conditions dictate that organic matter not be broken down by bottom-living organisms, such as worms, serpent stars, holothurians, and others, or by oxidation. These conditions exist in a few places when currents do not bring adequate oxygen to the bottom to ensure survival of the usual organisms. Certain anaerobic organisms capable of living in an oxygen-starved environment break down the organic matter somewhat, but the usual result is an organic rich black mud. Such areas are called *stagnant basins*. Several fjords of Norway, the Black Sea, Chesapeake Bay, and other localities contain sediments enriched in organic matter as

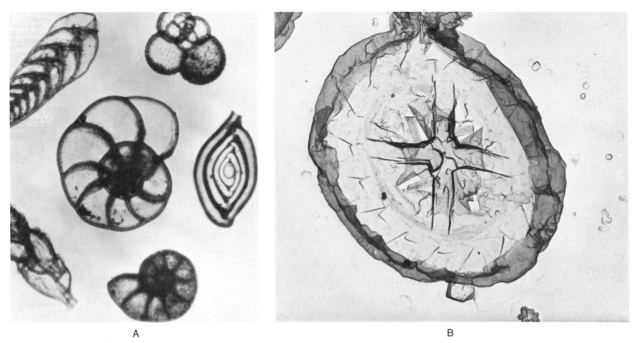

Figure 16–12. *A.* Photomicrograph of some foraminifera (Courtesy of Ward's Natural Science Establishment.) *B.* Electron photomicrograph of a coccolith from the Cretaceous of Alabama. Magnified 20,000X. (Courtesy of W. W. Hay, University of Miami.)

much as 5%. Normal deep-sea concentration of organic matter is about 1%.

THE OCEANIC CRUST

The crust beneath the sea is basaltic in composition. Much of this crust is covered by a veneer of sediment of varying thickness. Oceanic crust is much thinner than continental crust, but it is denser. Whereas continental crust may be as much as 24 mi (40 km) thick, oceanic crust averages only 1–3 mi (3–5 km) in thickness. Continental crust, whose composition is close to that of granite, has a density of about 2.7; the density of oceanic crust is about 3.0.

CONSTANCY AND MOTION OF SEA LEVEL

We generally think of sea level as being a constant. We relate elevations to mean sea level. Yet evidence from historical times indicates that sea level has changed in many parts of the world and over the whole world during much longer periods of time. Remains of ancient Greek architecture are being flooded by the Mediterranean Sea. Similar effects have been noted elsewhere (Figure 16–13). There is evidence also that land is rising or that sea level is being lowered (Figure 16–14). Quite often after an earthquake in a coastal area, changes in sea level will be notable. During the 1964 earthquake in Anchorage, Alaska, the land rose in some areas and fell in others along the coast, creating land where there had been none and inundating areas that had been previously above sea level. Changes of this type are called *tectonic changes* in sea level. They are generally local in extent, affecting only part of a continent, and are a result of forces operating within the continent.

Worldwide sea level changes have also

Figure 16–13. Portion of the Adriatic Coast which has been sinking during historical times. (U.S. Geological Survey.)

Figure 16–14. Portion of the Pacific Coast on San Benito Island, Baja California that has been exposed during recent geologic times. The flat areas are old elevated marine terraces. (Warren Hamilton, U.S. Geological Survey.)

taken place. These are termed *eustatic changes*. They have, for the most part, been attributed to waxing and waning of continental ice sheets. It has been calculated that sea level once dropped more than 300 ft (100 m) below its present level all over the world; it would rise an additional 100 ft (35 m) if the present icecaps were to melt (see Chapter 15).

It has also been suggested that eustatic changes could be produced by adding sediment to fill partially an ocean basin. This would require a great volume of sediment. Perhaps a more reasonable cause for eustatic sea level changes of other than glacial origin might be the change in shapes of ocean basins over millions of years (because of movement of continents) and changes in depths of ocean basins (because of plate tectonic processes; see Chapter 20).

Table 16–5. Comparison of discharges of major rivers and ocean currents.

	Cubic meters per second (millions)
Ocean currents	
Antarctic	150–200
Gulf Stream at Florida	25
Gulf Stream at Cape Hatteras	100
Kuroshio (Japan)	50
Major rivers	
Amazon	0.226
La Plata–Parana (Argentina)	0.0466
Congo (Africa)	0.0408
Yangtze (China)	0.0291
Ganges–Brahmaputra (India)	0.0233
Mississippi–Missouri (United States)	0.0181
Yenisei (Siberia)	0.0178
Orinoco (Venezuela)	0.0175
Mekong (Thailand–Vietnam)	0.0163
Lena (Siberia)	0.0157

SOURCE: Courtesy of Prof. Billy L. Edge, Clemson University.

OCEAN CURRENTS

Surface currents

Currents are like great rivers of water within oceans. However, the amount of water carried by different currents dwarfs the discharge of even the largest rivers. Table 16–5 compares the discharges of some of the larger rivers with estimates of discharges of major ocean currents. It is very difficult to estimate the discharge of an ocean current, because it is difficult to define its exact boundaries.

Surface currents control the climates of many coastal areas. The warm Gulf Stream keeps the British Isles from having a harshly cold climate. On the other hand, the cool Japan Current moving down the West Coast of the United States moderates the climate there and prevents it from being too warm, but does not make it too cold.

Surface currents are driven by winds (Figure 16–15). They follow prevailing wind directions within oceans. All equatorial currents are directed westward because of westward moving trade winds immediately

north and south of the equator. Farther north and south, from about 40° north and south latitude, winds move from west to east as the prevailing westerlies; the ocean currents do likewise. Circulation of air is governed partially by the east–west rotation of the earth – hence the east–west movements of many currents in the oceans.

Density currents

Currents also move along ocean floors (Figure 16–16). These are generally cold, more highly saline waters that descend in Arctic and Antarctic regions and flow toward the equator in the Atlantic. That from the Arctic flows near the bottom past the equator and finally approaches the surface in the Antarctic region. This current is thought to form from warm, more saline (because of evaporation) Gulf Stream water cooling and sinking along the east coast of Greenland. The cold Antarctic water is formed as sea ice forms, leaving more saline water that sinks to the bottom and flows northward.

Denser, though warm, saline water enters

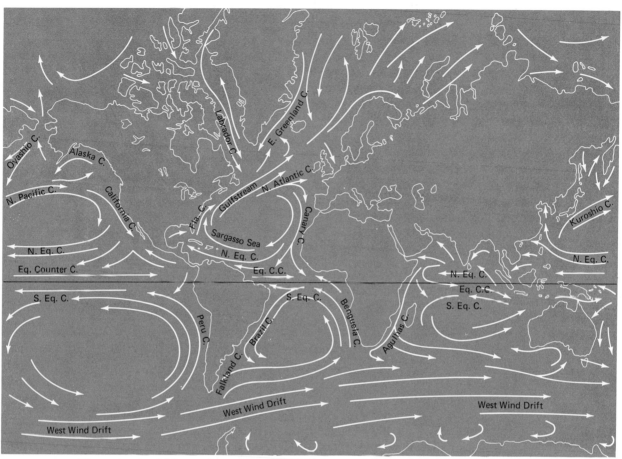

Figure 16–15. Major surface currents in the oceans of the world. (A. C. Duxbury, *The Earth and its Oceans*, Addison-Wesley Publishing Company. Used with permission.)

the Atlantic from the Mediterranean at a depth of about 1500 ft (500 m). This water is a thin, rapidly moving mass that spreads out and flows down the sea floor at measured velocities of up to 3 ft/sec (100 cm/sec). This current has scoured the Atlantic sea floor to a depth of 4500 ft (1500m). Beyond 150 mi (250 km) from Gibraltar it spreads out, moving northwestward as a thin layer. This layer has been recognized over much of the North Atlantic.

Turbidity currents

Currents of sediment that intermittently sweep down continental slopes and onto sea floors are *turbidity currents*. They are intermittent, because they involve the release of an unstable mass of sediment that has built up at some point; it is then set in motion by an earthquake, by a change in a bottom current, or simply by too much sediment built up on too steep a slope.

Once a turbidity current begins its journey, it gathers momentum quickly and may reach a maximum velocity of more than 75 ft/sec (25 m/sec) on continental slopes and as much as 20 ft/sec (7 m/sec) near the bases of continental rises. An earthquake in 1929 on

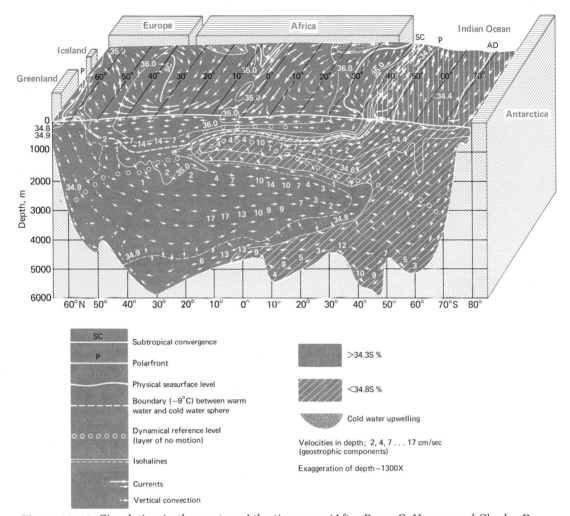

Figure 16–16. Circulation in the western Atlantic ocean. (After Bruce C. Heezen and Charles D. Hollister, *The Face of the Deep*, Oxford University Press, 1971. Taken from G. Wust, 1949, Block diagramme der Atlantischen Zirculation auf Grund der "Meteor" Erg. Killer meer es forschungen 7: Fig. 1. Used with permission.)

the Grand Banks off Newfoundland set off a turbidity current that severed several trans-Atlantic telephone cables. The velocity of the current is known from the positions and times that the cables were broken (Figure 16–17). However, that a turbidity current was responsible for the severing of the cables was not accepted for a number of years; later laboratory experiments and other data demonstrated that such currents indeed exist and do have the energy to cause damage of the

type that was witnessed on the Grand Banks.

Turbidity currents lose their energy and velocity as slope decreases. They spread out into a fan, which begins to deposit the sediment it is carrying. Large particles, mostly fine sands, are deposited first; then silts and finally clay come out as graded beds. These may be scoured and filled by other layers such that cross-bedding also results. But graded bedding in fine sediments is thought

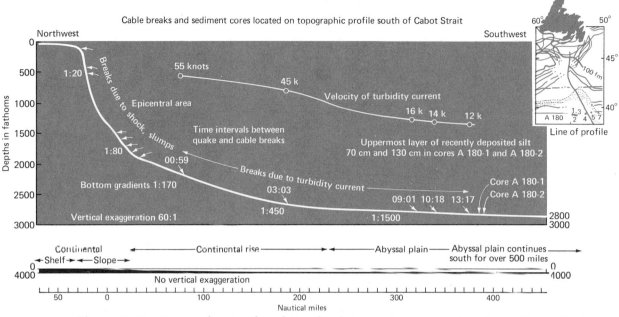

Figure 16–17. Diagram showing the relationship between the times that sections of transatlantic cable were broken by a single turbidity flow following the 1929 Grand Banks earthquake. (From *The Face of the Deep,* by Bruce C. Heezen and Charles D. Hollister. © 1971 by Oxford University Press. Used with permission.)

to be characteristic of *turbidites,* turbidity current deposits.

OCEAN WAVES AND TIDES

Waves and wave motion

Most of the waves we see in oceans are *wind-generated waves.* Submarine earthquakes produce waves that occasionally may cross an ocean, produce extensive damage in a coastal area, and even rebound to cross the ocean again but produce little or no damage on the return trip. These are seismic sea waves or tsunamis. Tides cause oceans to rise and fall periodically and may produce an effect of higher wind waves in an area when superimposed on a high tide.

In order to talk about waves, we must familiarize ourselves with their terminology (Figure 16–18). The highest point on a wave is its *crest.* The lowest point is its *trough.*

The vertical distance from crest to trough is the *amplitude* (*h*), and the horizontal distance from crest to crest or from trough to trough is the *wavelength.* So, if we say that a wave has a height of 15 ft. (5 m,) we are also saying that it has an amplitude of 15 ft (5 m). Just as the amplitude of waves can vary, so also can the wavelength. Wind waves with wavelengths of 2100 ft (700 m) have been observed.

Have you ever watched a floating leaf or a cork as waves pass it? A single wave does not transport the object very far and, if you

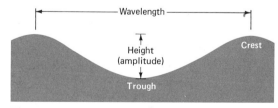

Figure 16–18. Anatomy of a wave.

288

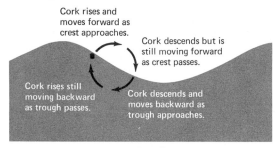

Figure 16–19. Motion of cork in a wave.

watch it carefully, you will see the object actually move back as the wave approaches. It will then appear to climb and move forward as the crest and backside of the wave pass and then will back up again for the next wave. The leaf or cork actually follows a circular path (Figure 16–19), but each time it does not return to its precise starting point. It moves forward a bit; thus it traces overlapping circles. The diameter of the top circle is equal to the wave height at the surface.

This circular motion of waves is not merely a surface phenomenon. It occurs below the surface, but the sizes of the circles decrease with depth (Figure 16–19). Generally the circles are so small at a depth that is slightly greater than one-half the wavelength that effects are negligible.

What happens when water depth is less than one-half the wavelength? The circles become flattened into ellipses, and circular wave motion becomes elliptical motion. Also, the shallower the water, the more drag and slowing down of the lower part of the

area affected by the wave. Yet the upper, near-surface section continues to move along at the same velocity. The top first becomes asymmetric; then water starts to spill from the crest. Finally, as the water becomes very shallow, the wave *breaks* and a zone of *surf* forms along the shore (Figure 16–1). Breakers may build gradually, or they may come up suddenly. This is determined by the slope of the bottom at any given point. In areas of gently sloping bottom, as along most of the East Coast of the United States, breakers will build up gradually. Along the Pacific Coast the slope is more steep, so breakers will build rapidly and appear suddenly from what appears to be an almost calm sea (Figure 16–20).

Wind waves

· Winds produce waves in the ocean, or on any body of water, and their size is directly related to the velocity of the wind. A gentle breeze may only ripple the water, whereas a 40-knot wind may produce waves that are 16 ft (5 m) high. The amount of time and the distance over which the wind blows also affect the height of waves.

Wind waves, once generated, may travel for great distances from the area where they originated. Waves that have formed in the South Pacific may break on the coast of California 5000 mi (8000 km) away.

The wind that produces one set of waves may die, or the waves may continue to travel great distances from their area of origin. Along the way winds may blow from other directions, superimposing other waves onto the first set. Several sets of waves of different amplitudes may be superimposed at times to produce interesting interference patterns (Figure 16–21).

Seismic sea waves

Coastal or submarine earthquakes frequently produce waves with very low amplitudes, 3 ft (1 m) or less, but with tremendous wavelengths. Wave crests may be

Breaking Waves

 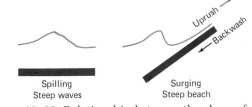

| Plunging
Long-low waves | Spilling
Steep waves | Surging
Steep beach |

Figure 16–20. Relationship between the slope of the sea bottom along the shore and the kinds of waves that break on the shore. (After D. L. Inman in F. P. Shepard, *Submarine Geology*, 3rd ed. Harper & Row, Publishers, 1973.)

Figure 16–21. Interference patterns produced by waves intersecting from different directions near Los Angeles, Calif. (U.S. Geological Survey).

350 mi (500 km) apart. These waves travel at high velocity, on the order of hundreds of miles (kilometers) per hour, and hence possess a large amount of energy. These misnamed tidal waves are properly called *tsunamis* or *seismic sea waves*.

A ship at sea would not note the passage of a seismic sea wave because of its low amplitude. Yet, when one of these waves encounters a land area, much of its great energy is expended there (Figure 16–22). Waves 45–65 ft (15–20 m) high have been recorded

in Hawaii as a result of seismic sea waves. On April 1, 1946, an earthquake off the Aleutian Islands produced seismic sea waves that reached the northern part of the Hawaiian Islands in 5 hr, traveling a distance of over 2500 mi (3700 km). Their average speed was close to 480 mi/hr (800 km/hr).

As one of these waves approaches, the sea retreats from its normal position, the surf zone moves seaward, and there is an eerie silence. Then the waves surge in, rapidly increasing to their maximum size.

Figure 16–22. Damage produced by a seismic sea wave near Seward, Alaska. (NOAA photo.)

Many people in coastal areas, particularly around and within the Pacific basin where major earthquakes are common, have been killed by seismic sea waves. Since World War II a system has been devised to detect long wavelength waves so that people from low-lying areas can be warned of the approach of seismic sea waves in time to evacuate. This, coupled with a worldwide network of seismic stations, has reduced the number of persons killed by the hazard to a very few.

Tides

The gravitational pull of the sun and the moon produces a bulge in the earth toward them. Although *ocean tides* are most noticeable, tides in the solid earth are also measurable. When the sun and the moon are aligned, abnormally high tides result. These are *spring tides*. When the sun and the moon are at right angles to each other, abnormally low tides called *neap tides* occur (Figure 16–23). The change from one to another takes place every two weeks.

Many variables affect the height of a tide (Figure 16–24). These include the relative positions of the sun and the moon, the Coriolis force, the position of the point of observation on the coast, and even the point at which ocean observations are being taken at a given time. Predictions of

Neap tides

Spring tides

Figure 16–23. Relationships between the position of the Earth, sun, and moon, and neap and spring tides.

high and low tides that are printed in daily newspapers are based on the solution of an equation involving all these factors. Solution today is made by computer. It may also be done mechanically with a *Kelvin tide-predicting machine,* an instrument that attempts to duplicate the up-and-down motion of each factor that influences tidal motion and then draws a curve based on this motion. Tides at a given point along a coast may be predicted to the nearest minute.

An interesting local phenomenon produced by tides is a *tidal bore.* This is a rapidly advancing wave that results as water attempts to flow into a narrow bay (Figure 16–25). In some areas such as the Bay of Fundy in Nova Scotia, the bore becomes an advancing wall of water 3 ft (1 m) high and more. The largest known bore, in the Chien-

A

B

Figure 16–24. Portion of the coast of South Carolina at both high (A) and low (B) tide. The tidal range here is less than 2 meters (about 5 feet).

Figure 16–25. A tidal bore (Courtesy of City of Moncton, N.B.)

Tang River in China, is 9–12 ft (3–4 m) high. Others occur in the Amazon and Seine rivers.

Although we know a great deal about occurrences and effects of tides in coastal areas, we know little about tides in open oceans and at great depths. Undoubtedly this will be an important area of study in the future.

SHORELINES

Oceans shape the edges of continents and islands by tearing them down in some areas and building them up in others. The rugged Pacific coast of the United States contrasts markedly with the Atlantic coast from New York City southward. The smooth, wide beaches of Florida, the Carolinas, and Virginia likewise contrast with the irregular coastline of Maine and maritime Canada. What causes this difference?

Keep in mind that sea level has fluctuated over the past 2 million years, so today's coastal area may be yesterday's continental shelf or some point inland. Also, realize that parts of the continents are actually rising, while others are subsiding. These points will influence the role of the sea in shaping the coast.

Additional factors that are very important include the type of terrain and rock types bordering the coast, whether it is mountainous or flat and whether it is composed of resistant or weak rocks. Whether rivers bring large volumes of sediment to the sea, and whether the sea can dispense the sediment or allow it to accumulate, also affect the shape of a coast. Another item to remember is that parts of coasts and even continental shelves have been glaciated. In places glaciers carved valleys along the coasts; while in others they pushed sediments in front of them onto continental shelves. A portion of Long Island is a good example of a glacial terminal moraine.

The coast of the United States

Consider the factors listed above, and think about the coastline of the United States. From Maine to New York City it is a glaciated coastline modified by wave action. The coast of Maine is dominated by glacial erosional features, and deposition was prominent along the coast from Massachusetts to New York.

South of New York City is the irregular coast along which river valleys have been flooded, forming *estuaries* such as Chesapeake Bay, Delaware Bay, and the barrier-island-protected estuaries off Albermarle Sound of Virginia and North Carolina. Further south, from South Carolina to Florida, the coastline becomes emergent and straight.

The Gulf Coast of Florida is smooth to the panhandle where it is again submergent; several estuaries exist, such as Appalachicola Bay. To the west is submergent Mobile Bay in Alabama. Westward to Texas the Gulf Coast is dominated by the great Mississippi

delta. In Texas the coastline returns to one of a smooth, submerged, gently inclined plain.

Mountains accompanied by a narrow continental shelf dominate the Pacific Coast. Moreover, no large deltas result from the accumulation of sediment, even from rivers like the Columbia. Strong currents here disperse sediment and prevent it from accumulating. The Pacific Coast is dominantly emergent, and the Canadian coast is a fjord coast that is continuous to Alaska. Some glaciers enter the sea along the coast of British Columbia and southeastern Alaska.

Erosional and depositional processes and features of waves

A rocky coastal area that has been inundated by the sea likely would present an irregular shoreline that waves would attempt

Figure 16–26. Refraction of waves around a headland, Brazil. (U.S. Geological Survey.)

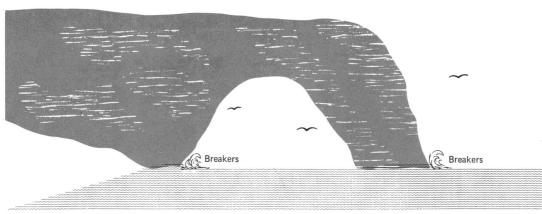

Figure 16–27. A sea arch.

to smooth by erosion and deposition. If waves approach the shore from the sea, irregularities such as a mass of rock extending seaward (a *headland*) would break up evenly spaced waves and cause them to bend or refract around the obstacle (Figure 16–26). This *wave refraction* becomes responsible for erosion along the shoreward extension of the headland and eventually causes it to be cut through, leaving an isolated mass of rock that is separated from the mainland. This is a *sea stack*. Occasionally waves will undercut the shoreward area of a headland leaving a *sea arch*, a bridgelike connection suspended above the sea (Figure 16–27). All of these features are eventually eroded away, as the sea smooths the irregularities of the coast.

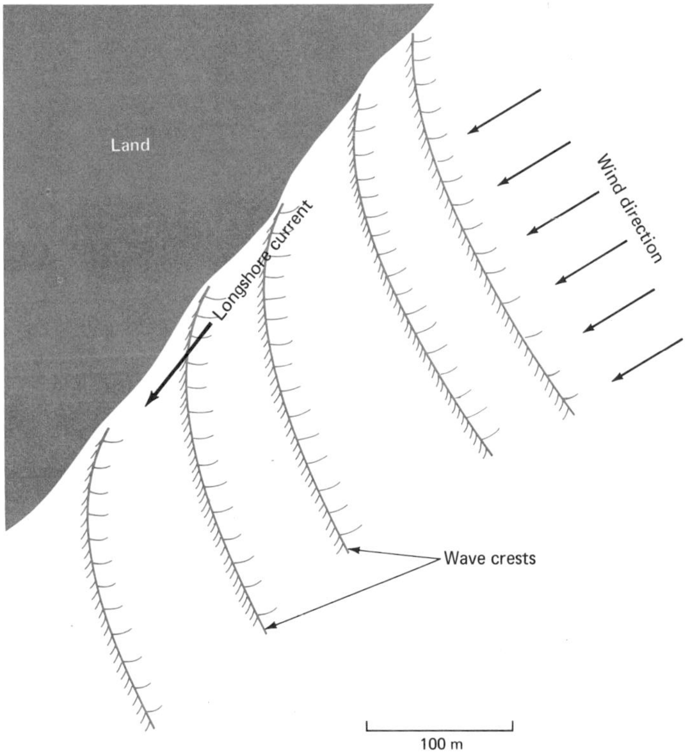

Material that is eroded from part of the coast or that is brought to the sea by streams may be winnowed and transported out to sea or deposited along the shore in shallow water. Waves meeting the shore at an angle rather than head-on may form *longshore currents*, which transport material parallel to shore along the coast (Figure 16–28).

Beaches may be built along the shore as sand is deposited and then may be thrown higher by storm waves. The highest part of a beach is the *berm*; it may serve as a source of sand for inland-migrating dunes (Figure 16–29). Beaches may form on shore or may be separated from it by a *lagoon*. The latter, a *barrier beach*, may be cut through by a *tidal inlet* (Figure 16–30). Deposition of sand across the mouth of a bay forms a *baymouth*

Figure 16–28. Development of longshore currents.

Figure 16–29. Beach area berm and inland migrating dunes, North Carolina.

Figure 16–30. Tidal inlet connecting the sea with a lagoon, North Carolina.

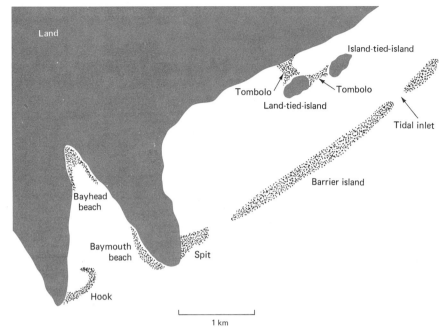

Figure 16–31. Depositional features along a shore.

bar or *baymouth beach*. Deposition at the head of a bay results in a *bayhead beach* or *bayhead bar*. A *spit* is a mass of sand extending from some point of land into the water. The term *hook* is used to describe a curved spit (Figure 16–31).

Sand is occasionally deposited between an island and the shore or between two islands; in each case it forms a connecting beach. The mass of sand connecting either is called a *tombolo*. The connected island there becomes an *island-tied island* or a *land-tied island* (Figure 16–31).

THE OCEANS: A GREAT EQUILIBRIUM SYSTEM

Oceans are remarkably uniform in some respects, the most notable of which is salinity. The salinity of oceans is very close to 35 ppt over most of their extent. Yet oceans are not a static system; they are constantly changing. Currents move about, waves travel across the surface, and tides move the great mass of oceans. In Chapter 4 we looked at the concept of equilibrium and the fact that the attempt to attain a state of equilibrium is a goal of all dynamic systems in nature. Oceans are no different. The World Ocean is a great system that is constantly attempting to maintain a state of equilibrium within itself and with its surroundings, principally the atmosphere.

Continuous interaction exists between oceans and atmosphere. All gases in the atmosphere are dissolved to a point of saturation in the near-surface portions of the seas. As oxygen is exhausted by organisms in a shallow sea area, more dissolves from the atmosphere. As carbon dioxide is taken from seawater to be used by organisms to make shells of calcium carbonate, more CO_2 is dissolved from the atmosphere into seawater. Conversely, if the atmosphere is depleted of CO_2 by plants in some area, that CO_2 in the oceans can return to the atmosphere, even to the point of dissolving $CaCO_3$ shells to maintain equilibrium.

The seas, or any large body of water, also serve as a great heat reservoir that moderates temperatures along many coasts. When air cools, heat is given up by oceans to warm the air. When air is warmer than the sea, some heat in the air is taken into the ocean. However, most of the heat in the oceans comes directly from absorbed sunlight.

THE FUTURE OF THE OCEANIC EQUILIBRIUM SYSTEM

We are changing the composition of our atmosphere. As we continue to add various pollutants to, and as winds carry them over, the oceans, this must affect the oceanic system. We have also been using the oceans for years as dumping grounds for garbage, industrial wastes, municipal wastes, and at one time even nuclear wastes. And we are considering mining the deep-sea floor for manganese nodules. Considering the vastness of the World Ocean, it seems inconceivable that we could significantly change it. Yet Thor Heyerdahl reports that the surface of the Sargasso Sea in the western Atlantic is already a mass of floating plastic bottles and other debris. He also says that oil slicks covering many square miles (kilometers) may be encountered in places in the Atlantic.

Yet in many places the sea is still clean, and we have not yet replaced order with pollution. Hopefully we can contain and diminish this pollution of the World Ocean in the future; we may have to depend on its delicate uppermost 300 ft (100 m) for future food supplies. Also, planktonic plants (diatoms) generate oxygen that helps to replenish that in the atmosphere. So, if we upset the equilibrium system of the oceans, we immediately cut off a necessary source of food and damage the atmosphere.

SUMMARY

The oceans cover almost three-fourths of the surface of the earth. We are learning more about them now and are realizing how important they are to our future.

We know today that sea floors are not flat, featureless plains but consist of a worldwide system of interconnected ridges separating smaller abyssal plains. Submarine volcanic seamounts disrupt the plains and have been formed along ridges, rifts, and continental margins. Very deep oceanic trenches occur near continents and island arcs. A continental margin may have a shallow continental shelf, a continental slope, and a continental rise above an abyssal plain. In places the continental slope continues downward into a trench.

Oceanic sediments may be continent derived, or they may be derived from within ocean basins. Continental margin sediments, as well as pelagic sediments, are mostly land derived. Both may in places contain a significant amount of inorganic, glacial, or biologically derived material. The oceanic crust on which pelagic sediments rest has a basaltic composition.

Sea level changes may be tectonic or eustatic. Tectonic changes involve movement of part or all of a continent or are changes in the crust that affect the shape or volume of an ocean basin. Eustatic changes are worldwide changes related to withdrawal or addition of water to the sea or to changes in the overall capacity of the ocean basin.

Ocean currents are of several types. Wind-generated surface currents are the most apparent. Density currents caused by differences in salinity or temperature are also important. Turbidity currents caused by a sudden release of an unstable mass of sediment down a slope may be locally significant.

Waves involve circular motion of particles they affect. This motion decreases downward and is nearly negligible at depths of greater than one-half the wavelength. Most waves are generated by winds. Seismic sea waves are produced by earthquakes. The gravitational pull of the sun and the moon produces a bulge in the oceans.

Wave erosion and deposition, and currents operating near and along shores, have shaped the coasts of continents. Many erosional and depositional features show these effects.

The oceans are an equilibrium system. As one part changes, another may also change to compensate and restore the effects of the first change. The atmosphere interacts with surfaces of oceans as part of this system. The oceans are thus in a delicate state of balance. Our future on earth depends on how wisely we can maintain this balance.

Questions for thought and review

1. Why should we learn as much as we can about oceans?
2. How does a continental rise differ from a continental slope?
3. Why are many seamounts flat-topped?
4. What is the most likely origin of submarine canyons?
5. Why is it difficult to predict the kinds and distribution of sediments of continental shelves?
6. Why do we find little or no calcium carbonate in sediments at depths greater than 13,500 ft (4,500 m) in oceans?
7. How may sea level change?
8. Why do currents exist in oceans?
9. How can turbidity currents break trans-Atlantic telephone cables?
10. Explain the circular nature of the motion of ocean waves.
11. How are seismic sea waves produced? Why are they so damaging to coastal areas and not to ships at sea?
12. How does a neap tide develop?
13. What evidence do we have that oceans are a great equilibrium system?

Selected readings

Bascom, Willard, 1959, Ocean Waves, *Scientific American*, vol. 201, no. 2, pp. 74–84.

———, 1960, Beaches, *Scientific American*, vol. 203, no. 2, pp. 80–94.

Emery, K. O., 1969, The Continental Shelves, *Scientific American*, vol. 221, no. 3, pp. 106–122.

Ewing, Maurice, and Engel, Leonard, 1962, Seismic Shooting at Sea, *Scientific American*, vol. 206, no. 5, pp. 116–126.

Fisher, R. L., and Revelle, Roger, 1955, Trenches of the Pacific, *Scientific American*, vol. 193, no. 1, pp. 36–41.

Goldreich, Peter, 1972, Tides and the Earth, *Scientific American*, vol. 226, no. 4, pp. 42–52.

Gregg, Michael, 1973, Microstructure of the Oceans, *Scientific American*, vol. 228, no. 2, pp. 64–77.

Heezen, B. C., 1956, The Origin of Submarine Canyons, *Scientific American*, vol. 195, no. 2, pp. 36–41.

———, 1960, The Rift in the Ocean Floor, *Scientific American*, vol. 203, no. 4, pp. 98–110.

Heezen, B. C., and Hollister, C. D., 1971, *The Face of the Deep*, Oxford, New York.

MacIntyre, Ferren, 1970, Why the Sea Is Salt, *Scientific American*, vol. 223, no. 5, pp. 104–115.

Marx, Wesley, 1967, *The Frail Oceans*, Ballantine, New York.

Stetson, H. C., 1955, The Continental Shelf, *Scientific American*, vol. 192, no. 3, pp. 82–86.

Turekian, K. K., 1968, *Oceans*, Prentice-Hall, Englewood Cliffs, N.J.

17 Wind and deserts

STUDENT OBJECTIVES

On completing this chapter, the student should:

1. formulate a definition of a desert
2. be able to identify characteristic topographic features in each of the six western deserts of the United States
3. recognize desert topography, including such features as pediments, bajadas, playas, pedestal rocks, and arroyos
4. understand the physical processes that work to produce desert features
5. distinguish between various types of sand dunes

How does the desert differ from any other land? Only in the matter of water—the lack of it. If southern France should receive no more than two inches of rain a year for twenty years it would, at the end of that time, look very like the Sahara. . . .

John C. Van Dyke

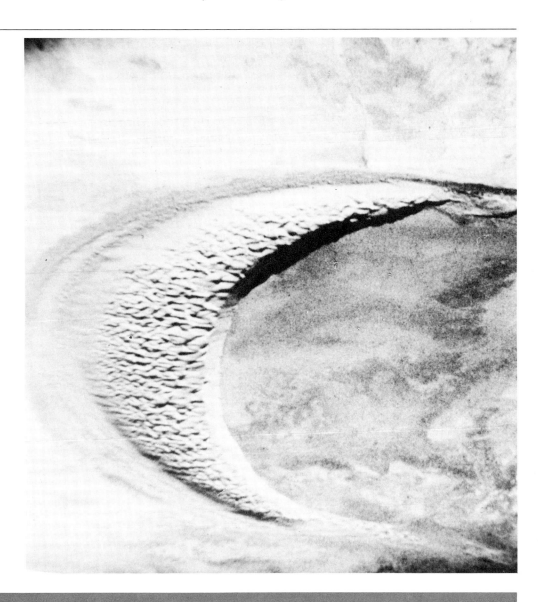

WHAT ARE DESERTS?

Most people see deserts as places with no vegetation, places of endless sand dunes reflecting extremely hot sun. Although it is true that relatively little vegetative cover is found in a desert, its chief characteristic is limited rainfall. Nor are deserts dominated by sand dunes. Because most of the world's population live in regions of normal rainfall (approximately 30 in. per year), many people believe that arid and semiarid regions are sparsely distributed around the earth. This is not true. Deserts and arid regions cover more surface area—more than 30%—of the earth than do humid regions (Figure 17–1).

One-fifth of the entire land surface of the United States can be classified as desert. If we add to the desert lands the semiarid regions of the Midwest we find that about one-half the United States has a limited amount of rainfall.

Deserts usually have less than 10 in. of rainfall annually; 10 in. of annual rainfall distributed evenly over an entire year can support a lush grassland. In deserts, however, rains fall in heavy showers and during only a few months of the year. The entire amount of annual rainfall may come down in three or four rainstorms. Frequently one-third to one-half the annual rainfall drops in a single cloudburst (Figure 17–2). Rainwater usually runs off through arroyos (channels of intermittent streams), or sinks into the ground before it can be used by plants. In hot deserts rain often evaporates before it reaches the ground.

In spite of conditions hostile to any form of desert life, a typical desert landscape re-

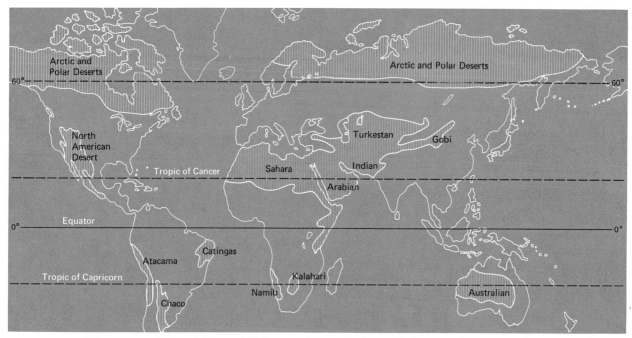

Figure 17–1. Most of the world's largest deserts lie in two belts near the Tropic of Cancer and the Tropic of Capricorn. In the Northern Hemisphere, the region north of the 60th parallel has an annual precipitation under 10 in. (25.4 mm), except for the West Coast belts. This includes polar and arctic deserts.

Figure 17–2. Although deserts receive less than 10 in. (25.4 mm) of precipitation annually, one-third to one-half of it may fall in a single cloudburst. What was formerly a dry creek bed may suddenly become a torrential river. (Courtesy of U.S. Soil Conservation Service.)

veals many kinds of plants and animals. All living things in the desert have developed modifications for survival and for the propagation of their species in desert environments.

TYPES OF DESERTS

In broad terms deserts may be classified, according to their geographic location, as: (1) tropical, (2) middle latitude, and (3) high latitude. *Tropical* deserts are hot or at least warm throughout the year. *Middle-latitude* deserts undergo seasonal changes and usually show extreme seasonal changes in temperature. *High-latitude* deserts are cool or cold most of the year. Even in tropical deserts daytime and nighttime temperatures fluctuate widely. Because deserts have little or no cloud cover, heat loss after sundown is fast and dramatic.

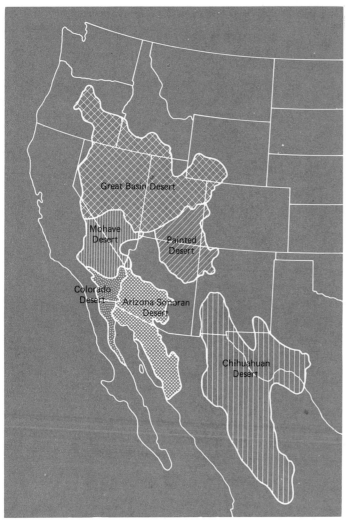

Figure 17–3. The great mountain chains of the West Coast prevent moist ocean winds from passing across them without first dropping much of their moisture. As a result, most North American deserts lie east of these high ranges.

NORTH AMERICAN DESERTS

Our discussion of deserts will concern itself primarily with those of North America, focusing on the western United States and northern Mexico (Figure 17–3). Six large desert areas are found in these regions. They stretch from northern Oregon to central Mexico. Included in this group are: (1) the Great Basin Desert, (2) the Mohave Desert, (3) the Arizona–Sonoran Desert, (4) the Colorado River Desert, (5) the Painted Desert Group (Piñon), and (6) the Chihuahuan Desert.

The Great Basin Desert

The Great Basin Desert is the bleakest of all American deserts. It encloses much of the land between the Sierra Nevada and the Rocky Mountains. Its landscape is not only the most monotonous, but it is almost completely void of trees except along a few of its watercourses. These may sport a few willows or cottonwoods. It is an arid upland of plateaus and basins. The high Sierras to the west (Figure 17–4) block moisture from the Pacific. Cacti for all practical purposes are absent except for a few low clumps growing in areas protected against cold winters and dry summers.

Lake Bonneville. When the Pleistocene glaciers retreated, much of the Great Basin consisted of streams and lakes receiving large amounts of water from melting ice. Remnants of this glacial lake system are Great Salt and Sevier lakes in Utah. These were formerly enclosed by ancient Lake Bonneville. As glacial waters receded, Bonneville shrank, leaving behind a string of lakes in the deeper basins. Evaporation of water from Lake Bonneville left a large, almost level desert, with a floor so solidly packed that it is often used in world-record speed-car trials (Figure 17–5). Lake Bonneville also left behind the most sterile of all American deserts, because evaporation of water left in its place large quantities of mineral salts.

A record of the retreat of Lake Bonneville can be traced by a series of seven terraces. People formerly believed that these were remnants of early Mormon agriculture. They are, instead, a series of previous shorelines cut by Lake Bonneville as it shrank in size

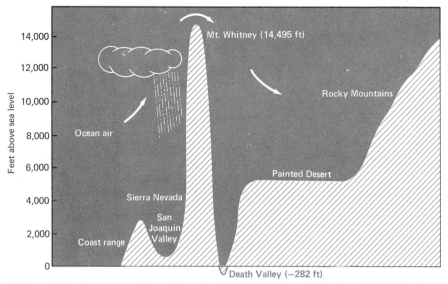

Figure 17–4. A cross section of western United States shows how West Coast mountain ranges produce the "rain shadow" effect. Ocean air rising up mountain slopes cools; its moisture condenses and falls as rain or snow on the ocean side of the slope. Very little moisture remains in the air after it crosses the mountain peaks. Cold ocean currents along lower California and northwestern Mexico have a drying effect, too, producing the desert.

(Figure 17–6). The ancient lake has left its imprint on the landscape in the forms of sandbars, beaches, wave-cut cliffs, and

Figure 17–5. Bonneville salt flats, Utah. (United Press International Photo.)

deltas similar to those created by modern lakes.

Lake Lahontan. Lake Lahontan in Nevada and California is another prehistoric lake that has left behind some interesting remnants. At its maximum the lake was the size of modern Lake Erie. The remains of the lake are found in Honey Lake in California and in a group of lakes—Pyramid Lake, Walker Lake, and Humboldt and Carson sinks—in Nevada. A number of the scattered Nevada pools and lakes belong to the Lahontan Lake system. Many of these lakes are only intermittently filled with water. They may be dry for as long as nine months a year. Some disappear entirely during years of low rainfall.

Playa lakes. Temporary lakes that form after heavy rains in arid regions are known as *playas*. (*Playa* is Spanish for beach and temporary lake.) A typical Great Basin playa

Figure 17–6. Beach terrace of ancient Lake Bonneville. (Courtesy of Buffalo Society of Natural Sciences.)

Figure 17–7. Playa Lake. (Courtesy of Buffalo Society of Natural Sciences.)

contains water only during the wet winter months. It sometimes covers a 450-mi² (1250-km²) area; yet its average depth is only a few inches (centimeters) (Figure 17–7).

In basins enclosed by mountains, drainage from all margins is toward the center. A nearly level basin is built up by a gradual influx of sediments from the fringing moun-

Figure 17–8. The Great Salt Lake, Utah, is a remnant of a glacial lake system. As glacial waters receded, the low spots became a string of lakes. Great Salt Lake is one of the larger of these lakes. During periods of drought, its salinity is greater than that of the Dead Sea. (Courtesy of Utah Travel Council.)

tains. This shallow basin spreads out water evenly, preventing pockets from accumulating. Because of this, playa lakes retain water for only a short time after a rainfall and lose it quickly through evaporation or percolation. As the lake water evaporates, sediments and dissolved minerals are deposited on the lake bed to form an *alkali flat*. If the concentration of salt is high, the flats are known as salinas.

Great Basin lakes. Lakes in the Great Basin Desert depend on rainfall from the mountains bordering the basin for their source of water. A basin lake has no outflow. With a high rate of evaporation and no outflow, minerals dissolved in the lake become more and more concentrated. If evaporation leaves behind large amounts of sodium chloride, the lake becomes a *salt lake*. Some desert lakes, known as *bitter lakes*, leave behind sulphates. Great Salt Lake is a combination salt lake and bitter lake (Figure 17–8). During dry spells Great Salt Lake is seven times as salty as the oceans. In prolonged periods of drought its salinity exceeds that of the Dead Sea. *Borax lakes* precipitate various borate minerals. Owens Lake in California contains rich deposits of bicarbonate of soda (Figure 17–9).

Figure 17–9. Owens Lake, California, contains rich deposits of bicarbonate of soda. (Courtesy of Ward's Natural Science Establishment, Inc.)

It would appear that most desert lakes are a cruel joke of nature. In a region where water means life, the lakes are more sterile than the desert sands surrounding them (Figure 17–10).

Figure 17–10. Does this sign designate a place, or is it a warning? The high mineral content of desert lakes makes the water unfit for drinking. (U.S. Geological Survey.)

Pediments, bajadas, and playas. Millions of years ago mountains in the Great Basin area were raised by upheavals within the earth and then reduced by wind and water, only to be raised again to fight these same elements of erosion over and over again. We find in the Great Basin Desert, as well as in other American deserts, the battered tops of mountains, half buried in soil that has eroded from their slopes. Centuries of erosion have filled the valleys between the worn-down mountains tops. Long, gentle slopes now reach out from all sides. These slopes are called *pediments* (after the Greek word for the low, slanting roofline of temple architecture—Figure 17–11.) Pediments are one of the more common landscape features in American deserts (Figure 17–12).

The pediment is believed by most geologists to be an erosional feature. As mountain slopes retreat from the basin, pediments are formed as gently sloping bedrock surfaces, which may be covered by a thin veneer of debris. Desert landform development is to a large degree related to pediment formation and the extension of the pediment, the *bajada*.

Bajadas are low deposits of alluvium. *Alluvium* is a general term applied to sediments dropped from streams as they lose

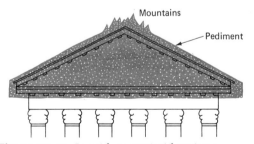

Figure 17–11. In arid or semiarid regions, weathering and erosion attack the steep faces of upraised mountain ranges, wearing backward the faces and thinly veneering the slopes with gravel. The erosional form so produced is called a pediment, since a cross section of it would resemble the pediment of a Greek temple.

307

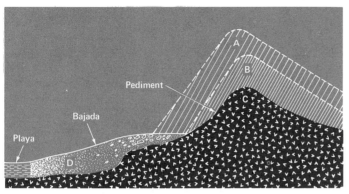

Figure 17–12. Cross sections of pediment, bajada, and playa landscape. Solid rock removed by erosion are shown in *A* and *B*. The remnants of the mountain are shown in *C*. Loose material grading from gravelly on the right, to sandy, to silty and clayey on the left, are shown in *D*.

their capacity to transport. The bajada is formed from a series of coalescing alluvial fans spreading their detritus radially outward from high ground from whence the streams originated. Materials found in the bajada consist of gravel and even large boulders that have been carried to lower levels by mudflows.

Pediment and bajada slopes are gentle, usually less than a 7° incline. When they abut against a mountain, there is a sharp break in slope, which increases rapidly from 15° upward to a maximum of 90°.

Sediments that reach the playa during rainstorms are silt or clay size. After the storm, runoff water from neighboring mountains evaporates quickly. As new deposits dry, they crack and curl, resembling an autumn lawn badly in need of raking (Figure 17–13). The relationship of the pediment, the bajada, and the playa are illustrated in Figure 17–12. The illustration shows that, as mountains retreat by erosion, a pediment is formed as a gently sloping bedrock surface. Coalescing alluvial fans (see Figure 18-39) build up a bajada below the pediment. Fine materials (silt and clay) are carried to the playa in the middle of the basin. Without an

outlet to carry water from the playa lake, the water evaporates, leaving behind a salt veneer. Ancient salt flats so formed were mined first by Indians and then by whites. The largest exposed salt deposit in the United States is Searles Lake, California; its upper layer alone is 70 ft (23 m) thick.

The Mohave Desert

North American deserts have much in common; yet each has its own community of plant life. In the Great Basin Desert the characteristic plant is sagebrush. When a traveler approaches the southwestern edge of the Great Basin Desert he encounters his first Joshua tree. This is a yucca that grows as high as 40 ft. From a distance its branches look like the arms of a giant. John C. Fremont in 1844 was probably the first European to see a Joshua tree. He described it as "the most repulsive tree in the vegetable kingdom." To the Mormon colony of California that set out across the Mohave to join the main group in Utah, the tree had a different meaning. The angular branches of the tree appeared to be arms of a man urging them to cross this unfriendly desert and go to the Promised Land. They named this strange tree Joshua after the leader of the children of Israel who led them into the land of Canaan.

The Mohave region had in the distant past numerous rivers and lakes that supported a wide range of plant and animal life. As recently as 10,000 years ago, a number of now-extinct animals roamed this area. They included mammoths, ground sloths, camels, and three-toed horses. As climate became arid, this life disappeared. Reminders that we find today of the earlier abundance of water in this region are ancient lake shorelines and wave-cut terraces on mountainsides. Numerous basins found in this desert, such as Panamint, Silver Soda, and Manix basins, were at one time filled with pleasant lakes.

The primary attraction of the Mohave is Death Valley (Figure 17–14). It consists of

Figure 17-13. Rapid evaporation or water in playa lakes causes the new deposits to curl and crack. These leaflike forms of dried silt eventually disintegrate, leaving the playa paved with a layer of salt and silt. (U.S. Geological Survey.)

two shallow basins surrounded by ranges of low hills. The southern basin contains extensive salt beds. Here the lowest spot in North America, Badwater, lies 282 ft (86 m) below sea level (Figure 17-15). Oddly enough the highest point on mainland United States, Mt. Whitney, with an elevation of 14,495 ft (4,418 m) above sea level is only 80 mi (129 km) to the west. The Panamint Range with an elevation of 6,000-11,000 ft (1,800-3,353 m) forms the western wall of Death Valley. The Amargosa Range, with elevations from 4,000-8,000 ft (1200-2400 m), forms the eastern boundary. The northern end of the valley is capped by the Last Chance Range; the southern end is closed by the Owlshead and Avawatz mountains.

Death Valley is only about 1 million years old in its present appearance, a short space in geologic time. Death Valley is not a true valley. Valleys are formed by running water. During the Pliocene and Pleistocene epochs a structural upheaval taking place in the southwest accentuated the region's topography. In some places block mountains were uplifted; in some places the basins moved downward. Death Valley is one of the basins that dropped. It has sediments of marine origin in depressions 200 ft (60 m) below sea level. If all the rocks, gravel, sand, and silt that have been carried into Death Valley

Figure 17-14. Aerial view of Death Valley. (U.S. Geological Survey.)

ture of 134°F (57°C) stood as a world record until a temperature of 136°F (58°C) was recorded in the Libyan Desert.

At one time Death Valley was occupied by large Lake Manley. The shores of the ancient lake cut terraces into mountain slopes. These are clearly seen at Shoreline Butte in Death Valley National Monument. Lake Manley was once 100 mi (161 km) long and 600 ft (180 m) deep. Today it is a temporary pond. The evaporation of Lake Manley left behind salt beds mixed with gravel 1000 ft (300 m) thick. In Devil's Golf Course, salt beds cracked into great blocks. These blocks have been eroded by wind and water into a landscape similar to that found in Bryce Canyon.

In spite of the apparent bleakness of Death Valley, plant and animal life is abundant. There are 608 different kinds of plants, and 230 species of birds found in Death Valley National Monument. Of commercial interest, the presence of borax in the valley was discovered in 1880, mines were developed, and, until recently, borax was taken from the valley.

The Arizona–Sonoran Desert

South of the Mohave Desert is the Arizona–Sonoran Desert, the best-known desert in North America. It spreads from southern California into western Arizona but lies mainly in the Mexican state of Sonora. It is the most magnificent desert in North America. The indicator plant is the saguaro cactus.

The western extension of the Arizona–Sonoran into California encounters the lower reaches of the Colorado River. This region of the Sonora is known as the *Colorado River Desert*.

Just as the Death Valley basin formed the heart of the Mohave Desert, the Salton Sink is the heart of the Colorado River Desert. Spanish explorers called the Salton Sink "the valley of torture." It is a depression 200 mi (322 km) long and approximately 50 mi

Figure 17–15. In the southern basin of Death Valley lies Badwater, the lowest in North America. It is about 280 ft (85 m) below sea level. Just 80 mi (110 km) west of Badwater is Mount Whitney, the highest point in mainland United States. Mount Whitney is 14,495 ft (4,418 m) above sea level. (Courtesy of Ward's Natural Science Establishment, Inc.)

since it was formed were removed, the floor of the valley would be 10,000 ft (3,000 m) below sea level.

The average annual rainfall in Death Valley is about 2 in., and annual days of sunshine number 351. Its maximum air tempera-

(80 km) wide. One-fourth of its 8,000-mi² (22,240-km²) area is below sea level.

Unlike Death Valley, Salton Sink does have a permanent lake, about 35 mi (56 km) long and approximately 20 mi (32 km) wide. Its surface is more than 200 ft (60 m) below sea level. Like Great Salt Lake, this California lake is highly saline and is known as the Salton Sea.

The history of the Salton Sink is curious. At one time the entire desert laid submerged below the Gulf of California. The gulf at that time extended northward to the San Bernardino Mountains. This arm of the gulf was cut off by sediments deposited at the mouth of the Colorado River. The delta so deposited stretched from mainland Mexico to the peninsula of Lower California. Water trapped behind the sediment dam became known as ancient Lake Cahuilla. The Colorado River replenished the lake periodically by pouring its floodwaters into it through what is now a dry channel called New River.

Lake Cahuilla occupied Salton Sink until approximately 1770, when it began losing water faster than it could be replenished. By 1900 it was reduced to a small saline lake. The present Salton Sea was formed in 1906 when the Colorado River broke from its channel during a high-water period to flood the sink. During some intervals of the flooding stage, water poured into the sink at a rate of 100,000 ft³ (2,913 m³) per second. It stabilized at approximately its present size in 1920.

The entire Colorado River Desert is an unstable area. Should the land subside about 50 ft (16 m) in the area bordering the Colorado River or the Gulf of California, it could once again be completely flooded.

The portion of the Arizona–Sonoran Desert that lies in southwestern Arizona is noted not only for its giant cacti, but it can boast most of the landforms commonly associated with desert landscapes. Within the boundaries of this desert are located Organ Pipe Cactus National Monument, Saguaro National Monument, and Antelope Peak. Each of these three places has an interesting story to tell.

The saguaro is the largest of all North American cacti. Some of these grow to heights of 50 ft (15 m) and weigh more than 10 tons (9060 kg). The saguaro was once described as "a tree designed by someone who had never seen a tree." It displays numerous adaptations for coping with limited desert rainfall. The gigantic fluted column is really an adjustable water reservoir. As water is taken into the cactus, the pleats expand like those in an accordian. Often after a heavy rain the plant absorbs a ton of water.

Organ Pipe Cactus National Monument is in southwestern Arizona on the Mexican border. Here on the floor of the desert or on the slopes of fine, rugged mountain ranges, we can see some of the most spectacular of all desert scenery. The organ pipe cactus, for which the monument is named, grows best on rocky slopes. The cactus has many ribbed branches that grow upward from ground level. Some may reach a height of 25 ft (8 m).

The Painted Desert

In east-central Arizona lies the Painted Desert, land of brilliant colors and amazing landscape shapes. It is one of the most popular vacation areas in the Southwest.

This is the land of mesas, flat-topped rocks rising above the desert floor (Figure 17–16). Cliffs and canyon walls of sandstone have bright bands of reds, oranges, yellows, and browns. Here we find the Navajo and Hopi Indian reservations. The Hopi build their communal dwellings like forts atop mesas.

In this high country—3,500 ft (1,070 m) above sea level—winters are cold and often snowy. Summers are hot, and thunderstorms are intense. Frequently at night are beautiful displays of sheet lightning on the horizon.

At the southeastern edge of the desert is Petrified Forest National Park. Here one can

Figure 17–16. A typical mesa. (U.S. Geological Survey.)

see fossilized tree trunks 150 million years old. To the north is Canyon de Chelly National Monument with its deeply cut canyons of red sandstone. Other attractions in the Painted Desert are Monument Valley, Rainbow Bridge, Meteor Crater, and Sunset Crater. Each of these is a study in the geologic work of nature.

The Chihuahuan Desert

The Chihuahuan Desert takes its name from the Mexican state in which it mainly lies. It includes a large portion of the Mexico plateau south of the Rio Grande and portions of southeastern Arizona, southwestern New Mexico, and western Texas.

Among the attractions of this desert in New Mexico are widespread lava flows, lost rivers, and extensive dunes of white sand, composed of almost pure gypsum. At the southwest end of the dunes is ephemeral Lake Lucero. Gypsum-bearing waters from nearby mountains flow into the lake. The water there evaporates during the summer, leaving behind large deposits of gypsum. These deposits are picked up by prevailing southwesterly winds and are deposited in the dune area. In White Sands National Monument gypsum assumes the shape of rounded hills; some rise to a height of 50 ft (15 m).

The landscape in this region is constantly changing as winds sweep over the dunes. Dunes and ripple marks are perpetually moving, forming patterns that are rarely duplicated. What is happening here is what happens in the Great Sahara Desert. Desert winds become a principal agent of landscape formation.

ALLUVIAL FANS

Although alluvial fans can be found in most North American deserts, some classic examples of alluvial fan deposition are found along the margins of Death Valley. The barren, steep mountain slopes in this region shed large quantities of debris when they are eroded by runoff from torrential rains. Floodwaters in mountain canyons have a large proportion of coarse bed load. This is carried swiftly downslope because of the steep gradient. When the water emerges upon a gently sloping basin below the mountain range, the stream can no longer transport its debris load and aggradation occurs. Free to shift from side to side, these temporary streams spread their excess load in the form of an *alluvial fan* (Figure 17–17).

An alluvial fan takes the form of a low cone with its apex at the canyon mouth. At

Figure 17–17. When a stream flows down a slope onto level terrain, its velocity is reduced. The water cannot carry its load of sediment, and a deltalike deposit forms. Subsequent floods produce similar deposits in the general area of the earlier deposits. Each new flood produces a new stream channel in the alluvium deposited at the base of the mountain. This process of shifting stream deposits continues, and an alluvial fan builds up into a conelike feature with its apex near the mouth of the stream.

its outer edge the fan's slope grades gently into the flat plain below. The size of sediment particles is larger near the mouth of the canyon, where boulder material can be found. The sediments grade down by progressively finer particles toward the outer edge of the fan.

Large fans in mountainous deserts are not uncommon. A fan may be several miles (kilometers) in radius from its apex to the outer edge. But most alluvial fans are small. Some measure only a few feet (meter) in radius. In larger desert fans mudflows are layered between alluvial deposits.

Alluvial fans are important in desert lands, because they act as groundwater reservoirs. Water that enters the fan at the apex moves downward and outward through the gravel and sand. Wells sunk into alluvium at the lower slopes will often have artesian flow, because water becomes trapped in permeable gravels deposited between impermeable mudflow layers.

The long-term supply of water from alluvial fans is limited by the slow rate of recharge. The frequency of torrential rains that feed the alluvial fans will determine the amount of groundwater availability. Extensive pumping for irrigation can quickly exhaust the groundwater supply.

PEDIPLANATION

In the Antelope Peak region of southwestern Arizona (Figure 17–18) are found a number of geologic features associated with *pediplanation*, the terminal stages of degradation of desert landforms. In addition to pediments, bajadas, playas, and alluvial fans, all of which have been previously described, pediment passes, inselbergs, and volcanic necks are indicators of an ancient landscape.

Pediment slopes on opposite sides of a mountain range may extend themselves backward through the mountain and join. The term applied to this feature is *pediment pass*. The road crossing Antelope Peak goes through a pediment pass.

The land surface east of Antelope Peak is flat except for a few sharply protruding structures. These *inselbergs* are steep-sided residual hills and mountains and rise abruptly from the plain (Figure 17–19). They are generally bare and rocky. An inselberg can stand alone as a sentry or can be a part of a mountain group.

A *volcanic neck* is solidified material filling a vent of a dead volcano. If a volcanic neck resists degradation, while the mass of the volcanic cone erodes away, the column or crag of igneous rock left standing is called a volcanic neck (Figure 17–20).

Another striking desert landscape feature

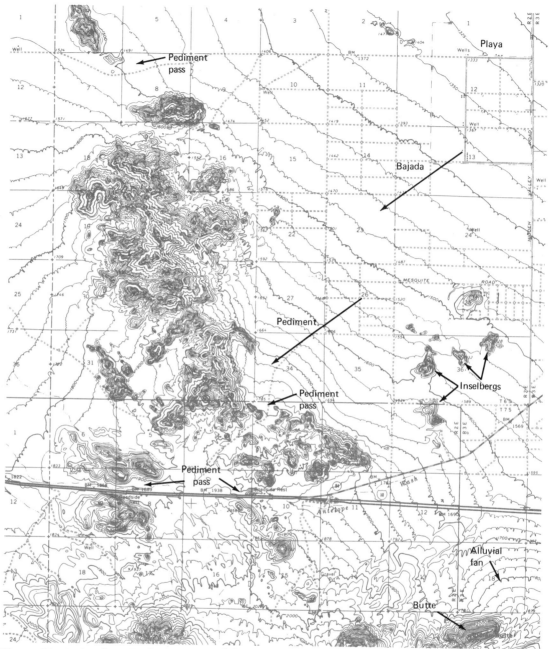

Figure 17–18. Antelope Peak, Arizona, exhibits geologic features associated with pediplanation.

is the *outlier*, part of any stratified sedimentary rock group that lies detached from the main body of the rock formation. Intervening rocks have been removed by denudation. A spectacular example of an outlier is Spider Rock in Canyon de Chelly National Monument (Figure 17–21).

Presence of these features is evidence that a region is in the old-age period of an erosional cycle.

Figure 17–19. Inselbergs are steep-sided residual hills that appear prominently above the general level of the desert floor. Pediment passes are narrow, flat rock-floored regions between inselbergs. (U.S. Geological Survey.)

FLASH FLOODING IN DESERTS

Tourism in recent years has lured into desert regions thousands of visitors annually. Most are not aware of sudden desert downpours and of dangers of flash flooding.

On September 16, 1974, a tragedy was reported in the Lake Mohave region of Nevada. A flash flood wiped out a trailer village and marina in Eldorado Canyon on the shores of the lake, killing two people and listing seven others as missing. The seven were presumed buried under mud or swept into the lake. The torrential downpour in the high ground above the village loosened tons of mud that were carried down the canyon by a flash flood. A dam of mud and silt 45 ft (14 m) deep was deposited on the shore of the lake at the entrance of the canyon (Figure 17–22).

This process of landscape alteration has been functioning for millions of years in desert regions. Such past activity has produced all the desert landscape features that we see today. The action itself is not a tragedy of nature. It becomes one when people encounter natural forces without an understanding of the great capacity of our planet to produce change.

WIND EROSION AND TRANSPORT

Although most desert landform features are the result of the action of running water

Figure 17–20. Only the neck remains of what was once a 1400-ft (427-m) volcano in Shiprock, New Mexico. (Courtesy of New Mexico Dept. of Development.)

Figure 17–21. Spider Rock, Canyon de Chelly, Arizona, is an outlier, a part of a stratified rock that is detached from the main rock mass. (Courtesy of American Airlines.)

Figure 17–22. Tragedy strikes quickly in the desert. A sudden rainstorm in neighboring mountains may produce flash floods in the basins below. As vacationers search out remote scenic spots to visit, a lack of knowledge about life in the desert may be fatal.

and of mass movement, effects of wind erosion are the most obvious. How important is wind erosion in shaping desert topography? It is very effective in shifting about debris of weathered rocks. Wind keeps land surfaces swept clear of this debris in some places, sifts it, then piles it up elsewhere as sand dunes, or distributes it evenly as sheets.

Some landforms are produced primarily through wind erosion. These include: (1) broad, shallow depressions in deserts where soil is blown out by repeated movement of wind currents flowing in one direction; (2) broad, shallow caves in sides of sandstone cliffs where impact and abrasion of sand sculpts the sides; and (3) mushroom, table, and pedestal rocks.

Wind erosional action may be classified as: (1) deflation (blowing away of materials), (2) corrosion and abrasion, and (3) impact. *Deflation* includes removal of dust and smaller particles and transporting them by saltation to some other place. Sand, silt, and even dust in high-velocity winds become effective tools for *abrasion*. When sand blows over rocks for long periods of time, the rocks become worn and their surfaces become polished from repeated abrasion. Such rocks are called *ventifacts* (Figure 17–23). *Impact* is more effective than abrasion and takes place when sand grains are blown against a rock

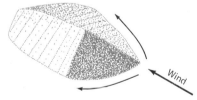

Figure 17–23. A common product of abrasion are pebbles, cobbles, and boulders (A) eroded in a peculiar way called ventifacts (Latin for wind-made). They usually have a high gloss and show a variety of facets or ridges. How does wind cut more than one facet on a rock surface? If the original rock is angular it may split the wind current so that two faces are cut simultaneously. Other explanations are: (1) The direction of the wind may have changed; and (2) the rock may have shifted its position because some external force acted on it.

surface. Momentum will dislodge other granular matter from the rock. Impact is the technique of sandblasting in today's industry. It is not the abrasive quality of sand that so efficiently cleans dirty exteriors of old stone buildings. Through impact dirty stone particles are removed, exposing clean stone underneath. Evidence of impact is found in pitted or frosted sand grains in deserts.

Wind does an excellent job of sorting sediments. It separates particles according to size, shape, and weight. Velocity of wind determines the maximum size of material it can transport. Wind transports matter in three ways: (1) in suspension, (2) by rolling, and (3) by saltation (Figure 17–24).

Smaller-sized particles are lifted, *suspended* in the wind, then carried by the wind to a new location. Heavier particles are pushed along the surface by the wind. Their shapes will determine how easily they will *roll* in the face of a strong wind. Some heavier particles are transported by *saltation*, a process resembling jumping (see Chapter 13 for a discussion of saltation in streams). One particle strikes another, forcing it to move. It, in turn, strikes a third particle, and so it goes on as long as the wind has sufficient velocity to activate the movement of small rock particles.

The sorting process begins initially as the wind begins transporting material. The sorting is further refined when wind velocity begins to drop. A *critical velocity* is required to sustain the movement of particles of various sizes (Table 17–1). When wind velocity falls below the minimum required to move a particle, that particle will settle. The result of this action is a well-sorted sediment.

WIND DEPOSITION

Wind deposition makes itself known in dunes and as sheets. Sand *dunes* are relatively rare surface features even in deserts, although they are not confined solely to them. Some of the larger sand dune areas in North America are found along the eastern coast and near the Great Lakes. Dunes are formed wherever there is a strong, constant

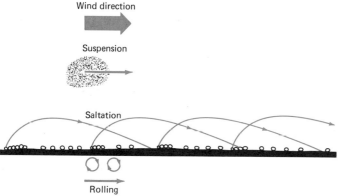

Figure 17–24. Wind is less effective than water as an erosion agent. Nevertheless, its action plays an important role in the transportation of earth materials in arid and semiarid regions. Wind picks up very fine particles and carries them aloft for hundreds or even thousands of feet into the air during a dust storm. These particles may be in suspension at these altitudes for long periods of time. Unlike water, wind normally cannot lift sand-sized grains from the ground. Sand particles are thrown into the air when a rolling sand grain collides with a stationary grain and propels that grain or itself into the air. This is called saltation. Some sand grains never rise into the air. They move from place to place by rolling.

Table 17–1. Wind velocity required to sustain a particle in air

Diameter (mm)	Critical velocity (m/sec)
0.01	0.01
0.05	0.25
0.10	0.50
0.20	1.00
0.25	1.50
0.50	3.50
0.75	5.70
1.00	8.00

or intermittent wind; a source of sand; and obstacles that can initiate the formation of a dune. Sand sources are varied. Along coasts and lakes the most common sand is quartz sand. In volcanic areas it can be olivine from volcanic rock and ash. In polar regions ice sand forms dunes.

Sheets are dust deposits spread over wide areas. Thickness varies within sheets. They are thicker near the source regions, thinning out farther away from the source. A good example of sheet deposits is the layer of *loess* that covers much of the central United States. Loess is a composite of small, angular particles of many rock types. The angularity of the particles stabilizes loess cliffs. For this reason, road cuts in loess deposits often have vertical sides. Examples can be seen in road cuts throughout the Mississippi River Valley.

Loess is semiconsolidated and contains small snail shells of species that live on land. It contains small tubes of calcite that filled holes left after the decay of tall grasses. Windblown dust and silt must have sifted down over fields of grass so slowly that the blades of the tall grasses were not forced down.

Composition of loess found in the United States is that which would have been produced by glacial grinding of rocks now exposed in the northern United States and in Canada. Rock material was carried by glacial meltwater to the outwash plains. After drying, winds picked it up and carried it southward to the Mississippi River Valley.

DUNE FORMATION

Sand dunes may accumulate whenever an obstruction blocks the path of moving sand. Highway departments in the northern United States annually erect snow fences near roads to form snow dunes away from roads and thus reduce the need for plowing. The obstruction temporarily decreases the velocity of wind, causing it to deposit the particles it carries in front or directly behind the windbreak. Sand moving by saltation hits the barrier or falls behind it as wind velocity falls off.

On the windward side of the obstruction a low slope forms. Sand moves up this slope by saltation. Just beyond the crest, wind velocity is reduced. Here the sand carried by the wind is dropped. A backslope to the streamlined pile of sand begins to form by the slumping and sliding of sand from the top of the dune. The angle of repose on the leeward side (slip face) is about 35°. The lee side of the dune is barely stable. Any addition of sand at the top will cause small slides on the lee slope. For this reason sand dunes are constantly migrating. A typical sand dune will move as much as 100 ft (30 m) in a year.

The shape of a sand dune is determined by the supply of sand in the region, by wind velocity, by the constancy of wind direction, and by the distribution of vegetative cover. Sand dunes fall into one of six general classifications: (1) fore (shoreline) dunes, (2) barchan dunes, (3) seif dunes, (4) longitudinal dunes, (5) transverse dunes, and (6) parabolic dunes (Figure 17–25).

Fore (shoreline) dunes

Along the shores of lakes and oceans, ridges of windblown sand called *fore (shoreline) dunes* are frequently built up. They can be found along the south and west shores of Lake Michigan, along the Atlantic from Massachusetts southward, along the coasts of Oregon and Washington, and down the southern coast of California.

Fore dunes are formed by strong onshore winds moving sand particles off the beach. Most coastlines have sufficient vegetation to check the inland advance of dunes. As a result of these obstructions, fore dunes are concentrated in a narrow zone that parallels the shoreline. They usually have an irregular surface and are at times pockmarked by *blowouts* (cup-shaped or trough-shaped hollows formed by wind erosion on a preexist-

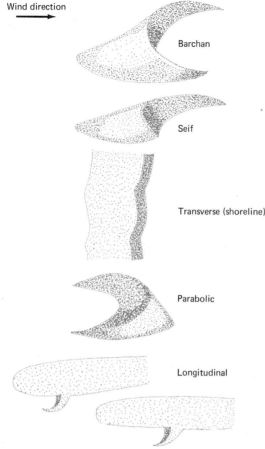

Wind direction

Barchan

Seif

Transverse (shoreline)

Parabolic

Longitudinal

Figure 17–25. Common types of sand dunes.

ing dune). Fore dunes are typically 5–6 feet (2 m) high.

Barchan dunes

Barchan dunes are crescent-shaped, and their horns point downwind (Figure 17–26). They are common when wind direction is constant and moderate, when the surface over which the sand must move is smooth, and when the supply of sand is limited. The steady wind blows the ends of the dune ahead of the rest of the dune, forming the crescent horns. Barchans move slowly. Smaller ones advance at a rate of about 50 ft (15 m) per year. Larger ones move only about one-half this rate. The maximum height of a barchan is about 100 ft (30 m). The maximum spread from horn to horn is about 1000 ft (305 m).

Seif dunes

Seifs are similar to barchans, but one wing is missing. This is caused by occasional shifts in wind direction with no shift in the direction from which the sand is supplied. Seifs may form large ridges as much as 700 ft (213 m) in height and many miles (kilometers) in length.

Longitudinal dunes

Longitudinal dunes are long ridges of sand that form in the general direction of wind movement. They form when sand is in short supply and the direction of the wind is constant. Smaller longitudinal dunes are about 10 ft (3 m) in height and are about 200 ft (60 m) long. In the Libyan Desert they commonly reach a height of 300 ft (90 m) and may extend for 60 mi (100 km).

Transverse dunes

Transverse dunes form ridges at right angles to the wind direction. Wind blowing inland from an ocean or a lake can produce long transverse dunes. The height of this dune rarely exceeds 15 ft (approximately 4½ m). Transverse dunes are common in arid and semiarid lands when sand is abundant but vegetation is scanty.

Parabolic dunes

Parabolic dunes are common in coastal areas where vegetation is present and partially covers the sand. They look like reverse barchans, their horns point into the wind rather than away from it. Blowouts (deflations) will develop in spots where vegetation is absent. Sand movement at either end of the parabolic dune is restricted from movement by vegetation. Ancient parabolic dunes, no longer active, can be found in the Upper Mississippi Valley.

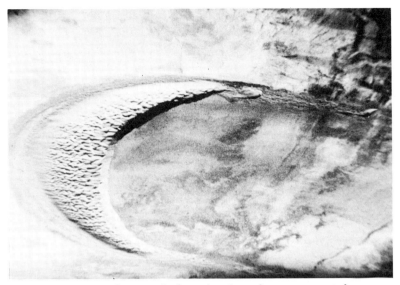

Figure 17–26. Aerial view of a large barchan shows crescent shape and developing horns. The height of this dune is about 100 ft (30.5m). Distance from horn to horn is about 1000 ft (305.5m). (Courtesy of Ward's Natural Science Establishment, Inc.)

DEFLATION

A form of wind erosion that produces no distinctive landform is *deflation*. Deflation is the lifting and entrainment of loose particles of clay- and silt-sized particles commonly referred to as dust. Turbulent eddies of wind lift dust grains vertically and diffuse them into the atmosphere to heights of from a few feet (meters) to several miles (kilometers).

Deflation will occur when clays and silts become thoroughly dried. These conditions exist in deserts, prairies, steppes, tidal flats, and dry lake beds. In the desert deflation removes light particles and leaves behind gravel and pebbles too large to be moved. These larger particles accumulate into a sheet that eventually covers fine-grained materials beneath and protects them from further deflation. This sheet of gravel and pebbles is called *desert pavement*. Exposed surfaces of pebbles may become coated with a dark, iridescent substance called *desert varnish*. In some regions the evaporation of capillary water that has risen to the surface leaves behind a deposit of calcium carbonate (caliche) or gypsum. These substances cement the desert pavement into a conglomeratelike slab resembling a paved surface.

SUMMARY

Deserts are regions that have less than 10 in. of rainfall annually, with most of the precipitation falling in a relatively short period of time. They are not confined to tropical areas. In addition to tropical deserts are middle-latitude and high-latitude deserts.

Six large desert areas are located in western United States. These include the Great Basin, the Mohave, the Arizona–Sonoran, the Colorado, the Painted, and the Chihuahuan deserts.

Structural features found in deserts include pediments, bajadas, playas, inselbergs, mesas, mushroom rocks, and pedestal rocks.

Drainage in western deserts is toward a central basin. A shallow temporary lake forms in the center. After water evaporates, mineral deposits form on the floor of the basin. Permanent lakes, such as Great Salt Lake and Salton Sea, are highly saline, because they do not have an outlet.

Death Valley in the Mohave Desert is not a true valley, because it was not formed by stream erosion. It is the product of structural upheaval. Originally its floor was 10,000 ft (3,050 m) below sea level. At present it is 200 ft (60 m) below sea level.

Although wind is not the most active agent of erosion in deserts, its effects are the most obvious. Landforms produced by winds include broad, shallow depressions; shallow caves in cliffs; and mushroom, table, or pedestal rocks. Rocks that are polished by windblown sand are called ventifacts. Wind transports particles by suspension, by rolling, and by saltation.

Two general forms of wind deposition include dunes and sheets. Principal dune forms are fore (shoreline), barchan, seif, longitudinal, transverse, and parabolic.

Sheet deposition by wind is illustrated by deposits of loess in the Mississippi River Valley. Loess is believed to be material from rocks crushed by Pleistocene glaciers. This material was carried to the outwash plain by meltwater, was dried, and then was blown toward the Mississippi River Valley by wind.

Deflation is the lifting and entrainment of loose particles of clay and silt. In the desert deflation removes light sediments from the surface, leaving behind gravel and pebbles. These larger particles accumulate into a gravel and pebble sheet called desert pavement. The pebbles become coated with a dark, iridescent substance called desert varnish.

Questions for thought and review

1. Describe some present-day problems resulting from sand encroachment.
2. Sand dunes are not stationary features. Describe how dunes form, how they move, and how dune types are influenced by such variables as amount of sand available, presence or absence of vegetation, and wind direction and velocity.
3. Pedestal rocks, mushroomlike residual masses of rocks found in southwestern deserts, have been described by geologists as products of wind abrasion. Formulate an alternate hypothesis that would adequately explain these features.
4. What is meant by pediplanation? What topographic features would be found in a pediplain?
5. If a region is to be classified as a desert, what criterion would best fit the definition?
6. Which of the western United States deserts has the largest land mass?
7. Describe desert pavement. How is it formed?
8. Geologists say that Death Valley is not a valley. Why shouldn't it be classified as a valley?
9. What features found in deserts indicate that precipitation comes in the form of infrequent torrential rainstorms?
10. Describe the origin of ancient Lake Bonneville. How can we trace shrinking (retreat) of this lake?
11. Describe the formation of an alluvial fan.
12. Explain the relationship between a pediment, a bajada, and a playa.

Selected readings

Bagnold, R. A., 1941, *The Physics of Blown Sand and Desert Dunes*, Methuen, London.

Blackwelder, Eliot, 1954, *Geomorphic Processes in the Desert*, California Division of Mines Bulletin 170, pp. 11–20.

Guthrie, R. D. and Zahl, P. A., 1972, "A Look at Alaska's Tundra," *National Geographic*, vol. 141, no. 3, pp. 293–337.

Sharp, R. P., 1949, Pleistocene Ventifacts East of the Big Horn Mountains, Wyoming, *Journal of Geology*, vol. 57, pp. 175–195.

Small, R. J., 1970, *The Study of Landforms, a Textbook of Geomorphology*, Cambridge, London.

Part

III

The mobile earth

18 Geologic structures

STUDENT OBJECTIVES

At the completion of this chapter the student should:

1. understand what structural geology and tectonics mean

2. know the various types of behavior of materials and how they relate to different types of geologic structures

3. understand dip and strike and how they are measured and used in understanding geologic structures

4. recognize differences among kinds of folds, and how and where they form

5. be able to relate differences among several fault types, how faults are recognized in the field, and how and where they form

6. understand what an unconformity is and know some differences between each kind of unconformity

Strange how much you've got to know
Before you know how little you know.

Anonymous

STRUCTURAL GEOLOGY

Have you noticed that in some places the layering in rocks is horizontal, in others it is tilted, and in still others it is crumpled and buckled (Figure 18–1)? Push against the loose edge of the pages of this book until the pages buckle. You have produced a *fold*.

A

B

Figure 18–1. *A.* Horizontal layering in sedimentary rocks in Tennessee that have not been deformed. *B.* Inclined layering in sedimentary rocks in an area in Alberta that has been deformed.

Take a jar that is not quite full of talcum powder, hit the bottom of the jar to compact the powder, and then rotate it slightly toward the empty part until cracks begin to appear. They will open up and slide toward the empty space (Figure 18–2). Cracks along which blocks of powder have moved are *faults* of one type. Rocks in the earth's crust have behaved similarly under different kinds of pressures over a much longer span of time to form folds, faults, and other structural features. The study of such features is undertaken as a part of *structural geology.* We call the branch of structural geology involved with study of large-scale features (including mountain systems, continents, and ocean basins) *tectonics.*

Knowledge of geologic structures enables us to locate and trace valuable mineral deposits, locate oil and gas fields, and evaluate the potentially hazardous nature of active faults. So understanding of structural geology is very important in evaluation of many problems of applied geology related to mineral and energy resources, engineering, and environmental planning.

BEHAVIOR OF ROCK MATERIALS

We can observe rocks that have been folded and others which have been broken under pressure. Obviously these represent two different kinds of behavior. Fractures represent *brittle behavior.* A plate dropping on the floor is an excellent example of brittle deformation. Many folds form during metamorphism when temperatures and pressures are high. Here rocks exhibit *plastic behavior.* It is easy to make plastic folds by bending and folding modeling clay. Folds also form in unmetamorphosed brittle rocks by slippage along bedding and/or cleavage planes.

Some materials spring back to their original shape after pressure is released. These materials exhibit *elastic behavior.* A rubber band or a thin sheet of plastic will spring back when it is deformed. However, if the

Figure 18–2. Faults formed in a box of powdered clay.

rubber band is stretched too far or the plastic sheet bent too much, both will exhibit brittle behavior as they break. So elastic behavior is therefore limited by the *time* the pressure is applied, because the material returns to its original shape after stress is removed, and also by the *elastic limit*, with too much pressure causing the material to enter into brittle or plastic behavior.

Rocks have an elastic limit beyond which they either break or fold under plastic deformation. The type of behavior beyond the elastic realm is determined by the rock type and by the amount of uniform pressure applied to all sides of the rock mass.

DIP AND STRIKE

Most sedimentary rocks are deposited with horizontally oriented bedding. However, when these and other rock types become deformed by folding or faulting, layers become tilted from a few degrees off horizontal, to many degrees, to nearly vertical (see Figure 18–1). When trying to recognize large structures in the field, geologists have found it necessary to measure how much the layering or foliation is inclined from the horizontal. This is called *dip*.

In addition to dip, another measurement is also taken. Because these layers of planes are inclined to the horizontal, they also trend in some direction. This trend is always measured at right angles to the dip, and measurement is made with a compass. The trend of an inclined surface is actually the direction of a horizontal line lying in the inclined plane and is called *strike* (Figure 18–3).

B

Figure 18–3. *A.* Measuring the dip of inclined layering using a Brunton compass. *B.* The strike of a bed is its trend measured as a compass direction.

A convenient way to measure strike is to find a place on a lake or a pool in a stream where the inclined layering or foliation enters the water (Figure 18–4). The compass measurement of the line of intersection of the water with the inclined rock surface, or any line parallel to this intersection, is

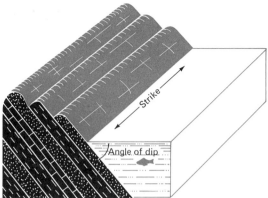

Figure 18–4. The strike of an inclined layer is easily measured when it intersects the smooth surface of a body of water.

strike. Dip direction may be obtained by splashing some water onto the surface and watching the water return directly down the slope. This route will be at right angles to the strike.

Dip–strike measurements are made with a *Brunton compass* (see Figure 18–3). This is a standard compass that contains a clinometer for measuring dip.

Measurements of dip and strike are made not only on bedding and foliation surfaces but also on joint and fault planes. They may actually be made on any two-dimensional (planar) surface. These measurements are commonly plotted on a map at the location where each measurement was made. In order to differentiate between various types of dip–strike measurements, each type has a different symbol (Figure 18–5).

FOLDS

Anticlines and synclines

Folds are bends, flexures, and crinkles of the layers in rocks. They may be formed in almost any kind of rocks. There are many kinds of folds, but two types serve as the basis for most others: anticlines and synclines.

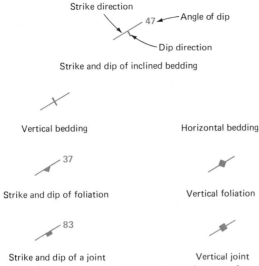

Strike direction

47 — Angle of dip

Dip direction

Strike and dip of inclined bedding

Vertical bedding   Horizontal bedding

37

Strike and dip of foliation   Vertical foliation

83

Strike and dip of a joint   Vertical joint

Figure 18–5. Standard symbols used on geologic maps to indicate different kinds of dip-strike measurements.

An *anticline* is a fold in which layers dip away from the center of the structure (Figure 18–6). The oldest rocks are found in its core, and younger rocks are located on its flanks. Anticlines are generally elongate features. A line may be drawn along the top or *crest* of an anticline, separating that part of a layer which dips in the opposite direction. This line is the *hinge* of the anticline. It is very similar to the fold *axis*. Fold axes in succeeding layers lie on an imaginary surface called the *axial surface* (Figure 18–6). The flanks of the fold are called *limbs*. A *dome* is a special type of anticline in which layers dip away from a central point (Figure 18–7).

A *syncline* is a fold in which layers dip toward each other (Figure 18–8). Youngest rocks are found in the center of the structure and older rocks are located in its flanks. The central part of a syncline is called the *trough*; the flanks are also called *limbs*. Axes and an axial surface may be recognized in a syncline as in an anticline (Figure 18–8). A special type of syncline in which layers dip toward a point rather than toward an axis is a *basin*.

Types of folds

The axes in many folds are not horizontal but are inclined. These are *plunging folds* (Figure 18–9). A fold will frequently have a horizontal axis along part of its extent and then will plunge at either end.

A fold whose axial surface is vertical and whose limbs dip similarly but in opposite directions on either limb is a *symmetrical fold* (Figure 18–10). One whose limbs do not dip similarly is an *asymmetrical fold* (Figure 18–11). Generally the axial surface in asymmetrical folds is not vertical, but it may be. Folds with gently dipping limbs are termed *open folds*; those in which limbs appear to be pressed together are *tight folds*.

Folds in which one limb has been folded beyond the vertical and is upside-down is an *overturned fold* (Figure 18–12). In some overturned folds limbs are parallel to the axial surface. These are *isoclinal folds* (Figure 18–13). If the axial surface of any fold is oriented parallel to the horizontal, we have a *recumbent fold* (Figure 18–14).

A uniform (usually gentle) dip of beds in the same direction for a distance of several miles (kilometers) is called a *homocline* (not really a fold). Local steepening of a homocline or otherwise uniform gentle dip results in a *monocline* (Figure 18–15).

Sizes of folds

Folds may be so small that they are best studied under a microscope (Figure 18–16). They range in size to many miles (kilometers) long and several miles (kilometers) wide. Folds of road-cut to hand-specimen size are probably easiest to observe and study.

How folds form

Folds may form in a variety of ways, the most common of which is by a mass of rocks that is compressed from one end (Figure 18–17). As pressure increases, rocks are pressed together until pressure must be relieved; they begin to buckle, usually upward (up-

A

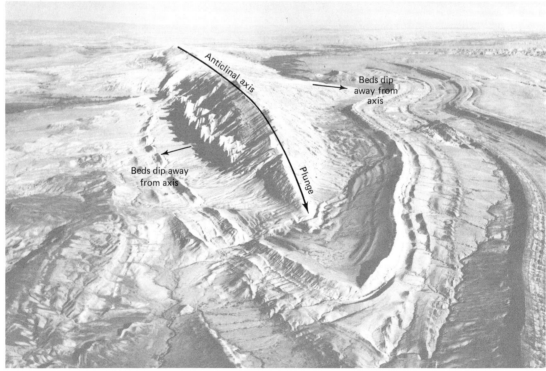

Anticlinal axis

Beds dip
away from
axis

Beds dip away
from axis

Plunge

B

Figure 18–6. *A.* Sheep Mountain anticline near Greybull, Wyoming. (Courtesy of John Shelton.) *B.* Sketch of the anticline showing its different parts.

ward is the easiest direction of relief). Low-amplitude symmetrical folds may form first, but, if pressure continues to be applied, folds may become asymmetrical or overturned as they lean away from the direction of applied pressure. If rocks being folded are in the outer part of the crust, they fold by slipping along bedding planes and may finally break, forming thrust faults (see below) along the common limb between anticlines and synclines. On the other hand, deeply buried rocks under conditions of high temperature and high pressure may be folded more severely and may behave plastically.

Tensional (pulling-apart) forces may occasionally produce folds (Figure 18–18) by causing the material above to sag as rocks below are faulted by being pulled apart. Folds produced in this way are mostly open folds. Monoclines are sometimes produced by this means. This type of folding may occur over a developing normal fault (see below). These structures may also be produced by vertical movements in the crust or may result from vertical movement of salt from beneath the area affected.

Sedimentation over a buried hill occasionally produces folds of low amplitude (Figure 18–19). These structures do not result from rock deformation but are part of sedimentation processes.

FAULTS

Anatomy of a fault

Faults are fractures in the outer skin of the earth along which a sizable amount of movement has taken place. Several kinds of faults have been recognized based on the kind of relative movement each has experienced. But first we must consider the makeup of a fault.

Valuable minerals have been deposited in crushed rock material along many faults. Many years ago, while these deposits were being mined, miners would walk along the lower block of the fault while the upper block rested above them (Figure 18–20). The underside of the fault zone being mined became known to the miners as the *footwall* and the block resting above as the *hanging wall*. These terms survive today as descriptive terms for the different parts of a fault in which the *fault plane*, the surface of movement, is tilted from the vertical.

Types of faults

In many instances it cannot be determined which side of a fault was actually moved. So our definitions of faults are based on *relative movement*, movement of one side relative to the other.

A fault in which the hanging wall has moved down relative to the footwall is a *normal fault* (Figure 18–21). The fault plane of most normal faults is relatively steep, averaging about 60° off horizontal. Movement on most normal faults is predominantly vertical (*dip-slip*), although a slight amount of lateral movement is not uncommon.

Faults with a low angle of dip (45° or less) in which the hanging wall has moved up relative to the footwall are *thrust faults* (Figure 18–21). Thrust faults may have a very low angle of dip, and in these most of the movement is in a horizontal sense. Friction developed along the fault surface during movement of thrust faults frequently results in *drag folds* that form next to the fault surface (Figure 18–22).

A third class of faults includes those which have almost no vertical movement. In these *strike-slip faults* motion is directed laterally (Figure 18–21). The fault plane of most strike-slip faults is oriented vertically. Two types of strike-slip faults are based on

Figure 18–7. *A & B.* Two domes northwest of Riverton, Wyoming. (Courtesy of John Shelton.)

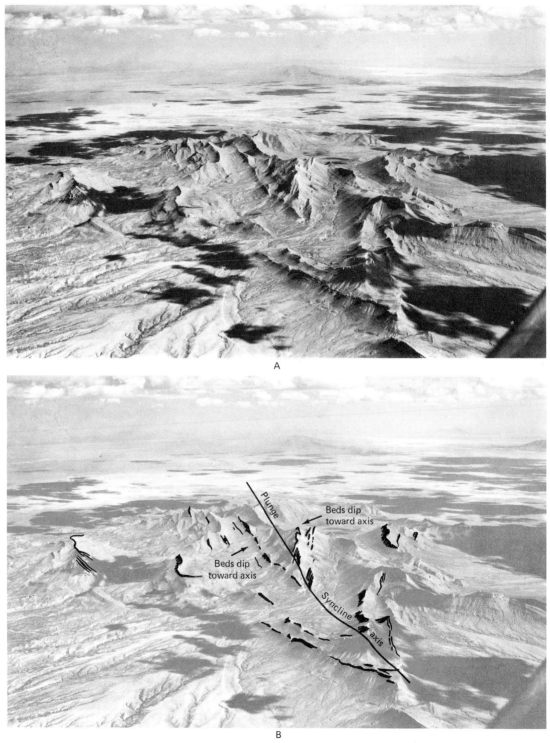

A

B

Figure 18–8. *A.* A syncline in Mexico southwest of El Paso, Texas. (Courtesy of John Shelton.) *B.* Sketch of the same syncline showing its morphology.

A

Figure 18–9. *A*. Plunging folds near the Salton Sea, California. (Courtesy of John Shelton.) *B*. Sketch of structure in A showing relationships between axes and limbs.

B

A

B

Figure 18–10. A. Symmetrical folds near Kingston, Tenn. B. The axial surfaces divide these folds into symmetrical parts.

Axial planes

A

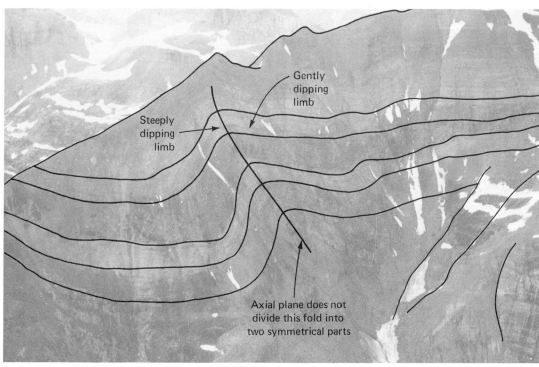

Gently
dipping
limb

Steeply
dipping
limb

Axial plane does not
divide this fold into
two symmetrical parts

B

Figure 18–11. *A & B*. An asymmetrical fold, British Columbia.

A

B

Figure 18–12. *A* & *B*. An overturned fold. One limb is upright but the other is upside-down.

Figure 18–13. *A & B.* A isoclinal fold, South Carolina. Both limbs are parallel to the axial surface in this type of fold. This is also a special type of overturned fold.

movement relative to an observer (Figure 18–22). If the left side of the fault has moved relatively toward (or the right side away from) an observer facing the trace of the fault, the fault is called a *left-lateral* or *sinistral* strike-slip fault. Strike-slip faults in which the right side has moved relatively toward the observer are called *right-lateral*, or *dextral*, strike-slip faults. The San Andreas Fault in California is a right-lateral strike-slip fault, while the Great Glen Fault in Scotland is a left-lateral strike-slip fault.

How faults form

Each of the three different kinds of faults is formed in a different way. Thrust and strike-slip faults are generally formed by segments of the earth's crust being pushed together by *compressional forces*. That thrust faults form in some cases and strike-slip faults in others is the result of a different orientation of forces.

Thrust faults often form by a fracture's following original weaknesses in a sequence of rocks. Certain rock types, such as shale,

Figure 18–14. *A & B.* An isoclinal recumbent fold, South Carolina.

A

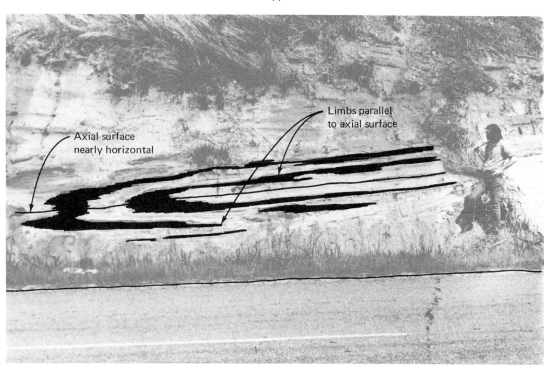

Axial surface
nearly horizontal

Limbs parallel
to axial surface

B

A

Flat beds

Flat beds

Steepened beds

B

Figure 18–15. *A* & *B*. The Comb Ridge monocline, Utah. (Courtesy of John Shelton.)
(*B*. Drawing of structure in *A*.)

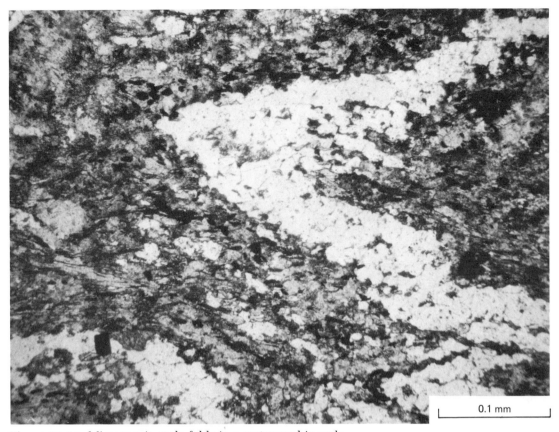

Figure 18–16. Microscopic-scale folds in a metamorphic rock.

0.1 mm

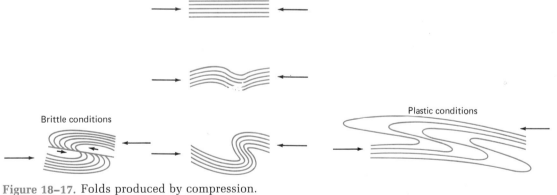

Brittle conditions

Plastic conditions

Figure 18–17. Folds produced by compression.

Figure 18–18. Folds and normal faults produced by tensional forces.

Figure 18–19. Folds produced by sedimentation over a buried hill.

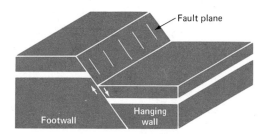

Figure 18–20. The parts of a fault.

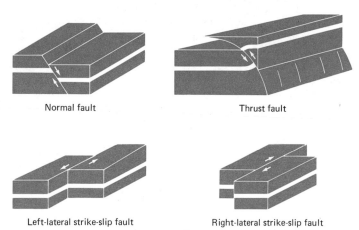

Normal fault

Thrust fault

Left-lateral strike-slip fault

Right-lateral strike-slip fault

Figure 18–21. Different kinds of faults.

rock salt, and gypsum, often serve as the weakness zone for fractures to follow. Because these faults follow beds or form parallel to bedding, they are called *bedding thrusts* (Figure 18–23).

Normal faults are formed by *tensional forces*, which try to pull things apart (see Figure 18–18). Not only are faults of this kind produced by actual tensional forces within the earth. They are also produced in un-

consolidated materials (soils, loose sediments) on a small scale, when their slope is too steep (see Chapter 12), as well as on a large scale, as in unconsolidated sediments of the Louisiana and Texas Gulf Coast (see Chapter 9). Normal faults of this type are not the result of movements in the earth's crust but of movement under the influence of gravity within masses of material ranging from a few feet (meters) to several miles (kilometers) in thickness. Because gravity is the prime mover of many normal faults, some geologists prefer to call them *gravity faults*.

How do we recognize faults?

Faults are fractures that exist in the earth. But there are other kinds of fractures. How can we distinguish them? By definition, we say that faults are "fractures that have experienced appreciable movement." If this movement is ongoing at present, there should be no problem in determining that this fracture is a fault: We could return to the same place over a period of time and measure the accumulated offset. Also, if it is an active fault, earthquakes may be produced by its movement, and seismic instruments may be used to locate the movement along it. But what if we suspect that the fracture in question is an inactive or dead fault?

Geologists use several means to demonstrate whether or not a fracture or even a boundary between two rock types is a fault. Some of these are:

1. repetition of one or more recognizable layers and omission of other layers
2. recognition of actual movement surfaces (*slickensides;* Figure 18–24)
3. observation of zones of crushed rocks (cataclastic rocks; see Chapter 10) along the fracture
4. offset of some easily recognized bed along the fracture
5. offset of topographic features, streams, ridges, and mountains along recent faults (Figure 18–25)

Criteria (1) through (4) may be used to

Figure 18–22. Photograph of a thrust fault in South Carolina showing drag folds developed near the fault plane.

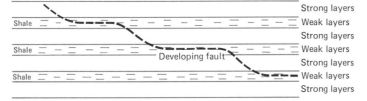

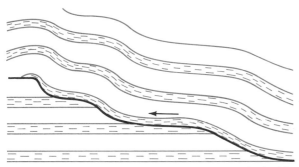

Figure 18–23. Development of a bedding thrust in a sequence of sedimentary rocks.

Figure 18–24. Photograph of a slickensided surface, Tennessee. The grooves and lines indicate the direction of movement.

How large or small are faults?

Faults may be very small features occupying a portion of a hand specimen (Figure 18–26). They may also be large structures that are traceable for hundreds of miles (kilometers) and have displacements measureable in many miles (kilometers). The still active San Andreas Fault in California (see Figure 19–2) is an example of a large fault. Ancient faults, such as the Great Glen Fault in Scotland and the Brevard Fault in the Appalachians of the southeastern United States, are examples of large faults that ceased to move hundreds of millions of years ago.

recognize ancient or recent faults. Using these criteria, many ancient faults that have been inactive for hundreds of millions or even billions of years have been found.

Figure 18–25. Offset of the course of a stream by the San Andreas fault, Carrizo Plain, California. (R. E. Wallace, U.S. Geological Survey.)

Figure 18–26. A small-scale thrust fault in a specimen of amphibolite.

WHERE DO FOLDS AND FAULTS FORM?

Folds and faults may be observed in some form in almost any part of any continent. They may also be found in ocean basins. However, most thrust faults, strike-slip faults, and tight folds, including most overturned, isoclinal, and recumbent folds, occur in mountain belts or their eroded roots in shield areas (see Chapter 21).

Large normal faults may also be found in mountain systems as well as in other regions, such as the Gulf Coastal Plain of Texas and Louisiana, where Mesozoic and Cenozoic sediments have been broken by large normal faults (Figure 18–27). Most of these developed as sediment was being deposited, and movement toward the Gulf occurred as the mass of sediment increased in size. Some of these are traceable for more

Figure 18–27. Development of large down-to-basin faults in the Gulf Coast of southern Louisiana.

than 100 mi (160 km) and have vertical displacements in excess of 3000 ft (1000 m). These normal faults have their downthrown side toward the Gulf of Mexico and are called *down-to-basin faults*. They, along with better-known salt domes (see Chapter 23), have been partly responsible for the accumulation of large petroleum resources in this area.

Gentle, open folds may occasionally be observed in the interiors of continents where tectonic activity has been minimal. A few faults also formed here.

OVERTURNED BEDS

Large overturned folds and occasional thrust faults present geologists with the problem of determining if beds are right-side up or upside down. In areas where we cannot see an entire structure, other means must be used to determine which beds are older or younger and to determine the tops of beds. The basis of these methods is the law of superposition, that the oldest beds in an *upright* sequence are on the bottom and the youngest are on top (see Chapter 1).

Various means, taken singly or together, may be used to determine tops and bottoms of beds and sequences. Features used in the methods outlined below are collectively known as *primary structures* and are the result of depositional, crystallization, or erosional processes (see Chapter 9).

1. *Fossils*, the remains of preexisting organisms, may be used to determine if a sequence is upright. Through careful study of fossils in rocks in areas that have not been deformed, we have learned that certain organisms — *index fossils* — are found only in layers of a particular age. Other index fossils occur in older and younger rocks. Thus if index fossils are present in the sequence in question, determination of the oldest and youngest beds and thus the tops in that group of rocks can likely be made.

2. Cross-beds (see Chapter 9) may be used to determine tops of beds. Three characteristics of crossbeds enable us to use them for this purpose (Figure 18–28):
   (a) They are commonly truncated at the top by normal bedding planes.
   (b) They are concave toward the tops of beds.
   (c) Cross-beds are tangent to normal bedding planes at the bottom of a bed containing cross-beds.

3. Rapid deposition of poorly sorted sediment in water frequently results in *graded beds* (see Chapter 9). During deposition heavier particles sink first; then successively smaller particles are deposited so that there is a gradual change of particle size from the bottom to the top of a bed (Figure 18–29).

4. *Ripple marks* may be used to determine tops of beds. *Current ripples* are asymmetric and appear the same whether they are upright or overturned (Figure 18–30). *Oscillatory ripples* formed in lakes or in the oceans by back-and-forth movement of water results in ripples with large and small crests (Figure 18–30). These do not appear the same when overturned and can therefore be used to determine tops of beds.

5. Other sedimentary features, including *mud cracks, rain imprints,* and *tracks and trails* of organisms that walked or crawled across soft sediment, are also useful in determining tops of beds.

6. Buried *lava flows* may be used to deter-

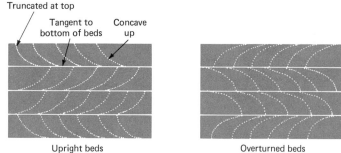

Figure 18–28. Characteristics of crossbeds.

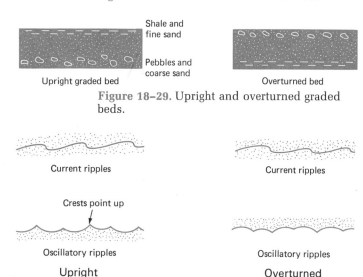

Figure 18–29. Upright and overturned graded beds.

Figure 18–30. Upright and overturned current and oscillatory ripple marks.

Figure 18–31. Joints in a massive quartzite in Georgia. Two joint sets (directions) are present here.

mine the top of a sequence if certain characteristics are present. A flow may have baked the rocks beneath and even picked up some fragments of the rocks over which it was passing. These fragments would still be present near the base of the flow after it cooled. Gas bubbles rise to the top of lava as it is cooling, and the tops of many flows are very bubbly (*vesicular*).

JOINTS

Fractures along which little or no movement has occurred are called *joints* (Figure 18–31). Most joints have an almost vertical orientation. Joints are present in most rock types in most places on earth. Some geologists think that joints form as a result of tensional forces, whereas other believe they are a product of shear. Most agree that they are the result of brittle behavior. It is likely that both tensional and shearing forces are responsible for jointing (Figure 18–32).

Joints of a particular type also form by shrinkage as lava cools or as mud dries. The polygonal pattern of mud cracks and the spectacular hexagonal columns formed in some lava flows are the result of tensional fracturing.

Sheeting is another special type of jointing that forms in very homogeneous rocks, such as granite. Sheeting is a set of fractures that parallel surface topography and is approximately horizontal (Figure 18–33). These fractures become more closely spaced toward the surface and more separated at depth. They are frequently wedged open instead of blasted when quarrying ornamental stone in order to minimize development of other fractures.

Joint patterns

Geologists have noticed in many areas that joints are arranged parallel to each other

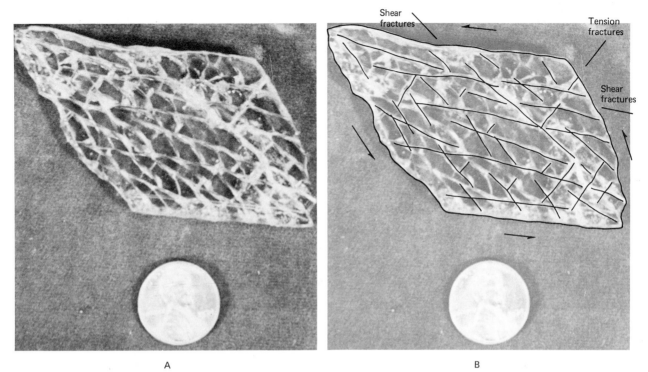

Figure 18–32. *A.* Fragment of fractured window glass involved in an automobile accident showing several sets of fractures. *B.* Sketch showing the different kinds of fractures in *A.*

Figure 18–33. Sheeting joints exposed in a granite quarry, South Carolina.

(Figure 18–31). All joints oriented in one direction make up a joint *set*. Two or more sets compose a *system* of joints. Joint sets and systems may be the result of regional uplift or of some other regional stress pattern.

Study of joint systems is important in construction of large engineering works. This is particularly true in dam construction where location and orientation of joints could affect the amount of leakage that the reservoir might experience once it is filled. If joints are open or if they cut soluble rocks, such as limestone, they must likely be filled during construction of the dam. Rocks that are intensely jointed or that contain large solution channels along joints might result in selection of a better site. Joints have also served as loci for formation of important mineral deposits.

UNCONFORMITIES

A surface of erosion or nondeposition in the geologic record that separates older from younger rocks is called an *unconformity*. Unconformities are not structures in the same sense as folds, faults, and joints. They have not formed directly as a result of forces operating within the earth. Yet they tell us that these forces have been operating, because they formed during a period of uplift of a portion of the earth's crust. They also represent a period in earth history that is not recorded in the rock record at that place: The unconformity was formed by erosion or by nondeposition.

A sequence of events may be related to the formation of any unconformity (Figure 18–34):

1. Oldest rocks present are formed; then all those that overlie the oldest rocks up to the break are formed.
2. These rocks are now uplifted and may be tilted, folded, faulted, metamorphosed, or left undeformed.
3. In either case erosion or a lack of deposition occurs after uplift for periods of thou-

sands or millions of years.
4. Younger rocks are laid down on the erosion surface.

Most unconformities represent old erosion surfaces on which younger rocks were later deposited. The old surface is nothing more than a landscape that is covered over. Hills, valleys, stream channels, old soils, sinkholes (ancient karst topography), and other topographic features have been recognized on unconformities. Considerable topographic relief has been measured in some areas on unconformities as well. More than 400 ft (150 m) of relief has been buried beneath Middle Ordovician rocks in eastern

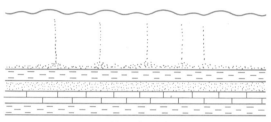

Deposition of sediment

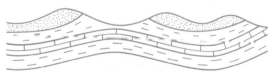

Folding, uplift, erosion

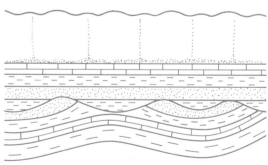

Subsidence and deposition of more sediment results in an angular unconformity

Figure 18–34. The formation of an angular unconformity.

Tennessee; more than 2000 ft (700 m) of relief exist on the Late Precambrian unconformity in Scotland, whereas that on the Precambrian erosion surface in the Grand Canyon is more than 800 ft (275 m).

Kinds of unconformities

Several types of unconformities have been recognized based on rock types occurring above and below the erosion surface as well as on the relative orientation of layering across the break (Figure 18–35).

A *disconformity* is an unconformity in which the rocks on both sides of the break are sedimentary, and there is no difference in the orientation of the bedding planes; that is, both sets are parallel. When this type was formed, there was simply uplift and erosion with later deposition of more sediment on the old surface.

An unconformity with a nonparallel relationship between the layering on either side is an *angular unconformity* (Figure 18–36). Tilting or folding occurred prior to erosion and to deposition of younger rocks. All rocks involved are commonly sedimentary.

A *nonconformity* is an unconformity in which rocks beneath are igneous or metamorphic, whereas those above are sedimentary or lower-grade metasedimentary. If rocks below are metamorphic, an angular relationship may exist across the boundary; thus some nonconformities may be considered angular unconformities as well.

TOPOGRAPHIC EXPRESSION OF STRUCTURES

It would be convenient for us if all synclines were valleys and all anticlines held up ridges, but this is seldom the case. The types of rocks involved in folding and the kind of weathering they experience (see Chapter 11) help to determine the effect on the way in which folds and other structures are or are not raised into relief on the landscape. For example, if it were not for sandstones that

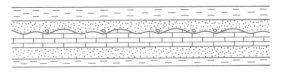

Disconformity

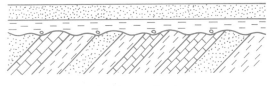

Angular unconformity

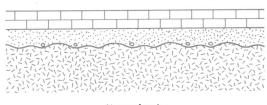

Nonconformity

Figure 18–35. Kinds of unconformities.

are resistant to the humid weathering conditions in Pennsylvania, we would not be able to see the folded structure expressed there as sinuous and zig-zag ridges (see Figure 11–6). In regions of sparse vegetation folds are much better exposed and are more easily observed (see Figure 18–6).

Factors that control the topographic expression of folds also determine whether a fault will be expressed in the landscape. In addition, active faults frequently offset recent modifications of the landscape by human activity (Figure 18–37). Some recent and ancient faults have a border zone of crushed rock that is more susceptible to weathering, so a valley may follow the trace of some faults (Figure 18–38). A fault may also form a series of flatirons, where several valleys trend perpendicular to the fault and the fault has truncated them (Figure 18–39).

Figure 18–36. An angular unconformity, Norris Lake, Tennessee.

Figure 18–37. Offset of the rows of trees in an orchard by movement along an active fault in California. (Courtesy of Clarence R. Allen, California Institute of Technology.)

Triangular flatirons are produced by erosion, leaving segments of the fault plane.

Jointing may be expressed in topography as a *rectangular drainage* pattern, that is, streams with right-angle bends (see Figure 13–28), and as fracture traces observable on aerial photographs (Figure 18–40).

SUMMARY

The study of geologic structures, including folds, faults, joints, and unconformities is undertaken as a part of structural geology.

Rock materials are deformed to produce different types of structures. Faults and joints are formed by brittle deformation. Cer-

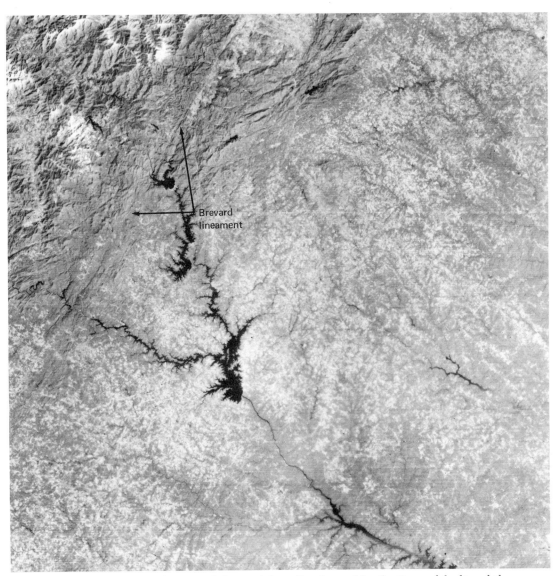

Figure 18–38. ERTS infrared image of the straight valley formed by the Brevard fault and the nearby Blue Ridge and Piedmont in the southern Appalachians. This segment lies in South Carolina and Georgia. Weathering of less resistant crushed rocks and carbonates in the fault zone produced the valley.(NASA photograph.)

Figure 18–39. Triangular facets (flatirons) showing recent movement on a fault, California (triangular facets can be seen in middle of photo). (U.S. Geological Survey.)

tain types of folds are formed by plastic deformation, and others form in brittle rocks.

Dip of a layer is its amount of inclination from the horizontal. Strike is the trend of an inclined layer measured as a compass direction. Measurement of dip and strike can be made on any inclined surface.

Folds are bends or flexures in rocks of the earth's crust. Anticlines are folds in which the oldest rocks are found in the center, and the layers dip away from the center of the structure. The layers dip toward the trough of a syncline, and the youngest beds are to be found in the center of this structure. Domes and basins are special types of anticlines and synclines. Plunging, symmetrical, asymmetrical, open, tight, overturned, isoclinal recumbent, and monocline are particular types

Figure 18–40. Large-scale fracture pattern in Wyoming produced by joints in a vertical aerial photograph. (U.S. Geological Survey).

of folds. Folds range in size from microscopic to many miles (kilometers). Compressional forces probably produce most folds but some form because of tensional forces.

Faults are fractures along which there has been appreciable movement. In normal faults the hanging wall has moved down relative to the footwall. In thrust faults the hanging wall has moved up relative to the footwall. Strike-slip faults involve lateral motion on the fault but almost no vertical movement. Normal faults are formed by tensional forces; compressional forces produce thrust and strike-slip faults. Faults may be recognized by several criteria. They range in size from a small hand specimen to large faults hundreds of miles (kilometers) long.

Most folds and faults are found in mountain belts and shield areas. Some folds and faults are found in almost any part of any continent and ocean basin.

Determination of tops and bottoms of beds and sequences may be made using primary structures including fossils, cross-beds, graded beds, oscillatory ripple marks, mud cracks, rain imprints, tracks and trails, and vesicular tops of lava flows.

Joints are fractures along which there has been no appreciable movement. Most joints form because of tensional and shearing forces. All parallel oriented joints in an area make up a set. Two or more sets compose a system.

Unconformities are erosional or nondepositional breaks in a sequence of rocks. Different types include disconformities, angular unconformities, and nonconformities.

Rock types and weathering conditions commonly control the degree to which structures are expressed in the landscape.

Questions for thought and review

1. What is meant by strike?
2. How do brittle and plastic behavior differ?
3. How would you recognize an anticline? A syncline?
4. Under what conditions do overturned folds form? Normal faults?
5. How would you recognize an active fault? An ancient thrust fault?
6. Where would you go if you wanted to see a thrust fault?
7. List two means by which you might determine a group of rocks as upright or overturned.
8. What is a joint system?
9. What is an unconformity? What does it represent?
10. What determines that a fold will be visible in the landscape?

Selected readings

Anderson, D. L., 1971, The San Andreas Fault, *Scientific American*, vol. 225, no. 5, pp. 52–68.

Beloussov, V. V., 1961, Experimental Geology, *Scientific American*, vol. 204, no. 2, pp. 96–106.

Cailleux, André, 1968, *Anatomy of the Earth*, McGraw-Hill, New York.

Calder, Nigel, 1972, *The Restless Earth: A Report on the New Geology*, Viking, New York.

Clark, S. P., Jr., 1971, *Structure of the Earth*, Prentice-Hall, Englewood Cliffs, N.J.

Summer, J. S., 1969, *Geophysics, Geologic Structures and Tectonics*, Brown, Dubuque, Iowa.

19 Earthquakes

STUDENT OBJECTIVES

At the conclusion of this chapter the student should:

1. know how and why earthquakes occur and where they are most likely to take place

2. be aware of effects produced by earthquakes on the landscape, at sea, and on human habitation

3. understand the meaning of the Richter scale and other ways of measuring magnitude of earthquakes

4. know the operation of the seismograph, types of seismic waves, and how they indicate location of an earthquake and nature of the earth's interior

5. be able to relate methods developed in forecasting earthquakes and preventive measures to be taken to lessen damage and injury

Everything in nature acts in
conformity with law.

Kant

A NIGHT OF TERROR

Tuesday evening was quiet, like many other evenings in Charleston, South Carolina. And, like many other evenings along the Carolina coast in August, this last day of the month was hot, oppressive, and sticky. Windows were open to receive whatever feeble breezes might stir. Many weary townspeople had already retired. Along the streets the hands of public clocks marched toward 10:00 P.M. The year was 1886.

Newspapermen on the second floor of the *News and Courier* Building were hard at work, sweating over Wednesday's edition. Along Tradd Street Dr. F. L. Parker, a physician, hurried toward home, his shadow dancing grotesquely in the light of gas lamps along the street. A short distance away J. K. Blackman was standing in his upstairs bedroom. Out of the dead stillness of the night Blackman heard a shaking noise and realized that the mirror over his bureau was quivering and rattling against the wall. It was 9:59 P.M.

At the same time in the *News and Courier* Building, newsmen became aware of a dull and distant roar, as though some heavy vehicle were rumbling by on the street below. A slight but perceptible vibration of the room was noticeable—nothing unusual. For a few seconds, at least, they paid little heed.

The roaring sound, increasingly louder, reached Dr. Parker on Tradd Street. It seemed to him that the noise was coming from the southwest. He turned in that direction, fully expecting to see a cyclone bearing down on the city. But the sky was clear. Then the vibrations of the earth itself came, increasing in intensity, and Parker felt the ground sway beneath him. He instantly realized the reason: This was an earthquake! Parker staggered across Tradd Street with the rising and falling ground. Under flickering gas lights he saw several undulations of ground, like waves on water, move across the street. He later estimated their height at 2 ft. A chimney crashed down in front of him.

Meanwhile at the newspaper office the dull and distant roar, instead of subsiding, swelled like a drum roll. The men leaped to their feet, baffled and dismayed as the walls around them swayed. One reporter later wrote:

> . . . there was no intermission in the vibration of the mighty subterranean engine . . . the tremor was now a rude, rapid quiver, that agitated the whole, lofty, strong-walled building as though it were being shaken—shaken by the hand of an immeasurable power, with intent to tear its joints asunder and scatter its stones and bricks abroad, as a tree casts its over-ripened fruit before the breath of the gale.

The newspapermen, acting instinctively, rushed for open air—down the stairs and into the streets—to be greeted by the choking dust of pulverized masonry rising in clouds to obscure the sickly yellow lights of the street lamps. The roar of the earthquake subsided momentarily, to be replaced by gathering shrieks of men, women, and children caught in the grip of panic. People groped their way outside, seeking explanation. Eight minutes after the first shock a weaker second shock was felt—weaker, yes, but a reminder that the earthquake was not over.

Half-clothed people congregated in the streets. Around them began the crackling of new danger—fire. The spreading flames at first triggered little human reaction. Fire was familiar; earthquakes were not. Most were listening intently and unconsciously bracing themselves for the ominous roar that would signal a new shock. And they came—three more severe shocks before midnight. More walls collapsed into the cluttered streets. Fires spread from building to building. More people were killed and injured.

Crowds jammed into Marion Square, a large, open area in Charleston, safe from tumbling walls. By midnight the entire aroused population had streamed into every available park or open area. Some escaped to boats anchored in the nearby harbor. There they held vigil, tense and listening, their fears heightened by fires around them and by the uneasy dread that the entire city, situated almost at sea level, might somehow be engulfed by giant waves sweeping in from outside the harbor. In the reality around them it seemed that any further imagined catastrophe might crystallize out of the night. Yet the waves held back. But the fires continued and the growling shocks as well.

All through the night the tremors pummeled Charleston, with bad shocks striking at 2:00 and 4:00 A.M. The morning light of September 1 brought no relief: Another major shock hit at 8:30 A.M. By daylight the extent of the damage was impressive. Few buildings had escaped destruction. The streets were piled with rubble. Overworked physicians, picking their way through debris, tended to the injured.

Throughout that day shocks continued to wrack the city—at 1:00, 5:00, and 8:00 P.M., causing further damage. Many people erected awnings and tents in open places. Few dared to return to what remained of their homes. Workmen doggedly began the task of removing rubble from the streets, which were lined with dangerously weakened walls that were liable to collapse at any moment.

The intense shocks liberated near Charleston, not recognizing state lines, had raced outward toward distant points like ripples in a pond. Four minutes after newsmen in Charleston beheld swaying walls, their fellow newsmen in the *Herald* Building in New York City felt the tremors. According to them:

> . . . the motion was so well defined that it was sufficient at each recurrence to press some of the writers against the tables. Electric wires suspended about the room swayed back and forth.

In Cleveland, Ohio, tremors were obvious enough to empty theatres, homes, and meeting places. Some men left bars, figuring that they had had enough to drink.

By the evening of September 1 the earth had worked out all of its kinks, and quiescence returned. The Charleston earthquake of 1886 was over. Repeated shocks and fires of a night and a day had destroyed or damaged 90% of the city. Seventy persons had died, and thousands had been injured (Figure 19–1).

CAUSES OF EARTHQUAKES

What happened within the earth to cause the destruction to Charleston in 1886? By what method did nature unleash powerful and mysterious forces capable of violently shaking the normally solid and immobile earth? Certainly an earthquake is not a rare event. During any week the odds are good that you will read or hear of an earthquake somewhere in the world. In fact, about 50,000 earthquakes every year are strong enough to be felt by someone.

Early ideas

Today we know with some degree of confidence the causes of earthquakes, but we can only surmise what primitive peoples thought about the shaking of the earth. An event of such terror might have conjured up visions of sinister demons or vengeful gods lurking deep within the earth. Even after early peoples turned from cave dwelling and nomadic life styles to agriculturally oriented settlements, the concept of an earthquake was one of fantasy. From legends and existing drawings we know that the earth was thought to be perched on the back of an animal. One group of North American Indians visualized that the earth was on the back of a turtle; to the Japanese the earth was supported by a giant spider; to some South Americans the earth rested on a whale. Whether the earth was depicted as flat or round, and no matter what the animal, the concept was basically the same: When the animal moved, the earth shook.

Oddly enough the Greeks of Aristotle's time (384–322 B.C.) were not much better off

Figure 19–1. Damage from the 1886 earthquake, Charleston, South Carolina. (U.S. Geological Survey.)

in their understanding of the origin of earthquakes in spite of their remarkable contributions to science in other areas. Democritus thought that water sloshed around inside a hollow earth and made it shake from time to time. Other Greeks, equally deceived that the earth was hollow, asserted that large masses of rock lying within the interior were subjected to alternate drying and wetting, causing them to break off and fall with a great thud, producing earthquakes.

Aristotle himself thought that these ideas were nonsense, but his own theories were no improvement:

> . . . not water nor earth is the cause of earthquakes but the wind — that is, the inrush of the external evaporation into the earth . . . The absence of the sun makes the evaporation return into the earth like a sort of ebb tide, corresponding to the outward flow; especially towards dawn, for [then] the quantity of wind in the earth is greater and a more violent earthquake results.

The hot sultry weather supposed to precede the rushing about of these subterranean, hurricane force winds was called "earthquake weather" by Aristotle. For centuries this term was accepted as factual, although we have no statistical support for the correlation of hot, humid weather and higher incidence of earthquakes.

Elastic rebound

In the late nineteenth and early twentieth centuries, scientists realized that an earthquake is the vibration of the earth as a result of a sudden, smashing blow. The smashing blow is caused by the bending and distortion of large masses of rock strained beyond the breaking point. When these rocks rupture, abrupt movement takes place along a fault as rocks grind and slide past each other, attempting to regain their original position.

This movement has been compared with the stretching and release of a rubber band. Like the rubber band, the rocks snap back with a sudden release of energy. This *elastic rebound* results in an earthquake. Behavior of rocks in this manner has been verified by observations along the San Andreas Fault both before and after the San Francisco earthquake of 1906 (Figure 19–2).

Figure 19–2. The San Andreas Fault, California. (Courtesy of John Shelton.)

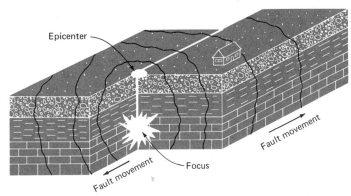

Figure 19–3. Location of focus and epicenter of an earthquake.

The center of the shattering and rebound of rocks is the *focus.* It is commonly several miles deep within the crust of the earth. Sometimes the focus is in the mantle, as deep as 400 mi (650 km) from the surface. The point on the earth's surface directly above the focus is the *epicenter.* Most destruction during a major earthquake takes place at the epicenter (Figure 19–3).

In studying faults, geologists became aware that the total amount that rocks shift along the break was sometimes thousands of feet. Along some big faults displacement could be measured in miles (kilometers). In the Appalachian and northern Rocky Mountains master faults involve rock masses in which lateral movement exceeded 75 mi (120 km; Figure 19–4).

Obviously such displacements did not happen all at once. Fault movement is often

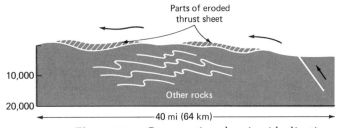

Figure 19–4. Cross section showing idealization of thrust faulting in the Appalachian Mountains.

described as a "jerky creep." Once a fault develops, periodic movements of 3–4 in. (7–8 cm) up to 3 ft (1 m) take place along the fault, relieving the strain that progressively builds up. Faults are the sites of earthquakes. Wherever active faults are known, it can be expected that earthquakes will occur in the area of the fault.

If you have experienced a severe earthquake, read about one, or seen a television report, you cannot help but wonder about the powerful earthquake-causing forces within the earth. For many years scientists were as baffled as laymen. They simply admitted that these great forces existed but that their cause was unknown. During the 1960s more and more evidence accumulated toward a unified theory to explain the dynamics of the earth's crust. This theory of *plate tectonics* is discussed more thoroughly in Chapter 20. Scientists realized that plate tectonics might explain not only the drifting apart of continents but also the cause of earthquakes.

Current theory holds that the earth's crust consists of six or seven relatively thin plates or slabs that can move or shift around independently of one another. Continents ride on these plates. Movement is slow—perhaps only 2 or 3 in. (5 or 7 cm) per year. Millions of years ago all the continents apparently were gathered together near the south pole. Subsequent breakup of this supercontinent about 200 million years ago inaugurated the shift or drifting of the continents to their present positions. This plate movement apparently has not yet stopped.

If we look at a map of the world on which the locations of earthquakes have been plotted (Figure 19–5), it will be obvious that quakes tend to be concentrated in certain regions. For example, one line of earthquakes extends from Alaska south to California; another concentration is located in the Mediterranean area. The Alaska–California line coincides with the contact between the American plate and the Pacific plate. Like-

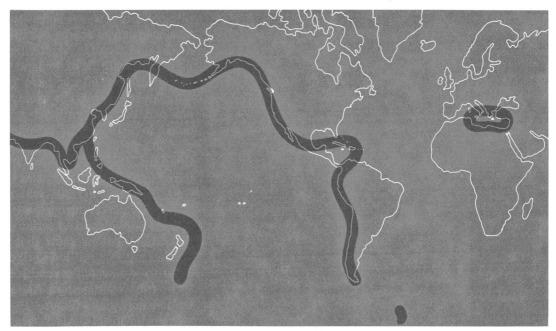

Figure 19–5. Distribution of earthquakes around the world.

wise the Mediterranean area is the site where the African plate nudges against the Eurasian plate.

When two plates contact one another, pressure builds up, because the plates are trying to move in opposite directions, or one plate is moving faster than the one it is in contact with. The pressures or stresses in the rocks that make up the plates can be borne for only so long. Fracturing and faulting occur, and earthquakes are recorded. Unrelenting pressure applied by one plate against another assures the repetition of one quake after another in the plate junction region.

At this point we can summarize that: (1) earthquakes are concentrated in certain areas of the earth; (2) these areas coincide with the edges of large plates that make up the earth's crust; (3) the edges of the plates push against each other; (4) rock masses finally rupture, creating faults along which strain is relieved; and (5) rock movement relieving this strain is so abrupt and intense that an earthquake results.

EFFECTS OF EARTHQUAKES

The cumulative effect of a multitude of small to moderate earthquakes over a period of thousands or perhaps millions of years is to change the topography or configuration of the land. Large areas may be elevated or depressed. Courses of rivers may be altered; lakes may form. During a major earthquake the process of landscape modification is accelerated. With the advent of human civilization, these changes have not always occurred in unpopulated areas. Thus the effects of earthquakes have broadened to include those involving human beings and their towns and cities. For convenience these various effects can be separated into direct or primary effects and secondary effects. A primary effect is a direct result of fault movement and of vibrations that radiate outward from the fault zone. A secondary effect is triggered by a primary effect.

Primary effects

The most vivid *primary effects* are rip-

Figure 19–6. Permanent shifting of the land along this highway occurred during the West Yellowstone earthquake of 1959. (Courtesy of Montana Highway Department.)

Figure 19–7. Earthquake damage near Anchorage, Alaska, in 1964. (Courtesy of National Oceanic and Atmospheric Administration.)

pling of actual waves across the ground surface, creation of scarps or cliffs, opening of cracks and fissures, shattering of rock, and permanent shifting of the land (Figure 19–6). Some primary effects have occurred over wide areas of land. For example, the southern part of India suffered a violent earthquake in 1819. During the quake an area 50 mi. (80 km) long and 10 mi (16 km) wide was lifted up about 10 ft (3 m). It was later referred to by its inhabitants as the "Mount of God." The coast of Chile, which experienced disastrous quakes during the 1960s, was subjected to a widespread series of temblors in 1835. Calculations show that more than .5 million mi² (1.3 million km²) of the Chilean coast were elevated several feet.

In 1964 southeastern Alaska was struck by a major quake (Figure 19–7). Sixty-five years prior to this, Alaska was the scene of an earthquake of comparable or even greater intensity. Fortunately the skimpy population in 1899 prevented any appreciable loss of life. Had a densely populated city been located along the western shore of Disenchantment Bay, the results would have been catastrophic. The area was upthrust 47 feet.

Secondary effects

In human terms *secondary effects* are often the most ruinous. The array of secondary effects includes fires, floods, landslides, breaking of water pipes and electrical lines, and epidemics. Of all secondary effects, fires seem to inspire the most dread. In the Charleston earthquake of 1886, fires destroyed 90% of the city. In the San Francisco earthquake of 1906, it was fire that caused the greater part of that city's destruction. Fire played a major role during the Tokyo–Yokohama earthquake of 1923. At that time most structures and private dwellings in those cities were constructed of highly flammable materials. In addition, cooking fires with open flames were maintained in homes. During the quake, thousands of such cooking fires were overturned. The holocaust that enveloped the Japanese cities burned more than a half million houses. Fleeing from the blaze, many people sought safety in pools of water, only to be boiled alive in the intense heat. Many others suffocated.

One of the best documented earthquakes is the one that struck Montana on August 17, 1959. Geologists point to this earthquake as an outstanding example of the variety of secondary effects that can result from a major quake. The scene of the disaster was West Yellowstone and vicinity, a region frequented by campers and other vacationers. The area of the earthquake's wrath was a 14-mi (22-km) stretch of valley that included Hebgen Lake, a 7 mi (11 km) body of water impounded behind a small dam, and narrow Madison Canyon, extending about 7 mi (11 km) below the dam (Figure 19–8).

On that Monday, at 11:37 P.M., masses of solid rock abruptly slipped 20 ft (6 m) along the Red Canyon Fault, releasing shock waves that converted a peaceful summer's evening into a nightmare for the 300 people trapped in the canyon. Racing shock waves ripped and tore apart roads and trails. Fissures split the ground surface, and masses of rock and soil poured across highways and inundated houses. The initial movements of the earth were over in a matter of seconds. But the worst was yet to come.

Almost immediately after the rocks snapped along the fault, the disturbance reached Hebgen Lake. A wall of water 20 ft (6 m) high formed and surged down the length of the lake. It swept over the top of the dam and boiled down Madison Canyon. Anything movable was caught up in the strong

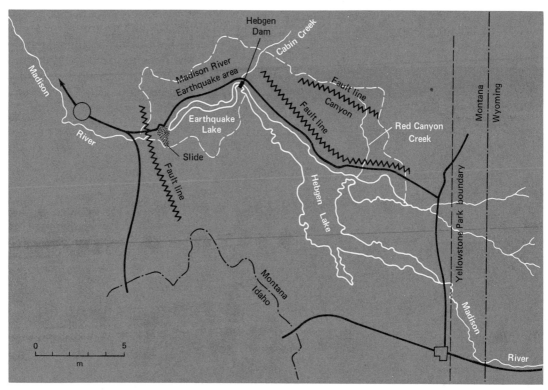

Figure 19–8. Sketch of West Yellowstone and vicinity.

current—tree trunks, planking, furniture, automobiles, and camping gear. Startled campers groped in the dark, struggling for higher ground. At the same time cottages along the lake shore were slumping into the water.

As the turbulent waters crashed down Madison Canyon, the earthquake shook loose 80 million tons of mountainside flanking the canyon. Within seconds this tremendous section of earth had slid into and across the valley and part way up the other side. The mass of soil and rock, 1.5 mi (2.4 km) wide effectively dammed the lower reaches of Madison Canyon. As the landslide descended, the air caught under its mass rushed outward with hurricane force. Trees were smashed to splinters, automobiles were accordioned, and several campers were crushed to death. Waters backed up behind the landslide, filling Madison Canyon. In this earthquake 28 people died (Figure 19–9). The lake, named Earthquake Lake, is still there.

Figure 19–9. Aerial view of the landslide that developed during the West Yellowstone earthquake of 1959. Several campers were buried under the slide. (Courtesy of Montana Highway Department.)

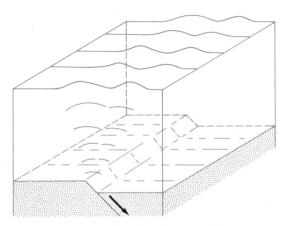

Figure 19–10. Production of tsunamis by faulting on the sea floor.

Among the most feared effects of an earthquake, especially for people residing in coastal areas, are *tsunamis*, or seismic sea waves. The tsunami manifests itself as a large wave or series of waves ranging in height up to 100 ft (30 m). When waves of such formidable height and volume crash into shoreline areas, property damage and loss of life can be considerable. These waves are not produced by the wind as are most other waves. Instead, they are caused by a sudden blow affecting the sea bottom. The resultant released energy is transmitted through the water. In most cases an undersea earthquake produces the sudden blow by abrupt up-or-down shifts along faults of portions of the sea floor (Figure 19–10). Tsunamis may also be produced, albeit rarely, by violent volcanic eruption in proximity to the sea or by submarine landslides. Even atomic bomb tests on remote Pacific islands have produced weak but detectable seismic sea waves.

EARTHQUAKE MEASUREMENT

Not all earthquakes are, of course, of equal intensity. Not only will intensities vary at the epicenter in separate quakes, but for one specific quake the effects will decrease outward with distance. A person at

Table 19–1. Kinds of damage or effects produced by varying earthquake intensity.

Mercalli scale	Typical effects	Richter equivalent
I	Instrumental. Detected only by seismograph	
III	Slight. Similar to vibrations from passing truck.	3
IV	Moderate. Causes shifting of small objects; larger objects tip back and forth.	4
VI	Strong. Overturning and falling of furniture, other objects	5
VIII	Destructive. Walls split and crack; chimneys collapse	6
X	Disastrous. General destruction of buildings, roads, overpasses; ground cracks open.	7
XI–XII	Catastrophic. Ground waves, shifting of land; floods; total destruction at epicenter.	8 or more

ranges from 1 (the weakest quakes) to ∞ (the strongest quakes). Actually an earthquake of 9 has never been recorded. The Richter scale is further graduated by decimals. The greatest earthquakes have reached about 8.7 on the Richter scale.

It should not be construed that a quake of 2 releases twice as much energy as an earthquake of 1, nor that a quake of 2 causes twice as much damage as a quake of 1. Furthermore a quake of 7 is *not* 7 times more powerful than a quake of 1. Instead, each number represents a quake 31 times more powerful than one of the previous number. For example, a quake of 6.5 on the Richter scale releases 31 times more energy than one of 5.5. An earthquake of 6 releases about one-third the force of the atomic bomb dropped on Hiroshima in 1945. However, a quake of 9 would be approximately equivalent to the energy release of 10,000 such atomic bombs.

the epicenter may see windows shatter and may be thrown to the floor; another person 1000 mi (1612 km) away would notice nothing unusual.

Initial attempts to categorize earthquakes were based on what individuals personally experienced and observed on damage around them. Thus the first types of earthquake intensity scales amounted to listings of earthquake effects in order of increasing seriousness. Table 19–1 shows one such kind of modified scale based on effects. Such scales are not completely satisfactory because of differences in construction of buildings and in the varying degrees of human reaction to seismic events.

A better way to quantify earthquakes is based on the amount of energy released at the focus. Continued improvements in instrumentation used to detect and measure earthquakes led to the use of the Richter scale in 1935. This scale is now commonly used by newspapers in reporting earthquakes. The Richter scale is logarithmic and

EARTHQUAKE DETECTION AND SEISMIC WAVES

The seismograph

The principal "improvement in instrumentation" that prepared the way for the development of the Richter scale was the *seismograph*. The invention of the seismograph, the earthquake detector, was the foundation for *seismology*, the systematic study of earthquakes.

The basic principle of seismograph operation can be simply illustrated. If you place your hand at one end of a long metal pipe and have someone strike the other end with a heavy hammer, you will feel the vibrations. The sharp blow of the hammer symbolizes the earthquake. The pipe is the earth itself, transmitting the vibrations to your hand. Your hand is like the seismograph picking up the vibrations from the earth. The seismograph's job is to record these oscillations in the earth and make them available to the seismologist for study.

In 1875 British geologist John Milne invented the seismograph while working for the Japanese government. He and a colleague, John Gray, constructed an instrument consisting of a weighted pendulum suspended by a string or wire. The framework that supported the pendulum was anchored to the ground. If the ground moved because of an earthquake, the framework of the seismograph also would move. However, the suspended pendulum would not move immediately because of inertia. A stylus attached to the pendulum would first record a series of wiggles on a moving strip of smoked paper called a *seismogram*. After these critical ground motions had been recorded the pendulum itself would move with a delayed response to the shifting ground.

Milne recognized that ground motion is complex during an earthquake. Not only were there up-and-down motions; the ground could also move horizontally east–west and north–south. He and Gray built a seismograph with three pendulums to account for the three different motions. From these beginnings other scientists designed other types of seismographs (Figure 19–11).

Seismic waves

As more and more seismograph records were studied and reviewed, it became apparent that each earthquake consisted of three kinds of distinct *seismic waves*, each of which arrived at the seismograph station at different times after the shock.

P (primary) waves reached the seismograph ahead of the others and made small but distinct wiggles. Within seconds or minutes *S (secondary) waves* arrived. They were always slower than P waves. *L (large or long) waves*, the slowest of all, arrived last but produced the largest tracings on the seismogram.

P and S waves travel outward from the focus directly through the earth and are known as *body waves*. P waves pass through rock with a vibratory motion comparable with that of an alternating electric current. S waves vibrate at right angles to their direction of propagation (Figure 19–12). P waves can pass through both solids and liquids; S waves can pass through solids only. The slower L waves are actually generated by P and S waves that strike the surface directly above the focus. P and S waves bounce back, while newly created L waves radiate outward on the surface like ripples in a pond.

Figure 19–11. A seismograph. (U.S. Geological Survey.)

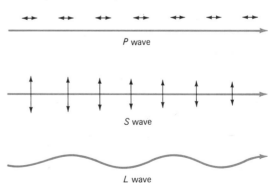

Figure 19–12. Mode of propagation of seismic waves.

Location of an earthquake

S waves travel at two-thirds the speed of P waves. The difference in velocities is fairly consistent. This fact opened the door to determine routinely the distance between the site of the earthquake and the seismograph station. The greater the time lag between P and S wave arrivals (P minus S), the further away the earthquake. Graphs were constructed so that distances could be determined quickly (Figure 19–13). Location of an earthquake epicenter, whether on land or in oceans, can be obtained readily by three cooperating seismograph stations (Figure 19–14).

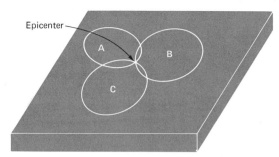

Figure 19–14. Stations A, B, and C know the distance to the earthquake but not the precise direction. Circles are drawn on a map around each station with a radius equal to the distance to the quake. The point of intersection of the three circles is the epicenter.

The earth's interior

Study of seismic wave behavior provides us with information regarding the nature of the earth's interior. As body waves pass through the earth, they are refracted and reflected at boundaries or *discontinuities* that indicate a change in the rigidity or density of the rocks. Reflected seismic waves arriving at seismograph stations indicate that the interior is made up of a series of concentric layers or shells. The earth does not simply consist of crust, mantle, and core. Several additional discontinuities exist. Furthermore, the core has a liquid outer part. The outer part appears to be liquid, because S waves are stopped when they try to pass. Seismic waves can tell us nothing directly about the composition of the interior, but we do know that material becomes increasingly dense with depth.

EARTHQUAKE PREDICTION

During the past few years scientists around the world, especially in Japan, the United States, and the Soviet Union, have made a concerted effort to develop criteria that would aid in forecasting the time, place, and magnitude of an earthquake. In 1972 the United States and the Soviet Union entered into a formal agreement to collaborate on this problem. Until recently such an effort (prediction) would have been regarded as something best relegated to fortune tellers and clairvoyants.

Some observers have claimed that the capability to predict earthquakes would be a bad thing and have even urged that all re-

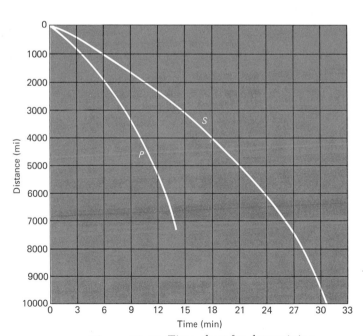

Figure 19–13. Time chart for determining distance to the site of an earthquake.

search on such projects be discontinued. They pointed out that if, say, Los Angeles were informed of an impending major earthquake, the paralysis of normal activity, panic, looting, drop in real estate values, and other attendant phenomena would be worse than the earthquake itself, if it ever came. Whereas these events are possible, it is doubtful that scientists would abandon their efforts in this direction. The only answer for humankind is to plan now on how to handle the situation should predictive capability be forthcoming. There is ample evidence now that such capability may indeed be forthcoming.

If moderate to great stresses build up along a fault for weeks, months, or years prior to an earthquake, most (if not all) geologists would agree that some kind of change in physical characteristics would be noticeable in rocks in the fault area. This was the basic assumption for investigation of premonitory phenomena in earthquake-prone areas. During the 1960s laboratory experiments showed that rocks subjected to pressure and high stress developed microfractures and other voids with a resultant increase in total volume. This was referred to as *dilatancy*.

Field studies involving hundreds of instruments were initiated to monitor seismic and kindred activity in zones of active faults. Old seismic records were reviewed. Changes in ground level, electrical conductivity, radon (a short-lived radioactive isotope) levels in water wells, and the ratio of *P* to *S* wave velocities were all observed and recorded. Taking these changes into account, especially the *P* and *S* wave velocities, the first fairly accurate predictions were made in the Garm District of the Soviet Union and at Blue Mountain Lake in the Adirondack Mountains of New York State. In the latter case an earthquake measuring 2.5–3.0 on the Richter scale was forecast two days before it actually happened.

These successes led to the formulation of what is now called the *dilatancy hypothesis* or *model*. It is still being tested, and there is no guarantee it is the ultimate answer in earthquake prognostication. But many scientists call dilatancy a "provocative" hypothesis.

Here is what seems to be happening: (1) rocks in the fault area develop cracks and voids as strain accumulates and porosity expands; (2) new or expanded voids are dry to begin with, or, if they contain water, there is loss of pressure; (3) after a time an influx of groundwater fills the voids, with increasing water pressure therein; and (4) increased pressure acts as a mechanism to trigger an earthquake.

This hypothesis satisfies all of the observed changes. Changes in ground level, as in Japan where the ground rose 2 in. (5 cm), were caused by the expansion of voids. Increase in electrical conductivity of the rocks was caused by the influx of water, an excellent conductor. Increases of radon would be predicted when groundwater is moving well rather than becoming stagnant or sluggish.

The most significant change was the decrease in the velocity ratio of *P/S* waves. *P* waves travel faster through rocks that have water-saturated voids than through rocks with dry voids. In test cases, the net loss of *P* wave velocity corresponded to the period of

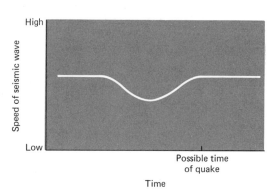

Figure 19–15. Graph showing schematically the behavior of *P/S* wave velocities prior to earthquake.

stress when new voids would be created or old voids expanded. The normal P/S velocity ratio was restored just prior to the earthquake, corresponding to the influx of water into rock openings (Figure 19–15). It seems also that, the greater the intensity of the earthquake, the earlier the change in the P/S ratio. Warning signs in this respect might manifest themselves three years or more before the earthquake itself. Only time and demonstrated repeatability of phenomena associated with the dilatancy hypothesis will attest to its viability.

PREVENTIVE MEASURES

While waiting for further developments in earthquake prophecy, there are a number of ways in which we can minimize damage and loss of life. First, we can build cities and towns away from known active faults. Even some so-called inactive faults should be avoided. In cities already situated within the zones of active faults, such as San Francisco, low structures of proper design and strength are preferable to high-rise buildings. Fragile and ornate appendages, such as balconies, should be eliminated. Schools and other buildings where large numbers of people congregate should be relocated as far from the fault itself as practicable.

It might be worthwhile also for us to review what to do or not to do if a major earthquake occurs. During the tremor, stand in doorways; if outside, stay away from walls. In the aftermath, stay off the streets, do not use telephones, and check for broken gas, water, and electrical lines. Additional information can be obtained from civil defense authorities.

Can earthquakes themselves ever be prevented? It is a tall order. A glimpse of this possibility was provided by a disposal well drilled by the U.S. Army near Denver, Colorado. Soon after fluids were pumped down the well, a series of small tremors was felt in the Denver area. Investigations indicated that the disposal well had been drilled through a fault. The fluids seemed to be "lubricating" the fault and causing slippage. When the disposal was discontinued, the earthquakes stopped.

Some scientists believe that if such wells were drilled along dangerous faults, they would move little by little rather than all at once. Thus the energy could be released in small, harmless amounts, and a major earthquake would be avoided. A great deal of care would need to be exercised in the application of this method. For example, to apply it to the northern section of the San Andreas Fault would be risky: There is a "lock" on the fault, and considerable energy is already stored up. Attempts to release this energy in small amounts might get out of control, and the major quake expected might come sooner rather than later.

SUMMARY

Earthquakes occur when rocks along faults, distorted by accumulated stresses, break and move in what is called elastic rebound. This movement takes place as much as several miles deep at a point within the earth known as the focus. The spot on the earth's surface directly above the focus is the epicenter. In major earthquakes most damage takes place at or close to the epicenter.

During severe quakes the ground itself moves in waves and may be permanently shifted up or down several feet. These movements may trigger landslides, fires, floods, or other secondary effects that often are more damaging than the earthquake itself. When earthquakes take place at sea, tsunamis, or large seismic sea waves, may be unleashed, traveling hundreds of miles (kilometers) at great speed to inundate and cause extensive damage and loss of life in coastal areas. The Pacific Ocean is notable for tsunami phenomena. The Richter scale, a logarithmic scale from 1 to ∞, is the most common measure of the intensity of an earthquake and is based on the amount of energy released.

Each earthquake releases vibratory energy that spreads outward from the focus. These vibrations can be detected even at great distances by the seismograph, which records the earth's vibrations—seismic waves—as a series of wiggles on a seismogram. Three kinds of seismic waves include P (primary), S (secondary), and L (large or long) waves, in order of their arrival times at the seismograph station. The time lag between arrival of P and S waves, which travel through the earth, permits calculation of the location of the quake. L waves travel on the surface and cause the most damage. Behavior of P and S waves reveals that the earth's interior is made up of several concentric layers, like an onion, and includes a liquid outer core.

Progress has been made in predicting earthquakes by measuring changes in physical properties of rocks in active fault zones prior to an earthquake. The most critical change involves reduction in speed of P waves as they pass through rocks whose porosity has increased because of applied stresses. Earthquakes may be prevented or rendered relatively harmless in the future, but in the meantime it is prudent to avoid population concentration and construction in zones of active faults. It is also wise to know what procedures to take in the event of a major earthquake.

Questions for thought and review

1. What specific effects would suggest that the Charleston earthquake of 1886 was a major quake?
2. Approximately how many earthquakes take place in the earth each year, including only those detected by the seismograph?
3. Defend the statement that "deep-seated rocks can behave elastically."
4. Is the focus of an earthquake only one specific point? Or can the focus be distributed along a fault?
5. About how many crustal plates exist? Do earthquakes only occur at plate junctions? If not, why not?
6. What is a deep-focus earthquake? How does it differ from a shallow-focus earthquake?
7. Cracks in the ground are known to result from major earthquakes. Would you guess that significant numbers of people are swallowed up by such cracks?
8. What types of faults would create scarps or cliffs during a major earthquake? Are there scarps or cliffs caused by movement along the San Andreas Fault?
9. In the San Francisco earthquake of 1906 and others, much damage to property was caused by liquefaction of the subsoil. What is liquefaction and what causes it?
10. Why did people "suffocate" during the 1923 Tokyo-Yokohama earthquake as fires raged around them?
11. Why do tsunamis steepen and grow higher as they enter the shallow water along coastlines?
12. As a tsunami approaches a coastal area, why does the water recede so dramatically, exposing parts of the sea bed normally covered by water at low tide?
13. Suppose an undersea earthquake occurs in the Pacific Ocean 1500 miles from Hawaii. About how many hours of warning would the people of Hawaii have that a possible tsunami was on the way?
14. What are the weaknesses in using the Mercalli scale to assess the intensity of an earthquake?
15. Is it possible for more than one P wave to arrive at the same seismograph station? How?
16. Under what circumstances would a seismograph station receive P waves but not S waves?
17. If the time interval between arrival of P waves and S waves at a seismograph station is 5 min, about how far away is the quake from the station?

18. What is a discontinuity?

19. At the time of the San Francisco earthquake of 1906, did seismographs exist to record it?

20. List all the things that might happen, including your own reactions, if you knew that a major earthquake would strike your city in 48 hours.

21. What is radon? What role does it play in earthquake prediction and why?

22. In what way did the pumping of fluids down a well in Denver, Colorado, support the dilatancy hypothesis?

23. A seismograph records small wiggles even when there is no earthquake. These are called microseisms. What do you think causes them?

24. What are foreshocks and aftershocks?

25. Formulate the dilatancy hypothesis in your own words.

Selected readings

Bullen, K. E., 1955, The Interior of the Earth, *Scientific American*, vol. 193, no. 3, pp. 56–61.

Grantz, Arthur, et al., 1964, *Alaska's Good Friday Earthquake, March 27, 1964*, U.S. Geological Survey Circular 491.

Hammond, A. L., 1973, Earthquake Predictions: Breakthrough in Theoretical Insight? *Science*, vol. 180, no. 4088 pp. 851–853.

Hodgson, J. H., 1964, *Earthquakes and Earth Structure*, Prentice-Hall, Englewood Cliffs, N.J.

Leet, L. D., and Leet, F., 1964, *Earthquake — Discoveries in Seismology*, Dell, New York.

Oliver, Jack, 1959, Long Earthquake Waves, *Scientific American*, vol. 200, no. 3, pp. 131–143.

20 Plate tectonics

STUDENT OBJECTIVES

At the end of this chapter the student should:

1. be able to follow the development of the concept of plate tectonics from its beginnings in continental drift theory to its present state

2. have examined evidence, both pro and con, concerning plate tectonics

3. recognize the impact of plate tectonics theory on the science of geology

*No great advance has ever been made
in science . . . without controversy.*

             Lyman Beecher

Suppose a time machine could take you back 200 million years and you could be suspended in a stationary satellite above the present position of the middle of the Atlantic Ocean. Instead of being above water, your satellite would be directly above the supercontinent of Pangaea (Figure 20–1). If you could remain in your satellite and travel through time toward the present, stopping occasionally to look down at the continents, you would see a magnificent spectacle of the separation of continents and their movement toward the positions they occupy today. You would also be a kind of "sidewalk supervisor" during the construction of the great modern mountain systems, such as the Alps, the Himalayas, the Andes, and the North American Cordillera.

This chapter will tie some of the modern concepts of the structure of the skin of the earth, the lithosphere (Figure 20–2), together with the evolution of the continents and ocean basins as well as with the nature and origin of mountain systems. According to J. Tuzo Wilson, plate tectonics theory may prove to be for geology what atomic theory is for chemistry and physics and evolution is for biology and paleontology: a theory that unifies many separate processes and accounts for their occurrences better than any previous theory. This will become more clear as we explore plate tectonics theory and its development. However, keep in mind that plate tectonics theory is in an early stage of development. Hypotheses are still evolving. They should be questioned at every

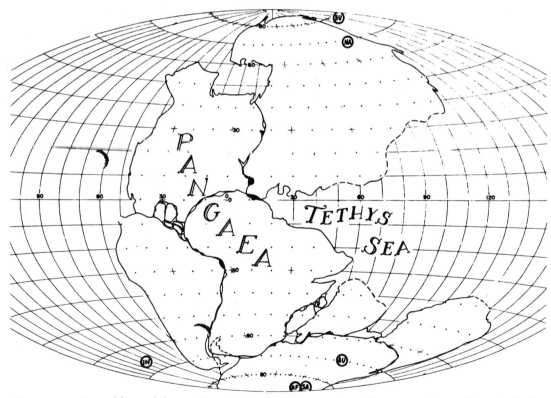

Figure 20–1. Assemblage of the continents into the supercontinent Pangaea as it may have looked 225 million years ago. This reconstruction was made on the 1000-fathom below sea level contour. (From Dietz and Holden, *Journal of Geophysical Research*, vol. 75 1970.)

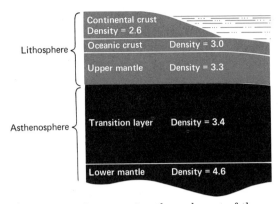

Lithosphere {
Continental crust
Density = 2.6
Oceanic crust   Density = 3.0
Upper mantle   Density = 3.3
}

Asthenosphere {
Transition layer   Density = 3.4
Lower mantle   Density = 4.6
}

Figure 20–2. Cross-section through part of the Earth showing the divisions of the crust and mantle.

stage and their value continually weighed. You, as a thinking student, should weigh and question the evidence and decide in your own mind if it is a far-reaching worthwhile idea or just an interesting mental exercise.

HISTORICAL DEVELOPMENT

Study a globe or a map showing the continents on either side of the Atlantic, (Figure 20–3) particularly South America and Africa. The protruding east side of South America would almost fit into the indentation on the west side of Africa if they were shoved together. This was probably first noted when the first reasonably accurate maps of the coasts of the continents surrounding the Atlantic were constructed. Perhaps it was here that the idea of *continental drift* was born.

The first known speculation on this subject was made by Francis Bacon in 1620. American naturalist Antonio Snider raised the possibility in 1858 that the continents may have drifted. However, it was not until shortly after the turn of the twentieth century that the concept gained any significant footing: O. Ampferer, an Austrian, and F. B.

Figure 20–3. The Atlantic Ocean floor and the continents on either side. (From *Atlantic Ocean Floor* by B. C. Heezen and Marie Tharp. Geological Society of America 1968.)

Taylor, an American, published articles in which the concept was presented. But Alfred Wegener, a German meteorologist, placed the idea before the eyes of most of the scientific world. The first statement of his ideas appeared in a technical article in 1912. His book, *The Origin of the Continents and Oceans*, published in 1915, set forth the theory of continental drift in as much detail as the state of geologic knowledge at that time permitted. Wegener held that each continent moved independently by plowing through oceanic crustal material and that drift was caused by drag produced by the rotation of the earth and by the gravitational attraction of the sun and moon.

Wegener's treatment and presentation of evidence were sufficient to convince a large number of geologists, principally in South America, southern Africa, Australia, and India. Most European and American geologists simply could not accept the theory without more concrete evidence, especially because there was so little at that time to indicate that any part of Europe and Africa was ever connected to North America. So Wegener's theory remained in the shadows of geologic thought in the Northern Hemisphere for many years.

Why did the geologists in South America, southern Africa, Australia, and India accept the idea so readily? The concept set forth by Wegener and others tied together things that they had known for some time. The *geographic fit* of South America and Africa is the most obvious piece of evidence. Geologists working in Brazil, South Africa, India, and Australia found Late Paleozoic *glacial deposits* and *pavements* (rock surfaces polished and striated by glacial movement) in each of these countries. Initially this would not seem too unusual until the distances that separate them today are considered. Also, India is north of the equator, the others south of it. Striations and grooves in the glacial pavements do not converge to a single ice nucleus or even to several in their present positions. However, if the southern continents and India are reassembled in a particular way into a supercontinent, the striations and grooves converge to a common center from which the ice sheet would have spread. Late Paleozoic glacial deposits have since been found in Antarctica, the Falkland Islands, and Malagasy (formerly Madagascar), a large island off the east coast of Africa.

Higher in the same sequence of rocks that contains the glacial deposits is an assemblage of fossil plants called the *Glossopteris* flora. The present distribution of this plant assemblage and the rocks that contain them are difficult to understand in terms of ancient climates and the great similarities of all the plants. Reassembling the southern continents in the same fashion as with the glacial striations provides a means by which similarities in the group of plants may be reconciled without the necessity of having them migrate over great distances to produce an almost identical assemblage on another continent. This supercontinent was called Gondwanaland by Wegener and was composed of South America, Africa, India, Australia, and Antarctica (Figure 20–4).

Fossil reptiles found in Late Paleozoic and Mesozoic rocks were judged by paleontologists who reconstructed their muscular arrangements to be unable to swim the great distances required for their present-day distribution. Shoving the southern continents and India back together seems to be a better solution here, too. However, this bit of evidence is more subject to question than some of that mentioned above.

Most Southern Hemisphere geologists and some Europeans were willing to accept the theory of continental drift with the evidence presented here, along with some additional conclusions and supporting data presented by Wegener. Yet it remained an idea of ill repute in North America and much of Europe until the late 1950s and 1960s, when other data began to accumulate as a result of new techniques and study of areas not pre-

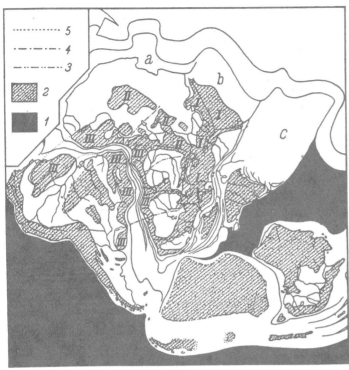

Figure 20-4. The supercontinent Gondwanaland. (From Alfred Wegener, *The Origin of the Continents and Oceans*, New York: (Dover Publications, 1929.)

viously accessible until recent years, namely, the ocean floors.

BIRTH AND REBIRTH OF A THEORY

Paleomagnetism

The revival of the older idea of continental drift and the beginnings of our present concept of plate tectonics came in the late 1950s, when British geophysicists P. M. S. Blackett, S. K. Runcorn, and Sir Edward Bullard discovered that certain minerals in rocks had locked-in the orientation of the earth's magnetic field at the time they had either crystallized or had been deposited. They collected oriented specimens of rock material whose ages were known and placed them, one at a time, into an instrument that would nullify the present magnetic field of the earth and measure the ancient field orientation.

This measurement of the ancient magnetic field is a measurement of *remanent magnetism*. Two types of remanent magnetism include *thermoremanent magnetism* and *depositional remanent magnetism*. Thermoremanent magnetism refers to the locking-in of the earth's magnetic field whenever an igneous body is crystallized and the temperature of the minerals that will interact strongly with a magnetic field falls below the *Curie temperature* for these minerals. Curie temperature is the temperature above which a material will no longer interact with a magnetic field; it is 578°C for magnetite, 670°C for hematite, 315°C for the iron sulfide pyrrhotite (FeS), and 770°C for iron. So, once the temperature of the cooling igneous body drops below this Curie temperature, the mineral aligns itself in the direction of the earth's magnetic field at that time.

When sediment containing magnetite or some other magnetically susceptible mineral is deposited, grains attempt to align themselves in the direction of the magnetic field of the earth as they settle through quiet water. Measurement of the remanent magnetism in specimens of this material may be made. This is depositional remanent magnetism, but measurements are not different from those made in rocks possessing thermoremanent magnetism.

Measurement of remanent magnetism in oriented igneous rock specimens enabled Blackett and others to determine the position of the earth's magnetic poles at different periods in the past. They found that pole positions were not the same during past times and that there was systematic migration of ancient magnetic poles toward their present positions. This migration of poles through time is called *polar wandering*. But determinations made in Europe and North America were offset by a particular amount

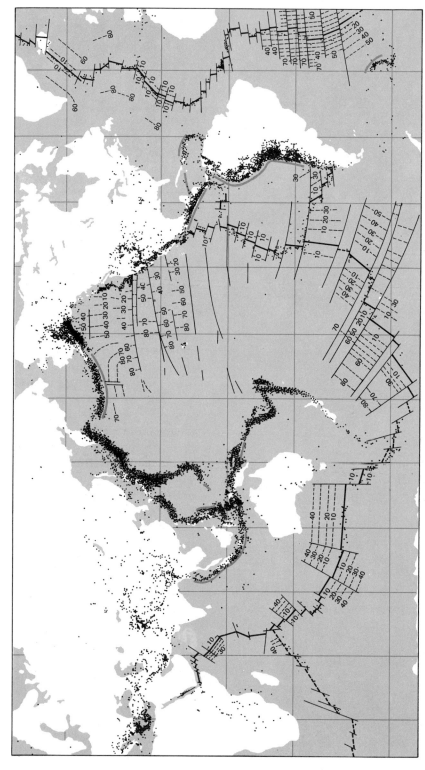

Figure 20–5. Distribution of the oceanic ridges, rifts and major occurrences of earthquakes. (From "The Origin of the Oceans," by Sir Edward Bullard. Copyright © 1969 by Scientific American, Inc. All rights reserved.)

that increases with increasing ages of the rocks. This difference disappears if the continents are shoved back together.

Study of the distribution of ancient climatic indicators, such as glacial deposits, tropical plant fossils, and bauxite deposits (which form in tropical regions), has suggested that the equator shifted its position in time. Wegener recognized that paleoclimatic data could be used as evidence for polar wandering and for continental drift.

Sea floor spreading

A second important set of data came from the ocean basins in the late 1950s. B. C. Heezen and Maurice Ewing, two American oceanographers who had been determining and studying ocean floor topography, concluded that oceanic ridges are all connected in a worldwide system (Figure 20–5). No reason for this was put forth initially, but in 1960 H. H. Hess suggested that the ridges represent areas where the sea floor is pulled apart and new crust is being formed as part of a worldwide rift system. The rift valley of Africa, the Red Sea, and the Gulf of California are features that are products of this system. Thus the idea of *sea floor spreading* was born.

Acceptance of the idea of sea floor spreading was not immediate. In the early 1960s several geophysicists discovered that in the geologic past the polarity of the earth's magnetic field had reversed itself. Although it could not be explained, reversal was found to have taken place 9 times in the last 4 million years and many more times from the Mesozoic until the present (Figure 20–6). In 1963 the magnetic pattern over a part of the Mid-Atlantic Ridge south of Iceland was published (Figure 20–7). The pattern consists of a series of parallel bands in which one stripe representing addition to the present magnetic field (called a *positive anomaly*) lay alongside one which subtracted from it (a *negative anomaly*). It had been discovered earlier that a positive anom-

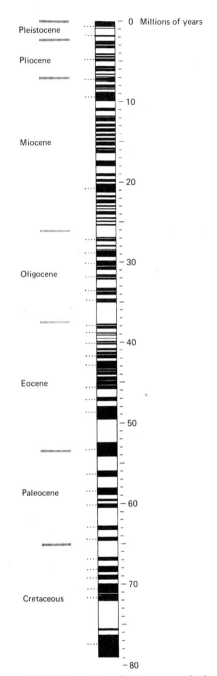

Figure 20–6. Magnetic polarity reversals from the Cretaceous Period to the present. This scale assumes a constant spreading rate in each ocean. Note that there is no systematic pattern of reversals. The dark areas indicate "normal" polarity, the light reversed. (From J. R. Heirtzler et al. *Journal of Geophysical Research* vol. 73, 1968.)

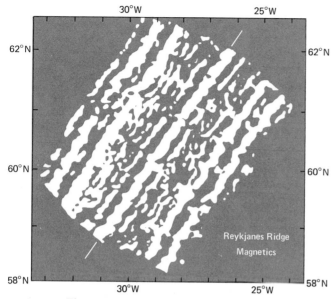

Figure 20-7. Magnetic stripe anomaly pattern along part of the Mid-Atlantic Ridge (Reykjanes Ridge) south of Iceland. (From J. R. Heirtzler et al. 1966, *Deep Sea Research*).

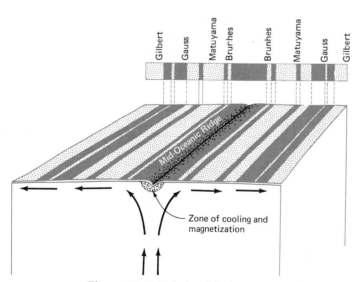

Figure 20-8. Relationship between sea-floor spreading the the development of the magnetic stripe pattern. (From Cox et al., "Research of the Earth's Magnetic Field." Copyright © 1967 by *Scientific American*, Inc. All rights reserved.)

aly exists on the ridge crest. It was then suggested that these stripes represent bands of normal and reversed polarity. F. J. Vine and D. H. Matthews interpreted this and similar magnetic stripe patterns observed elsewhere in the ocean floors as related to sea floor spreading. They concluded that, as material is added to the crust along a ridge, new material will lock-in magnetic polarity at the time it crystallizes. New material then slowly spreads away from the crest of the ridge, carrying its magnetic polarity with it. A reversal of the earth's polarity may occur later, and crustal material forming from that time until another reversal occurs will solidify with the polarity opposite to that which formed earlier (Figure 20-8). This, too, will spread from both sides of the ridge, and two parallel magnetic stripes will be produced. Subsequent reversals produce additional stripes. It was predicted that rocks on the ridges would yield younger radiometric ages than those away from the ridge crest. This was found to be true in general. But there are exceptions to this, so it is not so simple a relationship as might be expected.

The magnetic stripe pattern and the interpretation by Vine and Matthews opened the door to considerably greater acceptance of sea floor spreading and continental drift. Wegener's idea that continents plowed through a stationary ocean floor was one reason that continental drift had not gained wider acceptance. He had no evidence that continents had plowed through the ocean floor; yet there is considerable evidence that they would not be able to do so. The concept of sea floor spreading largely removed this objection, but as yet there was not a complete integration of the two concepts.

Subduction zones

Intermediate- and deep-focus earthquakes, associated with oceanic trenches, have been known and have been studied for many years. Also, land masses next to trenches have long been known as areas of intense earthquake activity (Figure

20–5). Hugo Benioff in his research on these earthquakes noted that they occur along zones inclined into the mantle and are always inclined toward the fringe of the ocean basin. This usually makes them dip beneath the edge of a continent, such as South America, or an island arc, such as Japan or the Aleutians (Figure 20–9). These inclined seismic zones associated with trenches are known as *Benioff zones*. R. S. Deitz proposed that these zones are areas in which oceanic crust is subsiding into the mantle and being reabsorbed at the same time that it is created and is spreading from oceanic ridges. Zones of resorption of oceanic crust were later named *subduction zones*. Major volcanic activity is also concentrated along or near ridges and subduction zones.

Other evidence

During the middle to late 1960s additional information was brought to light. P. M. Hurley and his colleagues determined the ages of some rocks from Brazil and others from western Africa using the rubidium–strontium and potassium–argon methods (see Chapter 4). They located a boundary separating rocks of about 2000 million years old from rocks about 600 million years old. This boundary exists in both Brazil and Africa and may be aligned if the continents are pushed back together.

G. O. Allard, at about the same time as Hurley, traced the roots of an ancient east–west-trending mountain system, consisting of a series of folds, faults, and particular rock types to the eastern edge of Brazil. He then went to Gabon in western Africa and found the continuation of this system (Figure 20–10). Both these latter pieces of evidence strongly suggest that Africa and South America had been joined and later drifted apart.

The fit of the continents across the Atlantic is something that appears obvious from a map. But sea level has not always been where it is today. In the mid-1960s British geophysicists Sir Edward Bullard, J. E. Everett, and A. G. Smith used a computer to obtain the best fit of the continents on either side of the Atlantic, based on the topography at a depth of 500 fathoms (3000 ft, 1000 m) along the edges of the continents. Their results are remarkable (Figure 20–11).

Oceanic ridges are features extending for many thousands of miles (kilometers). As spreading centers where new crustal material is created, the rate of spreading will doubtless vary along different segments of the ridge. Rates of spreading have been calculated along major ridges from determining the ages of magnetic stripes on either side; they vary considerably. Spreading rates vary from about 1–3 in. (2–7 cm) per year on the East Pacific Ridge, whereas on the Mid-Atlantic Ridge rates vary from less than .5 in. (1 cm) per year to greater than 1 in. (2.5 cm) per year. The Mid-Indian Ridge is spreading at rates from .5–1.25 in. (1–3 cm) per year.

Transform faults

How are different spreading rates compensated along a ridge? On a map of one of the ridge systems, fractures running across ridges and offsetting them are almost as prominent as ridges themselves (Figure 20–3). These fractures were at first thought to be strike-slip faults (see Chapter 18). J. Tuzo Wilson first proposed that these fractures are not strike-slip faults but a previously unrecognized type of fault produced by different spreading rates and called them *transform faults* (Figure 20–12). Transforms may compensate for different rates of movement between segments of ridges, ridges and island arcs, two island arcs, or other features. Seismic studies have shown that movement on transforms is confined to the connecting interval between the two ridge segments or whatever is separated by the fault. Controversy has arisen over whether the San Andreas Fault is a transform fault that connects one end of the East Pacific Ridge, where the ridge disappears beneath North America into the Gulf of California, and

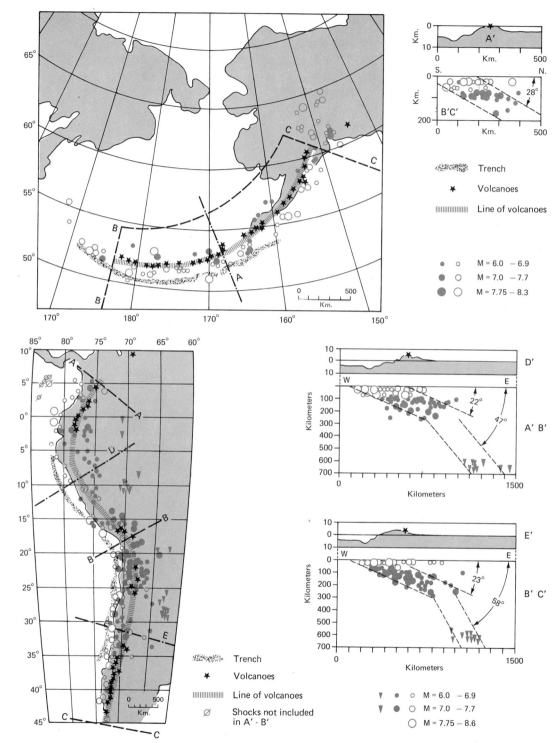

Figure 20–9. Cross-section through part of the crust and mantle showing the distribution of earthquake foci along a Benioff zone. *A.* Section through the Aleutian Islands. *B.* Sections through the Andes of South America. M = earthquake magnitude. (From Hugo Benioff, *Geological Society of America Bulletin*, vol. 65, 1954.)

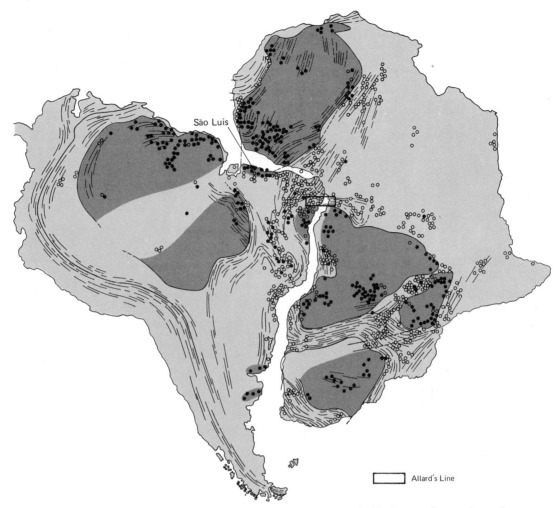

Figure 20–10. Map showing how Africa and South America probably fit together with Hurley's age boundaries and Allard's foldbelt as the basis for the fit. The solid dots represent rocks which are 2000 million years old and older. The open circles are younger rocks. The lines are trends in foldbelts. (After P. M. Hurley, "The Confirmation of Continental Drift." Copyright © 1968 by *Scientific American*, and G. O. Allard and V. J. Hurst, *Science*, vol. 163, 1969.)

another ridge segment that appears off the coast of Oregon.

Conclusion

So by the middle to late 1960s several separate sets of ideas were waiting to be united into a single concept. These were: (1) polar wandering, (2) knowledge of ocean floor topography, (3) sea floor spreading and magnetic reversals, (4) Benioff zones, (5) transform faults, and (6) a better fit of the continents around the Atlantic with features that could be traced from South America to Africa. Certainly the idea of continental drift had been established, but data from ocean basins implied that there is something more than continents involved, that continents may really be no more than passive spectators and scorekeepers in the evolution of the lithosphere.

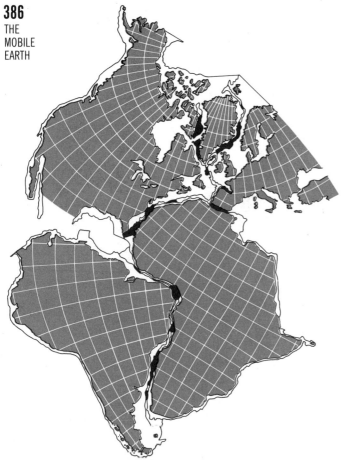

Figure 20–11. Reassembling the continents on either side of the Atlantic along the 500-fathom depth contour (Sir Edward Bullard et al., *Royal Society of London Philosophical Transactions*, 1965.) The dark areas are areas of overlap.

PLATE TECTONICS THEORY

The separate ideas and data that had been brought forth over the past several years were pulled together into a single concept in 1968 by American geophysicists Bryan Isacks, Jack Oliver, and L. R. Sykes. They divided the lithosphere into seven major slabs or "plates," each of which is bordered by oceanic ridges, subduction zones, and/or transform faults (Figure 20–13). They reasoned that plates are being generated along ridge systems and destroyed in subduction zones, as oceanic crustal material subsides back into the mantle and is reabsorbed. This is the essence of *plate tectonics theory*. Its importance is considerable, for it unifies our thinking about the evolution of continents and ocean basins and of their contained features. It has caused "a revolution in the earth sciences."

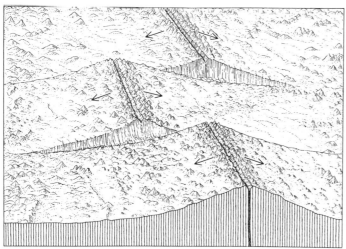

A

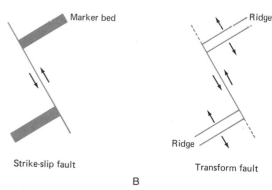

Figure 20–12. *A.* Ridge-ridge transform faults. Movement occurs to compensate for differences in spreading rates and movement in different parts of a plate. (From H. W. Menard, "The Deep-Ocean Floor." Copyright © 1969 by Scientific American, Inc. All rights reserved.) *B.* Differences in sense of movement of a transform and a strike-slip fault.

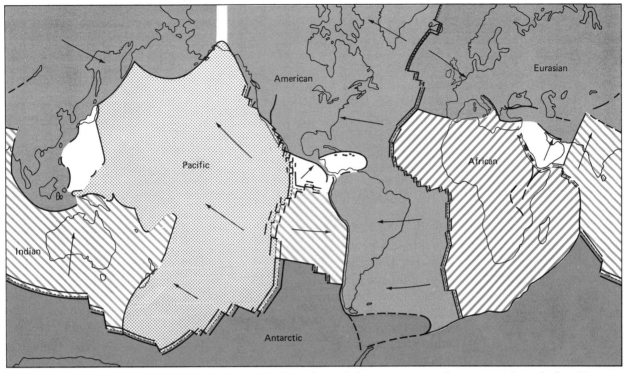

Figure 20–13. Major plates bounded by either subduction zones, ridges or transform faults. (From Sir Edward Bullard, "The Origin of the Oceans." Copyright © 1969 by *Scientific American*, Inc. All rights reserved.)

Role of the continents

Continents are simply passengers on plates and drift about in accordance with the movement directions of the plates on which they reside at a particular time. A continent may move in some direction for millions or tens of millions of years; then the plate system of which it is a part may stop moving for some reason and then move in another direction as a new plate system is formed.

Continents are passengers because they effectively float, (they are lighter) on the denser lower crust or mantle below (Figure 20–14). Continental crust has an average density of 2.6; that of oceanic crust is 3.0.

A paradox that is solved by this concept

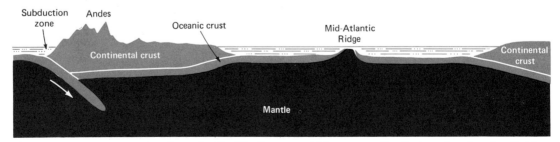

Figure 20–14. Cross-section from the Pacific Ocean to Africa across South America showing lighter continental crust floating on denser oceanic crust and mantle material.

is the age discrepancy between rocks of continents and ocean basins. Whereas it was first thought that ocean basins contained the oldest rocks on earth, it was discovered, when sufficient rocks had been dated, that the oldest rocks in ocean basins are about 200 million years old. Yet the oldest rocks on continents are about 3.7 billion years old. The paradox is eliminated when continents are thought to have been floating for eons on a substratum of transient oceanic crust that is generated, exists for a time, and is then consumed by the process of subduction.

Possible mechanisms

Any process existing on a global scale must have something that makes it operate, just as we need gasoline to make a car's engine perform. Wegener used the earth's rotation and the gravitational drag of the sun and moon to motivate the process of continental drift. Another suggestion is that lithospheric plates are effectively pulled away from spreading centers by the weight of the cooler mass sliding into a subduction zone. The process is operated by gravity. This theory has not gained widespread support.

The mechanism that receives the greatest support is mantle convection. We are reasonably certain that the earth becomes hotter from the outer crust toward the core. If you place a beaker of water over a heat source, convection currents will form and flow upward from hotter to cooler areas and return downward to be heated again (Figure 20–15).

Proponents of the convection mechanism maintain that slowly moving convection currents exist over part of the entire thickness of the mantle and that drag effects along the base of the lithosphere cause the plates to be moved (Figure 20–16).

Measurements of the rate of heat loss from the earth's interior indicate that oceanic ridges are in fact zones of abnormally high heat flow, whereas trenches are zones of below average heat flow. In terms of a convection mechanism, ridges are areas of upwelling in the convection cell, and subduction zones are places where the cell has cooled and turns down into the mantle again.

Advocates of mantle convection have coupled this with effects of the earth's rotation so that convection cells may be oriented by rotation. Spreading centers are oriented dominantly north–south; most plate motion is east–west. Actual spreading can be related to a spreading axis, which is located near the earth's rotation axis (Figure 20–17).

Contrary criticism and evidence

Any theory that is new and is part of a dynamic science must carry and endure the yoke of criticism toward either refinement or eventual demise. Such is the case with plate tectonics. Since the late 1960s, when the concept was first brought forth, it has been subjected to a barrage of criticism, much of it reasonable and justified. But some is easily answered and put aside. We will examine here those objections to the theory that remain significant.

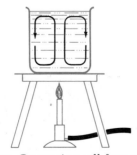

Figure 20–15. Convection cell formed in a beaker of water heated over a Bunsen burner.

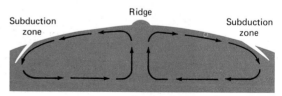

Figure 20–16. Convection in the mantle influencing and directing plate motion.

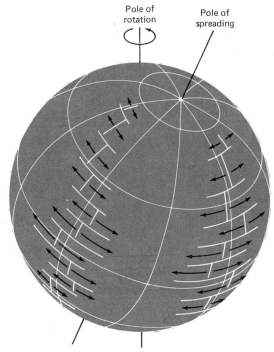

Pole of rotation

Pole of spreading

Figure 20–17. Relationships between the earth's rotational axis and the present spreading axis. (After J. R. Heirtzler, "Sea Floor Spreading." Copyright © 1968 by *Scientific American*. All rights reserved.)

Some of the early evidence cited by Wegener, the *Glossopteris* flora, and the fossil marine reptiles have been subjected to question as genuine evidence favoring continental drift. *Glossopteris* fossils also occur in the central Soviet Union. This detracts considerably from the idea that the southern continents needed to be in contact for this assemblage to develop. The ability of mesosaurs (marine dinosaurs) to disperse themselves by swimming is simply not known. That they were not able to do so is a conclusion drawn from anatomical reconstructions of their muscular structure from skeletal remains. Moreover, even many of the same kinds of land dinosaurs became dispersed on all the continents during the Mesozoic Era.

Several geologists claim that the magnetic stripe pattern parallel to the oceanic ridges may be explained in several ways other than being related to polarity reversals and sea floor spreading. One suggests that the stripes consist of a deep crustal pattern that has existed since the Precambrian and is expressed through younger rocks near the surface. Another believes that the pattern is related to different rock types having different magnetic properties dipping away from ridge crests. If either of these alternative interpretations is correct, the possibility that the sea floor has spread apart is diminished.

It has also been pointed out that the simple pattern of younger rocks on ridges to older rocks in adjacent ocean basins may be oversimplified. There are older rocks on ocean floors. This may indicate that either sea floor spreading did not take place or that the geology of ocean floors is more complex than has been assumed.

Most oceanic trenches contain sizable thicknesses, as much as several miles (kilometers), of sediment. Yet in several this sediment remains undeformed. Trenches are supposed to be zones of compression, and sediments should be deformed unless the sea floor in those regions has not moved since sediment was deposited. Geologist W. F. Tanner uses this and several other lines of evidence to conclude that trenches are really tensional rather than compressional features. If trench areas are tensional features, a major part of plate tectonics theory must be revised considerably.

The occurrence of igneous activity and active faults around plate margins, near subduction zones, and near spreading centers is easily explained in terms of plate tectonics theory. However, igneous bodies and active faults occur within continental areas that either are not part of a plate boundary situation at present or were not at the time that the features in question were formed (Figure 20–18). The active New Madrid Fault in southeastern Missouri produced the largest earthquake experienced in historical times

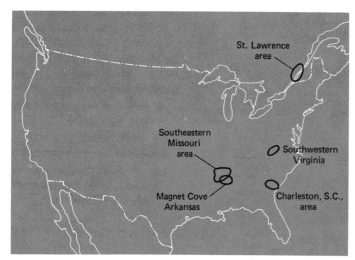

Figure 20–18. Locations of igneous activity and active faults in "stable areas" in North America.

in North America. It is located in what is commonly considered a "stable" area. The active Bowman Fault in South Carolina, which produced the large and destructive Charleston earthquake of 1886 (see Chapter 19), is likewise in another "stable" area. Igneous plutons in the middle of the continent are also out of place in the plate tectonics scheme. All of these things might be thought of as occurring as readjustments within a plate or taken as exceptions to the theory.

The active San Andreas Fault offers no difficulty in placing it near a plate boundary. However, there is considerable argument concerning its actual role in the plate tectonics regime of western North America and the eastern Pacific. As mentioned earlier, some geologists studying the sea floor off California maintain that the San Andreas is a transform fault. The same geologists can account for the total amount of movement on the San Andreas in a few millions of years. Yet others can demonstrate a classical strike-slip movement history spanning at least 100 million years and maintain that it is not a transform fault.

Refitting of the continents around the Atlantic has led to some questioning. Because the fit is based on the present shapes of the continents at a depth of 500 fathoms (3000 ft, 1000 m), this must be in error: The continents on either side of the Atlantic were separated from 180–65 million years ago. The continental edges have become covered by hundreds and thousands of feet (meters) of sediment derived from continental erosion. James Gilluly maintains that the computer fit was just a coincidence, and, if a fit were made on the old continental edge, it would not be as good.

A serious difficulty arises in attempting to explain the assemblage of old cratons in the great shield areas of the continents. The stable portions of shields are separated by orogenic belts that appear to have developed without large-scale plate tectonics movements. In fact, most of these old circular or elliptical nuclei are surrounded on all sides by orogenic belts (see Figure 20–10). How can plate tectonics processes, involving essentially straight-line movement, produce multiple orogenic belts encircling these cratonal areas? Plate tectonics processes may not have been operative throughout the 4–5 billion years of earth history, perhaps because only about 2 billion years ago were continents assembled into masses large enough for plate tectonics processes to affect them or to allow plate tectonics to become operative.

Evidence favoring plate tectonics

The theory of plate tectonics appears to explain many things about continents and oceans that we could not adequately explain before. Some evidence is questionable for one reason or another. But other evidence favoring continental drift and plate tectonics is very difficult to refute or to explain otherwise.

1. age boundaries that have been correlated from Africa to South America
2. truncated Precambrian mountain system (fold belt) in Brazil whose continuation has been found in Africa
3. Late Paleozoic glacial events that left deposits in widely separated places which today are both north and south of the equator
4. fit of the continents, particularly Africa and South America, on either side of the Atlantic
5. determination of ancient magnetic pole positions on different continents, indicating different paths of polar wandering, which may be reconciled by refitting continents
6. data on the greater amount of heat flow from the interior on ridge areas and lower rates of heat flow in the trenches
7. distribution of earthquake foci at depth along Benioff zones

Conclusion

Whereas other basic sciences have for many years had central theories around which to build their ideas, plate tectonics promises to be such a theory for geology. As a new theory it will undergo modification and refinement before it becomes as well accepted in geology as atomic theory and evolution are in other sciences.

Plate tectonics has already given us several practical dividends. It has provided a framework within which research toward a solution of a major environmental problem can be undertaken—that of understanding and predicting earthquakes. Also, exploration for petroleum along the edges of continents has already been assisted by assuming that continental drift has occurred. Yet in petroleum exploration it has also limited prospective areas. The match of Africa and South America by Allard has enabled an African nation to discover oil in rocks similar to those that have previously produced oil in Brazil.

SUMMARY

The idea of drifting continents had its origin in the nineteenth century, although it was not until the early part of the twentieth century and Alfred Wegener that most of the scientific world learned of it. There was little initial acceptance of the idea in North America and Europe, but the location of the best evidence in South America, southern Africa, India, and Australia led to better acceptance there. Late Paleozoic glacial deposits, *Glossopteris* flora, and the best fit of the continents were there.

Exploration of the deep-ocean floor, discovery of the worldwide ridge system, and study of paleomagnetism led to a rebirth of the continental drift theory and greater acceptance in Europe and North America. The idea of sea floor spreading was proposed and aided by magnetic polarity reversal patterns on sea floors. Study of the distribution and localization of large earthquakes near oceanic trenches provided the idea that trenches may be sites where oceanic crustal material is being reabsorbed into the mantle. The theory of continental drift achieved acceptance in most places with the discoveries of Hurley and Allard of age boundaries and of an ancient mountain system that can be correlated from Brazil to Africa.

The discovery of transform faults accounted for different rates of spreading along ridges. This information enabled a complete plate tectonics theory to be formulated. The lithosphere was divided into seven major plates and several minor ones. These plates are bounded by trenches (subduction zones), ridges (spreading centers), or transform faults. The continents are passive passengers on the plates. Rocks much older on continents than in oceans indicate that continents have survived for billions of years because of their lesser densities, whereas oceanic material has been continuously created and destroyed. Plate tectonics processes may be driven by convection currents in the mantle. Convection cells may be oriented by the earth's rotation.

Some of the evidence favoring plate tectonics and continental drift is questionable. *Glossopteris* fossils occur in Late Paleozoic rocks outside the area of former Gondwanaland, and it is not known to what degree certain Mesozoic marine reptiles could disperse themselves across ocean barriers. The magnetic stripe pattern may not record polarity reversals, and the simple pattern of younger rocks occurring on the ridges with older rocks toward the flanks is not so simple. Sediment in several supposed active trenches is not deformed, as it should be if trenches were active.

Correlation of age boundaries and a fold-belt from Brazil to Africa, Late Paleozoic glaciation, fit of the continents across the Atlantic, and polar wandering are all pieces of data that are difficult to refute.

Questions for thought and review

1. Why were Wegener's ideas not accepted by most European and North American geologists?
2. What major evidence did Wegener use to support his continental drift theory?
3. List additional evidence that led to for-mulation of the plate tectonics theory.
4. How does a transform fault differ from a strike-slip fault?
5. What is remanent magnetism? How does it originate?
6. What is sea floor spreading?
7. What is a subduction zone?
8. Why is continental crust not reabsorbed by the process of subduction?
9. How is one plate defined and set apart from other plates?
10. What is the role of continents in plate tectonics?
11. Explain the age discrepancy between rocks of continents and oceans.
12. What appears to be the most likely driving mechanism for plate tectonics?
13. List evidence contrary to the theory of plate tectonics.
14. Give three lines of evidence, difficult to refute, that favor continental drift.

Selected readings

Beloussov, V. V., 1968, An Open Letter to J. Tuzo Wilson, *Geotimes*, vol. 13, pp. 17–19.

Bullard, Edward, 1969, The Origin of the Oceans, *Scientific American*, vol. 221, no. 3, pp. 30–45.

Cox, Allan, Dalrymple, G. B., and Doell, R. R., 1967, Reversals of the Earth's Magnetic Field, *Scientific American*, vol. 216, no. 2, pp. 44–54.

Dewey, J. F., 1972, Plate Tectonics, *Scientific American*, vol. 226, no. 5, pp. 56–68.

Dietz, R. S., 1972, Geosynclines, Mountains, and Continent Building, *Scientific American*, vol. 226, no. 3, pp. 30–45.

Dietz, R. S., and Holden, J. C., 1970, The Breakup of Pangaea, *Scientific American*, vol. 223, no. 4, pp. 30–41.

Hallam, A., 1975, Alfred Wegener and the Hypothesis of Continental Drift, *Scientific American*, vol. 232, no. 2, pp. 88–97.

Heirtzler, J. R., 1968, Sea-Floor Spreading, *Scientific American*, vol. 219, no. 6, pp. 60–70.

Hsu, K. J., 1972, When the Mediterranean Dried Up, *Scientific American*, vol. 227, no. 6, pp. 26–36.

Hurley, P. M., 1968, The Confirmation of Continental Drift, *Scientific American*, vol. 218, no. 4, pp. 52–68.

Matthews, S. W., 1973, This Changing Earth, *National Geographic*, vol. 143, no. 1, pp. 1–37.

McKenzie, D. P., and Sclater, J. G., 1973, Evolution of the Indian Ocean, *Scientific American*, vol. 228, no. 5, pp. 62–74.

Menard, H. W., 1955, Fractures of the Pacific Floor, *Scientific American*, vol. 193, no. 1, pp. 36–41.

Menard, H. W., 1969, The Deep-Ocean Floor, *Scientific American*, vol. 221, no. 3, pp. 126–145.

Press, Frank, and Siever, Raymond, 1974, *Planet Earth*, Freeman, San Francisco.

Raff, A. D., 1961, The Magnetism of the Ocean Floor, *Scientific American*, vol. 205, no. 4, pp. 146–156.

Tazieff, Haroun, 1970, The Afar Triangle, *Scientific American*, vol. 222, no. 2, pp. 32–40.

Wilson, J. Tuzo, 1963, Continental Drift, *Scientific American*, vol. 208, no. 4, pp. 86–100.

———, 1968, A Revolution in Earth Science, *Geotimes*, vol. 13, no. 10, pp. 10–16.

———, 1968, Reply to V. V. Beloussov, *Geotimes*, vol. 13, no. 10, pp. 20–22.

21 Mountains and the growth of continents

STUDENT OBJECTIVES

At the conclusion of this chapter the student should:

1. know the differences between various types of mountains

2. understand common characteristics of mountain systems

3. be aware of modern ideas on the origin of mountain systems and the tectonic cycle, and their relations to the growth of continents

4. have explored the history of mountain chains after their compressional history has ended

5. understand the theory of isostasy and its bearing on the elevation of mountains and continents

May we not be looking into the
womb of Nature, and not her grave?

Lyell

Figure 21-1. Photograph of part of the Selkirk Mountains, British Columbia. These mountains are the result of folding, intrusive activity, vertical uplift, stream and glacial erosion.

The word *mountain* elicits different pictures in the minds of different people. Some think of lofty peaks devoid of vegetation and capped with snow; others might think of high rounded hills covered with trees. To still others a mountain may be a volcanic cone. They are all correct. These are all mountains, but each has had a different history since it was formed and perhaps a different mode of formation.

A mountain is a topographic feature that stands at least 1000 ft (300 km) in relief above its base. Mountains may be formed by volcanic activity (see Chapter 7), by compression (producing fold belt mountains; Figure 21-1), by tension (producing block fault mountains), and by vertical uplift (producing erosional mountains). Several types of mountains may be found in the same area. For example, block fault mountains are found in part of the southwestern United States, but the tensional forces that produced large normal faults in block fault ranges were formed millions of years after the compressional mountains developed (Figure 21-2). Also, erosion of a high plateau may leave mountainous erosional remnants of more resistant rock.

Compressional mountains occur in the elongate chains that are found in many parts of the world. These *fold belts, orogenic belts,* or *orogens* are mountain systems produced by compressional folding, by faulting, and by metamorphism of a complex assemblage of sedimentary and volcanic rocks. This section will deal principally with fold belts, their origin, and how they relate to the growth of continents.

CHARACTERISTICS OF FOLD BELTS

The great mountain chains of the world (Figure 21–3), the Andes, the Alps, the Himalayas, the North American Cordillera, and the Appalachians, are each somewhat unique. But they all have certain features in common (Figure 21–4).

Mountain systems are generally formed near the edge of a continent. The most notable exception to this is the Urals in the Soviet Union. However, other great mountain chains are all located near the edge of the continent on which they reside.

In general a belt of thrust faults and folds dies out toward the continent on the side of the chain closest to the continental interior (Figure 21–5). Examples of these *thrust belts* include the Canadian Rockies and their extension into the United States, the Jura Mountains of France and Switzerland, and

Figure 21–2. Part of the Basin and Range Province near Las Vegas, Nevada. The present mountains were produced by Cenozoic block faulting. At the edge of each range is a large normal fault. Paleozoic folding and thrust faulting occurred in this region as well. Remnants of the thrust sheets are present in the ranges. (Courtesy of John Shelton.)

the Valley and Ridge Province of the Appalachians. The thrusts become larger and have a greater amount of horizontal offset deeper in the mountain chain. Some of these faults are responsible for as much as several tens of miles (kilometers) of movement.

Proceeding farther into the mountain chain, we find thrust faults giving way to large folds called *nappes* (Figure 21–6). Deformational style changes from brittle to plastic as rocks have been metamorphosed. Metamorphism is slight in the thrust belt and then becomes intense where large plastic folds have formed. Actually heat and pressure of metamorphism create conditions for plastic deformation. Although faults may occur here, they generally have formed after rocks have cooled considerably. The zone of most intense metamorphism, frequently intense enough to involve partial melting of

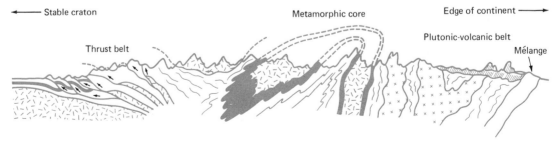

Figure 21–4. Generalized cross-section through a hypothetical mountain chain showing features common to most foldbelts.

399
MOUNTAINS
AND
THE
GROWTH
OF
CONTINENTS

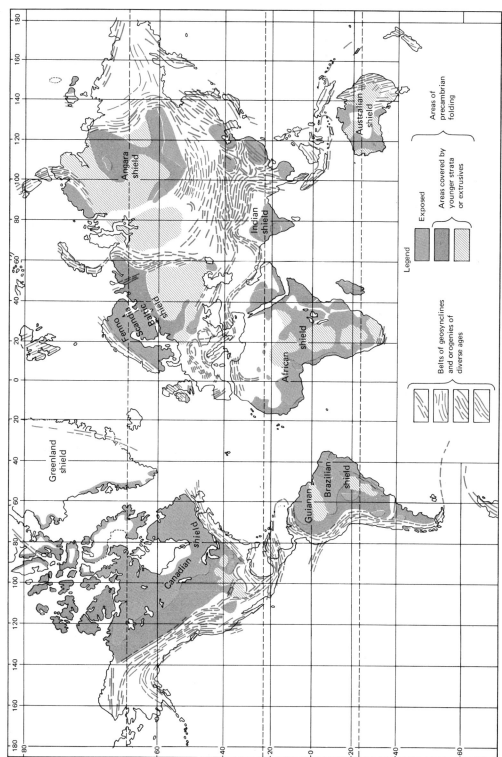

Figure 21–3. Locations of post-Precambrian mountain chains (From L. G. Weeks, 1952, American Association of Petroleum Geologists Bulletin, vol. 36.

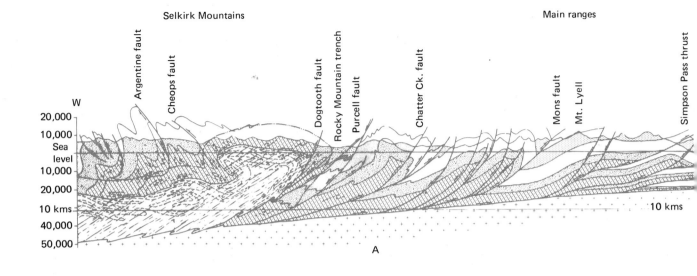

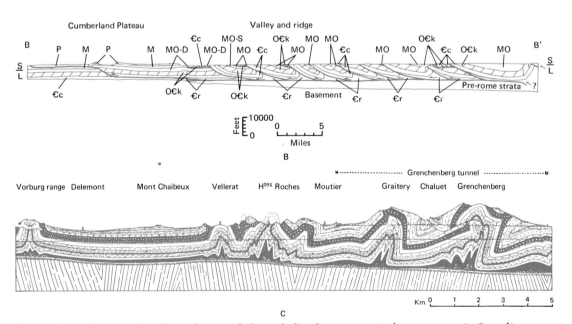

Figure 21–5. Cross-section through several thrust belts showing typical structures. *A.* Canadian Rockies. (After R. A. Price and E. W. Mountjoy, Geological Association of Canada Spec. Paper 6). *B.* Southern Appalachian Valley and Ridge. (After R. C. Milici, American Journal of Science, vol. 268, 1970, by permission.) *C.* Jura Mountains. (After Buxtorf, 1916, from L. W. Collet, *Structure of the Alps*, London: Edward Arnold Publishers. Ltd., 1927.)

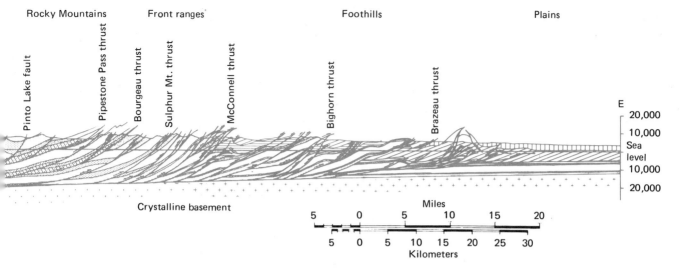

Rocky Mountains   Front ranges   Foothills   Plains

Pinto Lake fault | Pipestone Pass thrust | Bourgeau thrust | Sulphur Mt. thrust | McConnell thrust | Bighorn thrust | Brazeau thrust

E
20,000
10,000
Sea level
10,000
20,000

Crystalline basement

Miles
5   0   5   10   15   20

Kilometers
5   0   5   10   15   20   25   30

rocks, is the core of the mountain system—this is the *metamorphic core*. These areas are exemplified by the Shuswaps metamorphic complex in British Columbia in the Canadian Cordillera, the Pennine Alps, and the Inner Piedmont and part of the Blue Ridge of the southern Appalachians (Figure 21–7).

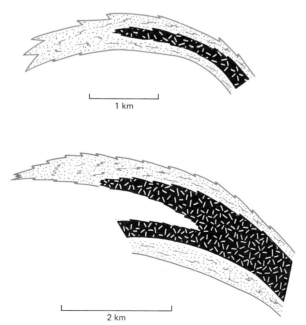

1 km

2 km

Figure 21–6. A nappe is a large fold which in some cases has been broken by a thrust.

Closer to the edge of the continent in some mountain systems is a zone in which many smaller or a few larger granitic plutons occur. Intensity of metamorphism decreases here. Such a belt exists in the Appalachians, the Andes, and the North American Cordillera but not in the Alps.

Toward the edge of the continent from the plutonic belt may be a zone of volcanic rocks produced from early volcanism or later volcanism after the major part of the orogenic belt has been built. Volcanoes in the Cascade Range in Washington and Oregon and the great expanse of volcanic rocks in the Andes were formed after most of the folding and faulting had taken place.

Just beyond the continental edge in some of the youngest mountains is an active oceanic trench. The best example of this is off the coast of South America, next to the Andes. An inactive filled trench exists off part of the coast of western North America, and trench deposits marking the location of an ancient trench can be found onshore along the coast of California. Rocks in this and other ancient trenches have been subjected to high-pressure and low-temperature metamorphism. They have also been crumpled, contorted, and otherwise deformed in such a way that the entire mass is in a rather

401

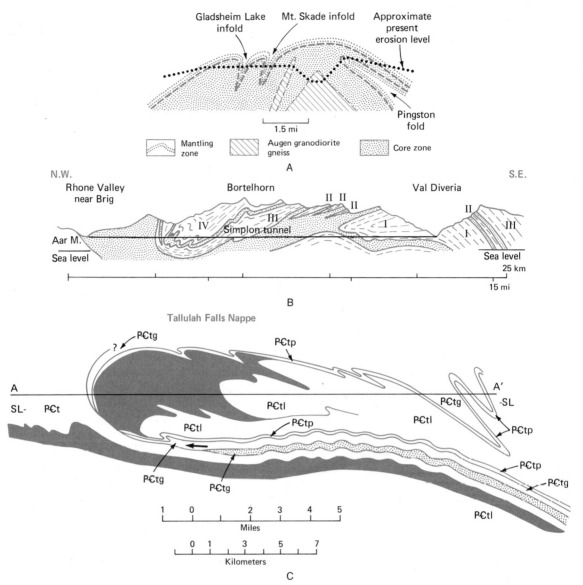

Figure 21–7. Cross-sections through portions of metamorphic core zones of several major mountain systems. *A.* Shuswaps metamorphic complex, Columbia Mountains, British Columbia. (After J. O. Reesor and J. M. Moore, *Geological Survey of Canada Bulletin*, 1965). *B.* Pennine Alps. (After Schardt, 1898, in E. B. Bailey, *Tectonic Essays*, 1935). The Roman numerals refer to different nappes. *C.* Southern Appalachian Blue Ridge. (After R. D. Hatcher, Jr., *American Journal of Science*, 1973.)

chaotic state. The term *mélange* (French, mixture) has been applied to such assemblages of rocks that are probably deformed former trench deposits. *Mélanges* have been recognized in every modern mountain system in the world (Figure 21–8).

One notable observation about mountain systems is their linear nature. All are much longer than they are wide, and thrusts and folds that are a part of fold belts are oriented parallel to the length of the chains.

So mountain systems contain (1) thrust belts along the inner, or continental, side, a complexly deformed metamorphic core, (2)

403
MOUNTAINS
AND
THE
GROWTH
OF
CONTINENTS

Figure 21-8. Disoriented, sheared, and jumbled rocks making up part of a mélange at Cache Creek, British Columbia.

mélange structures, and (3) belts of plutons. They are characteristically linear features.

ORIGIN OF MOUNTAIN SYSTEMS

Mountain systems display certain similarities, enough to suggest that they have been produced by similar processes. During the last century American geologists James Hall and J. D. Dana both recognized that, in areas that exhibit deformational and metamorphic characteristics of fold belts, there also has been a thickening of the sedimentary section, as well as a change in the nature of sediments that are found there. The term *geosyncline* was introduced to describe areas in which these extraordinary thicknesses of sediments accumulated. The *geosynclinal theory* grew out of the descriptive term and related thick accumulations of sediments to eventual destruction of the geosyncline by construction of a mountain system on that site by deformation and uplift of rocks deposited in the geosyncline.

The geosynclinal theory states that sections of the crust experience accumulations of large amounts of sediments that after a time were subjected to compression to produce folds, faults, metamorphism, and a mountain chain. The compressional mountain-building phase that terminates the history of the geosyncline is called an *orogeny*

—hence the term *orogenic belt*, a region that has undergone orogeny and that is a synonym for a fold belt or mountain system.

The concept of the geosyncline has been with us for over a century. It has been modified, and various features have been classified as geosynclines. The basic concept, as described above, was modified in the 1930s by German geologist Hans Stille. He subdivided continents into stable *cratons* and *orthogeosynclines* that flank cratons. He further divided geosynclines into two subdivisions, *miogeosyncline* and *eugeosyncline,* based on the nature of rocks found in each subdivision. The miogeosyncline contains a sequence of sedimentary rocks of the same type as those on the craton, including limestone, dolostone, shale, and sandstone; but the thicknesses of these rock units are much greater, increasing from about a mile (kilometer) on the craton to as much as 4 mi (6 km) or more in the miogeosyncline. The eugeosyncline consists of a mixture of volcanic flows, ash, and rapidly deposited sediments of slight to considerable thickness. As much as 9–10 mi (12–14 km) of volcanic and sedimentary rocks have been estimated to occur in some eugeosynclinal areas. Generally miogeosynclinal sequences are thought to accumulate on continental crust, whereas eugeosynclinal sequences may accumulate on oceanic crust and in some instances on continental crust.

The concept of the geosyncline and its eventual demise has been accepted for many years, and it has been recognized that the recurrent formation of geosynclines and orogeny through time is part of a *tectonic cycle.* But why does this process take place? What is the source of compression that produces mountain systems? Dutch geophysicist F. A. Vening-Meinesz proposed in the 1930s that an area of the earth's crust may be downbuckled and may receive an enormous thickness of sediment (Figure 21–9). As downbuckling continues, sediments on the bottom of the pile become heated and com-

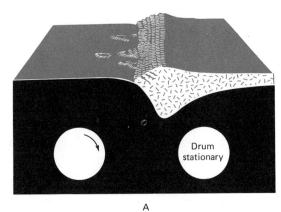

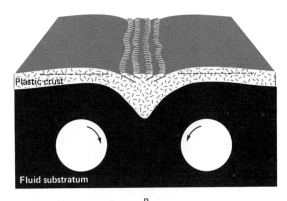

Figure 21–9. Two tectogene models driven by mantle convection. (From Griggs, *American Journal of Science,* 1938.)

pressed. Some begin to melt and expand, pushing those above them upward and outward. As the mass is squeezed tighter, rocks in the center are pushed out as well. Expansion on melting and crowding cause the mountain system to be uplifted and the thrust belt along the flanks to be formed. H. H. Hess later called this downbuckled zone a *tectogene.* F. A. Vening-Meinesz proposed that the downbuckling process was driven by convection currents in the mantle.

Gravity has been considered for many years to be important in the origin of mountain systems. Alpine geologists for many years thought that many of the great folds

405

MOUNTAINS
AND
THE
GROWTH
OF
CONTINENTS

A

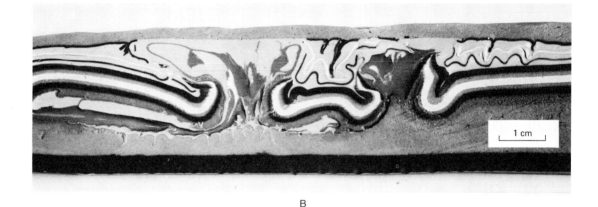

B

Figure 21–10. *A & B.* Centrifuged clay models showing structures produced after different amounts of time in a centrifuge. (From H. Ramberg, and Sojöström *Tectonophysics*, vol. 19, 1973.)

and thrusts in the Alps were the result of movement away from the uplifted core of mountain chains. Studies of centrifuged masses of colored layers of clay by Norwegian geologist Hans Ramberg have shown that structures of mountain chains can be produced by this means (Figure 21–10). The centrifugal force of the earth's rotation could conceivably produce the deep structures, and uplift could allow movement toward the flanks of the mountain chain. This movement would be motivated by gravity.

In his continental drift hypothesis Wegener proposed that mountains are formed by crumpling along the leading edge

of a drifting continent as it plows its way through the sea floor (see Chapter 20). This idea was originally scoffed at, but we see certain similarities to Wegener's proposal in our modern plate tectonics scheme.

Mountain building may be considered as a corollary to plate tectonics theory. In 1970 J. F. Dewey, a British geologist, applied plate tectonics theory to the process of mountain building. He concluded that mountains are built in two plate tectonics environments: by collision of two continents (*collisional mountains*) and by crumpling as one plate passes beneath another in a subduction zone (*cordilleran mountains*; Figures 21–11 and

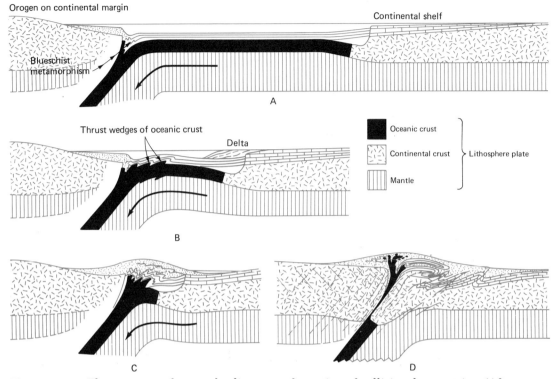

Figure 21–11. The sequence of events leading up to formation of collisional mountains. (After Dewey and Bird, *Journal of Geophysical Research*, vol. 75, 1970.)

21–12). The Himalayas are an example of a collisional mountain chain: Collision of India with Asia produced the Himalayas. The Andes are an example of a cordilleran mountain system.

Does the Dewey concept of mountain building conflict with the older geosynclinal theory? No — it probably makes the concept of the geosyncline more realistic. The miogeosyncline is a continental shelf/slope not unlike that which lies along the Atlantic coast of the United States today. The eugeosyncline is the lower continental slope or a deep-water area adjacent to a volcanic arc, such as Japan. Rock types observed in these sites are identical to those described earlier.

The early history of both the Appalachians and the North American Cordillera probably involved a stable shelf like today's

Atlantic coast. Their subsequent histories included development of island arcs seaward from the continent as trench subduction zones were formed and compression began to take place. Arcs of this type exist off the coast of Asia as the Japanese Islands, the Philippines, and others. However, this type of activity began in the Late Paleozoic in western North America. Orogenic activity continues today over part of that area, just as it began in the Early Paleozoic in the Appalachians and was finished there by the end of the Paleozoic.

We also know today that orogeny is not a one-step process. In the course of mountain building several pulses of compression may occur, while deposition is taking place and part of the chain formed before another. For example, parts of the Appalachians were folded, faulted, and even metamorphosed

407
MOUNTAINS
AND
THE
GROWTH
OF
CONTINENTS

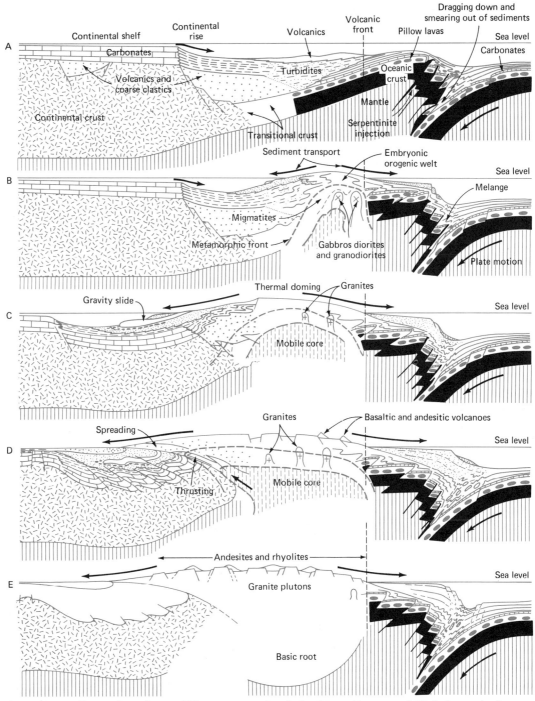

Figure 21–12. Formation of a cordilleran mountain chain. (From Dewey and Bird, *Journal of Geophysical Research*, vol. 75, 1970.)

early in the Paleozoic, while deposition was taking place elsewhere. Other parts of the chain were deformed and uplifted later.

TECTONIC CYCLES AND CONTINENTAL GROWTH

Orogenic activity in eastern North America occurred as the ancient Atlantic closed during the Paleozoic Era, when the Appalachians were formed. The present Atlantic was opened in the Mesozoic Era, and North America began its westward movement toward its present position to form the Cordillera along its western margin. It seems likely that in the future a volcanic island arc-trench-subduction zone will again develop off the Atlantic seaboard and orogenic activity will begin again in eastern North America. This will be the result of North America's being carried in another direction as a passenger on a different plate from the one on which it is riding today.

Here then is a *tectonic cycle*, which ends with collision and construction of a mountain system that is in turn eroded to produce another cycle of mountain building along the continental edge. This revised tectonic cycle was suggested by J. Tuzo Wilson in the 1960s. It has been suggested that this tectonic cycle be called the Wilson Cycle. Like other geologic cycles, this is based on the principle of equilibrium related to maintaining a balance between materials of the lithosphere and the interior and energy supplied from within and possibly from outside.

Plates probably move about in response to the way in which heat arrives at the base of the lithosphere and perhaps also because of the faster rate of rotation of the earth in the geologic past. It has been predicted that this process operated more rapidly in the past and will operate more slowly in the future, in response to a slowing of the earth's rotation, which would be attributable to the gravitational pull of the sun and the moon and also to the lesser amount of heat arriving at the surface from the interior. So presumably the entire plate tectonics process would cease to operate many millions of years from now.

As the tectonic cycle adds material to continents, the sizes of continents increase. It was known, before the concept of plate tectonics was developed, that the oldest rocks on all continents are found in shield areas and that shields seem to consist of a nucleus of very old rocks with younger rocks appearing in succession away from the old nuclei (Figure 21–13). These belts of different-aged rocks are the igneous and metamorphic cores of ancient mountain systems that were apparently added to the continents through the tectonic cycle. This process of continental growth is called *continental accretion;* it appears to have been operating since earliest geologic times. Plate tectonics has made this concept more easily explained.

Continents have thus become scorekeepers in the lengthy history of the earth. Even though they maintain a passive role in the plate tectonics scheme, they have recorded events taking place as oceanic plate systems were created, then destroyed, and ultimately formed into new ones. However, the process of plate tectonics may have been operative for only the last 2 billion years of earth history. Before that, other processes of continental growth and welding together of smaller continental blocks and vertical movement may have taken place.

POST-OROGENIC HISTORY OF MOUNTAIN SYSTEMS

Once a mountain system is built and compressional orogenic processes have ceased to operate, there exists a mass of crustal material that stands high above the remainder of the continent. As soon as it is raised above base level, the mountain system is subjected to processes of erosion, processes that attempt to level the mountain system. Yet a mountain chain may remain

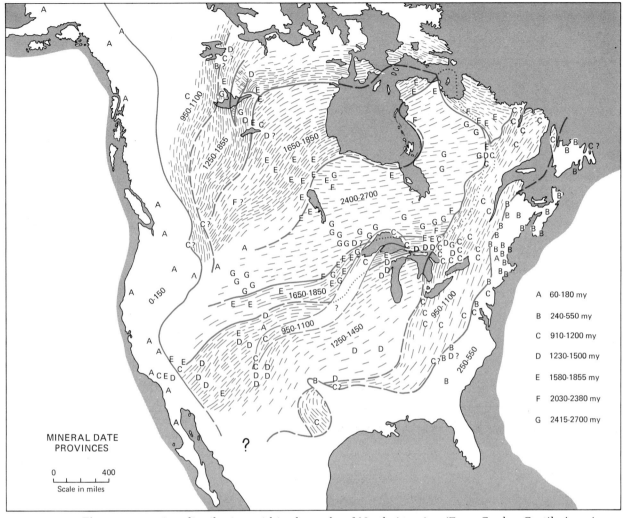

Figure 21–13. Age distribution within the rocks of North America. (From Gordon Gastil, *American Journal of Science*, vol. 258, 1960.)

A	60-180 my
B	240-550 my
C	910-1200 my
D	1230-1500 my
E	1580-1855 my
F	2030-2380 my
G	2415-2700 my

MINERAL DATE PROVINCES

0   400
Scale in miles

for millions of years and still not be completely eroded to the level of the surrounding continent. The Appalachians still remain over a mile (kilometer) above sea level over large areas. Orogenic forces ceased to operate there about 200 million years ago. What enables a mountain system to remain high in the face of powerful erosional processes that constantly operate to reduce it? This problem and its solution have been with us for some time.

During the nineteenth century a British government survey of India was conducted. It was known at that time that a correction had to be made for the deflection of the surveyor's plumb bob when an area near a mountain range, such as the Himalayas, was being surveyed. However, the calculated correction proved to be greater than the actual deflection (Figure 21–14). Geophysicists J. H. Pratt and G. B. Airy pondered this problem for some time and independently

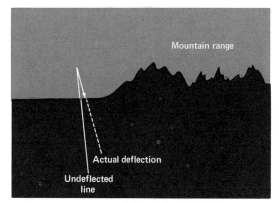

Figure 21–14. Deflection of the surveyor's plumb bob near a large mountain range.

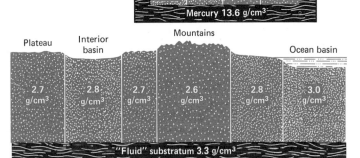

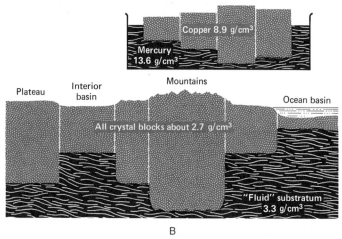

Figure 21–15. *A* & *B*. Pratt's and Airy's interpretations of the relative densities and volumes of crystal blocks. (From L. U. DeSitter, *Structural Geology*, 2nd ed., New York, McGraw-Hill Book Co., and G. Gilluly, J. Waters, and A. O. Woodford, *Principles of Geology*, San Francisco, W. H. Freeman, 1969.)

concluded in 1955 that crustal blocks of contrasting properties are involved in the deflection problem (Figure 21–15). The mountain mass was thought to be either a larger volume of material of the same density as that in the plains to the south or a mass of lesser density than the Indian subcontinent. In either case the Himalayas would "float" higher than the remainder of India. It was also realized that crustal blocks probably float on a plastic sublayer. This idea became known as the principle of *isostasy*.

The concept of isostasy embodies some of these assumptions originally put forth by Pratt and Airy and later refined by Vening-Meinesz and others (Figure 21–16). Today we know that isostasy is a condition of balance, or equilibrium, between parts of the earth's crust of different densities or volumes. Lightest parts of the crust float higher than heavier parts.

One of the most interesting confirming lines of evidence is the measurable uplift of northern Europe and North America following the melting of the last great continental glaciers 8,000–10,000 years ago (Figure 21–17). As glaciers formed and advanced, they added weight to continents where there was none. To readjust (reestablish equilibrium) to this added weight portions of continents supporting the ice sank. After melting, these areas began to reemerge. This process is like the loading and unloading of a barge. The unloaded areas are still rising at a measurable rate. The areas that supported the most ice sank the most and therefore have the fast-

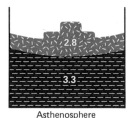 

Asthenosphere   Asthenosphere

Figure 21–16. A more recent interpretation of the relations between floating crustal blocks. (From L. U. DeSitter, *Structural Geology*, New York, McGraw-Hill Book Co., 1964.)

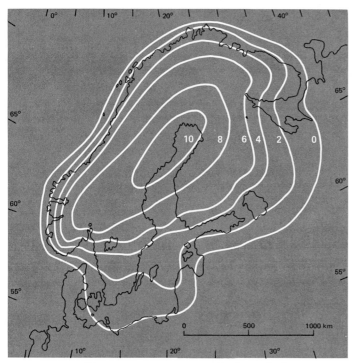

Figure 21–17. Amount of rebound of Scandinavia in millimeters per year following melting of glacial ice. (From R. A. Daly, *The Changing World of the Ice Age*, New Haven, Yale University Press, 1934.)

est rebound rate. We measure the slowest rates around the peripheral extents of the ice sheets.

We know from continuous measurement of shoreline elevations that the south end of Hudson Bay is being tilted northward and emptied into Baffin Bay by more rapid rebound. Conversely Lake Michigan is being tilted southward and is presently being partially drained southward through the Chicago and Illinois rivers, as well as through the Great Lakes to the northeast. However, as tilting continues, we should see increased drainage to the south and conceivably inundation of parts of Chicago in the next 20,000 years.

So isostasy accounts for vertical uplift of mountain ranges that are being eroded away by surface processes. As the mountain range is eroded down, its light root raises it up

again. This type of vertical movement is called *epeirogeny* or *epeirogenic movement*. It is the opposite of orogeny (orogeny involves horizontal forces). Orogeny tends to be more localized to a smaller part of the crust than does epeirogeny. As the range continues to be uplifted, the root becomes smaller and less able to elevate the range to heights as great as before. This helps to explain why older mountain ranges, such as the Appalachians or the Urals, are not so lofty as the Andes, the Himalayas, or the Alps, which are much more recent mountains still being uplifted in part by original mountain-building forces as well as by isostatic forces.

Erosion finally wins the battle over the mountain range when there is no more root to uplift the range by isostasy. The crust beneath the old mountain range will have the same thickness as the remainder of the continental interior, a more or less equilibrium thickness. At this point the old mountain system becomes a part of the stable craton, and the size of the continent is permanently increased.

SUMMARY

Several kinds of mountains exist: volcanic mountains, mountain chains produced by compression, block fault mountains, and erosional mountains. Most mountain chains share certain common features: location near the continental edge, a thrust belt along the side of the chain closest to the continent, a complexly deformed metamorphic core, *mélange* structures, a belt of plutons, and an overall linear aspect.

It was recognized many years ago that mountain chains are formed in a zone that is experiencing accumulation of greater thicknesses of sediments than is the continental interior—these are called geosynclines. The tectogene concept was proposed to account for orogeny before plate tectonics theory was formulated. Mountain systems may result from the plate tectonics processes of collision and subduction. Mountain building is probably a cyclic process related to continental growth. Evidence for this is the ages and nature of rocks in continental shields. These rocks appear to have been part of ancient mountain chains that were eroded and younger mountain chains formed on their flanks, thus increasing the sizes of the continents.

Isostatic epeirogenic uplift after orogeny is responsible for adjustment of the thickness of the continent to that of its continental interior by removal, by erosion from the top, of the thickened root beneath the mountain chain. This is the final step in the addition of material to growing continents.

Questions for thought and review

1. How do compressional mountain chains differ from block fault mountains?
2. Summarize common characteristics of mountain systems.
3. What is a geosyncline?
4. What is the tectogene concept?
5. How does a miogeosyncline differ from a eugeosyncline?
6. What is the tectonic cycle?
7. How do collisional mountains originate? Cordilleran mountains?
8. What is continental accretion?
9. How do orogeny and epeirogeny differ?
10. What modern-day measurements tell us that isostasy is operating?

Selected readings

Cailleux, André, 1968, Anatomy of the Earth, McGraw-Hill, New York.

Calder, Nigel, 1972, The Restless Earth, Viking, New York.

Dietz, R. S., 1972, Geosynclines, Mountains, and Continent Building. In Wilson, J. Tuzo, ed., Continents Adrift: Readings from Scientific American, pp. 124–132.

James, D. E., 1973, The Evolution of the Andes, Scientific American, vol. 229, no. 2, pp. 60–69.

Part
IV

Environments here
and beyond the earth

22 Planetology

STUDENT OBJECTIVES

At the conclusion of this chapter the student should:

1. understand the general objectives of planetology

2. be familiar with the sequence of events that led to manned landings on the moon

3. understand methodologies used to study lunar geology

4. understand how lunar rocks and surface features are used to interpret the history of the moon

5. perceive similarities and differences between lunar and Martian geology

Moon, lovely Moon, with a beautiful face,
Careering throughout the boundaries of space,
As I see thee I often reflect in my mind
Shall I ever, oh ever, behold thy behind?

Keats

416

ENVIRONMENTS
HERE
AND
BEYOND
THE
EARTH

In Chapter 3 we presented a brief survey of the universe without mentioning much about our nearest neighbors in space, especially the moon and Mars, which have received so much attention during the space programs of the 1960s and 1970s. The reasons for this are (1) there is now so much information available that a full chapter is necessary for discussion, and (2) many aspects of lunar and Martian geology are better appreciated after you comprehend the background of the previous chapters of this book.

The terms *astrogeology* and *planetology* are both used for the science that deals with the nature, origin, and history of planetary bodies. However, planetology places emphasis on comparative studies of planets, so models can be developed to explain planetary phenomena. To do this, of course, one must have something to compare. Prior to Mariner 9, which sent back enormous amounts of information from Mars, lunar

features were mostly rationalized in terms of earth analogs. Now meaningful comparative studies can be made between the moon and Mars as well as between the moon and the earth, and the earth and Mars. Additional data being accumulated on the sun, Venus, Mercury, and outer planets, such as Jupiter, assure an important role for the young science of planetology in mankind's attempt to understand the solar system.

THE MOON

The moon is a sphere 2,160 mi (3,484 km) in diameter (Figure 22–1). It has no water or air, although there is a very thin atmosphere of helium and argon. Day and night last nearly 15 days. Even during direct sunlight the sky is absolutely black. Temperatures range from 214°F to as low as −243°F. The moon has a density of 3.34 and thus is a lighter body than the earth, whose density is 5.5. As seen from the moon, the earth fills an area of sky 13 times larger than the full moon fills in our sky and shines with a brightness 74 times that of the full moon. (This is equivalent to the light of a 75-watt bulb at a distance of 6 ft.)

Early telescopic observations

Galileo was the first, in 1610, to point a telescope at the moon and record what he saw. He noted numerous circular depressions, large dark regions, and long shadows cast by mountains. Up to this time dark areas were imagined to be oceans and were called *maria*. The lighter areas were called *terrae*, or land. During the remaining 90 years of the seventeenth century further observations served to dispel this notion, but the terms are still used. Looking at the moon and constructing lunar maps became popular. More than 25 such maps of varying quality were made by scientists lucky enough to have telescopes. Probably the best of these maps is that of Cassini. His map shows hundreds of craters, rays extending from conspicuous

Figure 22–1. The full moon. (Courtesy of NASA.)

craters, and central peaks within some craters.

During the eighteenth and nineteenth centuries little substantive knowledge of the moon was achieved except to add more craters and other topographic details to existing maps. By the 1870s more than 33,000 craters had been mapped. Lunar observers of the mid-nineteenth century welcomed the advent of photography. The first photo of the moon was taken in 1840. Although there was great potential in photography, it added little to what was already known about the moon. Visual observation remained superior to distant photos of inferior resolving power. Utilization of photos would not be of importance until cameras could be brought into proximity to the lunar surface.

Unmanned probes

In the spring and fall of 1960, satellites Tiros I and II photographed the earth from orbit. These launchings foreshadowed the Ranger, Lunar Orbiter, and Surveyor programs, designed to obtain detailed pictures of the moon, not only simply to increase our knowledge of surface characteristics but also to select likely sites for later manned landings.

The Ranger program involved a hard landing on the moon. During the rocket's descent and prior to crashing, photos would be sent back to earth. This program provided the first glimpse of the moon with detail better than a telescope (Figure 22–2). On July 28, 1964, Ranger VII crashed into the Sea of Clouds. During its descent it obtained 4316 high-resolution vertical pictures of the moon, including definition of objects only 36 in. (91 cm) in diameter. Additional flights obtained several thousand more pictures from the Sea of Tranquillity (Ranger VIII) and near the crater Alphonsus (Ranger IX).

The Lunar Orbiter program consisted of a series of five spacecraft placed into orbit around the moon. The orbiters took detailed vertical and oblique photos. These photos were developed on board the spaceship, converted to an electronic signal, transmitted back to earth, and there reconverted using scanning systems (Figure 22–3). The Orbiter program lasted from August 1966 to August 1967.

Orbiter and Surveyor projects overlapped in the late 1960s, with the first Surveyor making a successful soft landing in the Ocean of Storms on May 30, 1966 (Figure 22–4). Surveyor I took more than 11,000 pictures, providing very detailed and panoramic views of the moon's surface (Figure 22–5). The soft landing allayed fears that the mare areas consisted of soft sands into which the surveyors (or later astronauts) would sink out of sight. Later surveyors were designed to carry out functions other than picture taking. Besides trenching into the lunar surface to observe soil characteristics (Surveyor III), they had the capability to instruct

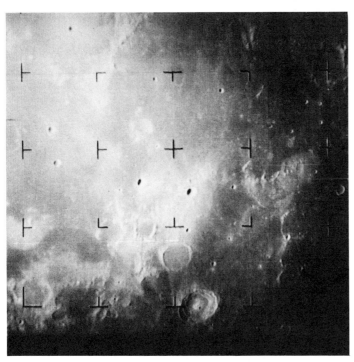

Figure 22–2. View of lunar surface taken during Ranger program (courtesy of National Space Science Data Center).

Figure 22-3. Marius Hills region. (Courtesy of National Space Science Data Center.)

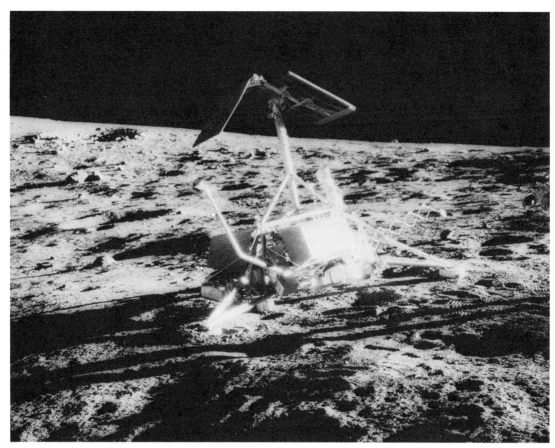

Figure 22–4. Surveyer spacecraft. (Courtesy of National Space Science Data Center.)

Figure 22–5. Composite view of lunar surface taken from Surveyer I. (Courtesy of National Space Science Data Center.)

420

ENVIRONMENTS
HERE
AND
BEYOND
THE
EARTH

the surveyor to pick up lunar rocks and hold them in front of the television camera so that observers on earth could have a better look at them. The last Surveyor (VII) conducted a lunar exploration in January 1968 in the vicinity of the crater Tycho.

Manned landings

Manned landings on the moon began with Apollo 11's giant step for mankind (Armstrong, Aldrin, and Collins) on July 20, 1969. Apollo 11 was preceded by 20 manned space trips in the vicinity of the earth (Mercury and Gemini programs) and the moon (Table 22–1). Landings on the moon involved 6 missions (Apollo 13 was aborted) that saw 12 astronauts explore the lunar surface, take several thousand photographs, and collect several hundred pounds (kilograms) of rocks and soil from a variety of lunar environments (Figure 22–6). These rock materials were subdivided and distributed to more than 200 research institutes around the world for analysis and evaluation. The flood

of data from all these activities caused almost daily reevaluations about the nature of the moon during the late 1960s and early 1970s. Lunar data is still being studied.

Rather than individualize the findings of each lunar landing (Table 22–2), we will summarize the collective results of the Apollo and other space programs as they apply to various lunar features and to the overall history of the moon.

Summation of lunar exploration

Approach to lunar geology. Data used to evaluate the nature of the moon and its history come from the following sources:

1. photographs and geologic maps
2. on-site observations by astronauts
3. instruments left behind
4. rock and soil samples
5. laboratory experiments
6. earth analogs

Geologic field work of the astronauts, al-

Table 22–1. Pre-Apollo 11 manned space flights.

Date	Mission	Astronauts	Remarks
5–5–61	Mercury	Shepard	Suborbital flight
7–21–61	Mercury	Grissom	Suborbital flight
7–20–62	Mercury	Glenn	3 earth orbits
5–24–62	Centaur	Carpenter	3 earth orbits
10–3–62	Sigma 7	Schirra	6 earth orbits
5–15–63	Mercury	Cooper	22 earth orbits
3–23–65	Gemini III	Grissom, Young	3 earth orbits
6–3–65	Gemini IV	White, McDivitt	62 earth orbits
8–21–65	Gemini V	Cooper, Conrad	120 earth orbits
12–4–65	Gemini VII	Borman, Lovell	206 earth orbits
12–15–65	Gemini VIA	Schirra, Stafford	Rendevous in orbit
3–16–66	Gemini VIII	Armstrong, Scott	First moon man in orbit
7–18–66	Gemini X	Young, Collins	43 orbits
9–12–66	Gemini XI	Conrad, Gordon	44 orbits
11–11–66	Gemini XII	Lovell, Aldrin	59 orbits
10–11–68	Apollo VII	Schirra, Cunningham, Eisele	10 days in orbit
12–21–68	Apollo VIII	Borman, Lovell, Anders	Orbited moon
3–3–69	Apollo IX	Scott, Schweikert, McDivitt	Test of lunar module
5–18–69	Apollo X	Cernan, Young, Stafford	LEM tested over lunar surface

Figure 22–6. Landing of Apollo 11. (Courtesy of National Space Science Data Center.)

Table 22–2. Manned landings on the moon.

Mission	Date	Astronauts	Landing area	Remarks
Apollo 11	7–20–69	Armstrong, Aldrin, Collins	Sea of Tranquillity	On surface 21.5 hr; brought back 48 lb (22 kg) of samples
Apollo 12	11–19–69	Conrad, Bean, Cooper	Ocean of Storms	Set up nuclear-powered station; seismograph records of moonquakes
Apollo 13	4–11–70	Lovell, Haise, Schweikert		Aborted
Apollo 14	2–5–71	Shepard, Mitchell, Roosa	Fra Mauro area	
Apollo 15	7–31–71	Scott, Erwin, Worden	Sea of Serenity	First motorized trips in Rover; obtained cores with percussion drill
Apollo 16	4–72	Young, Duke, Mattingly	Sea of Rains (Descartes area)	Studied and sampled lunar highlands
Apollo 17	12–72	Cernan, Schmitt, Evans	Taurus-Littrow Valley	Travelled 24 mi (39 km) during 22-hr exploration; collected 242 lb (110 kg) of samples

422
ENVIRONMENTS
HERE
AND
BEYOND
THE
EARTH

Figure 22–7. Fresh and degraded lunar craters. (Courtesy of National Space Science Data Center.)

though invaluable, was necessarily restricted in time; therefore other ways of conducting lunar geology had to be implemented. The photographic record is, of course, a prime resource for analyzing surface features and stratigraphy. Crater frequency, for example, can be used for relative age dating by counting the number of craters per unit area. More heavily cratered areas are older. Sharp crater rims and fresh *ejecta* (material thrown out during crater impact) signify more recent cratering than rounded and subdued crater rims (Figure 22–7). Successive lava flows interbedded with layers of ejected material permit stratigraphic analysis based on the law of superposition.

On a small scale, cores taken during the Apollo 15 mission showed more than 50 different layers with thicknesses ranging between 0.2–8 in. (0.5–21 cm). Geologic maps, carefully constructed from photos, provide considerable data on age relations. In this connection color and albedo (reflectivity) are an added means of dating. Higher albedo suggests younger material or material recently moved.

Instruments left behind on the moon to operate after the astronauts' departure provide information, among other things, about the moon's interior. Seismic stations pick up seismic energy just as on earth. The seismograph set up by Apollo 12 recorded 208 natural events or moonquakes during its first 8 months of operation. All moonquakes turned out to be less than 2 on the Richter scale and were caused by meteor impact or by internally induced tidal stresses in the lunar crust. Sensors placed in shallow drill holes recorded heat flow data. These data show a cold point approximately 27 in. (70 cm) below the surface. With increasing depth beyond this, temperature rises; however, heat flow appears to be only about one-half that of earth. *Solar wind* (particles issuing from the sun) was also measured. Solar wind particles were found later in lunar soil.

These may help show us the history of the sun.

Maria and terrae. Mare regions are large circular dark areas created by megameteorite impact. They are floored with numerous layers of iron-rich basalt and are veneered with fine, loose, medium to dark gray soil. In the southwestern part of the Sea of Tranquillity this soil, or *regolith*, ranges in thickness from 8–16 ft (3–6 m). Age dating shows the mare basalts to be 3.5–3.7 billion years old. The basalts apparently represent upwelling of molten lavas from the moon's interior following deep gouging caused by impact. The size of these meteorites must have been impressive although considerably smaller than the areas of mare that they created. Some scientists have suggested that asteroids of comparable size impacted against the earth early in its history and point to the Gulf of Mexico and Hudson Bay as scars of these encounters.

Heavily cratered mountainous terrain marks the highlands (terrae) areas. They have a higher albedo than the maria and stand higher above them topographically. The far side of the moon is dominated by highlands topography (Figure 22–8). Common rock types in highlands are anorthosite and breccias of various shades of gray containing fragments of plagioclase, olivine, and pyroxene, all of which occur in terrestrial rocks. Breccias may represent ejecta from medium- to large crater impacts. Highland rocks were dated at 4.5 billion years and are thus older than mare rocks.

Craters and crater formation. The most common and conspicuous features of the lunar landscape are the craters. They range in size from tiny features no larger than the diameter of a small pebble up to several hundred miles (kilometers). Prior to detailed space exploration it was argued that these craters had been caused either by meteoric impact or by volcanic activity. Although

424
ENVIRONMENTS
HERE
AND
BEYOND
THE
EARTH

Figure 22–8. Far side of the moon. (Courtesy of National Space Science Data Center.)

Figure 22–9. Meteor Crater, Arizona. (Courtesy of Flagstaff Chamber of Commerce.)

there are indeed volcanoes and volcanic craters on the moon (for example, Hyginus Rille), the dominant process of crater formation is impact.

Many craters have central peaks and slump blocks along rims. One interesting question concerns the mechanics of crater formation. One approach to understanding how craters are formed has been in scaled-down laboratory experiments. A gas gun is used to fire projectiles at various angles into a vacuum chamber containing a target of sand layers that simulate the lunar surface. The velocities of projectiles averaged about 4.6 mi (7.5 km) per second. Results showed craters with rims adjacent to curled or up-

turned sediment layers, shear features, and radiating ejecta. The most interesting comparable feature that we have on earth is Meteor Crater, Arizona (Figure 22–9). This crater is about 3900 ft (1200 m) in diameter. It also shows upturned strata along the rim. Coesite and stishovite, two forms of high-pressure silica found in the crater, bespeak its impact origin.

Laboratory experiments on angle of entry also explain the nearly perfect circular craters we see on the moon. If the angle of entry is greater than 15°, circular craters are always formed. Only at very low angles are elongate craters and "skip marks" formed. The latter types are not unknown on the

426
ENVIRONMENTS
HERE
AND
BEYOND
THE
EARTH

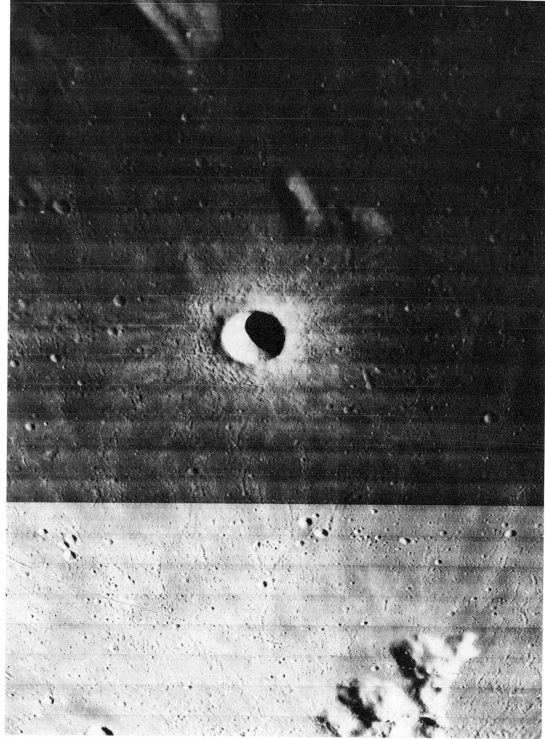

Figure 22–10. Elongate craters and skip marks produced by ejecta can be seen in this Lunar Orbitor photo. (Courtesy of National Space Science Data Center.)

moon. They seem to occur most often as secondary craters produced by low-flying ejecta from primary impacts (Figure 22–10; LOP III 113M).

The sequence of development of lunar craters involves, first, original impact, causing a shock wave that compresses, shatters, and faults the rocks. The meteor itself is probably vaporized in whole or in part at that moment. The crust, at first yielding to the shock, undergoes isostatic adjustment and rebound takes place. If the shock is great enough, central peaks are formed (Figure 22–11; LOP II 162H). The central peaks are analogous to the plume of water rising after the splash of a pebble into water. Ejecta is sprayed outward, and subsequent slumping along the rim walls takes place.

Figure 22–11. Crater with central peaks. (Courtesy of National Space Science Data Center.)

428

ENVIRONMENTS
HERE
AND
BEYOND
THE
EARTH

Erosional processes. Certainly there are no erosional processes operative on the moon that can compare with those on earth. In the absence of water and atmosphere chemical weathering is essentially absent. Nor can there be great valleys that are carved by rivers, abrasion of landscape by strong winds, or erosion along shorelines. Yet the surface of the moon is not immutable. Physical changes have taken place over millions of years. This is especially observable in the craters. The craters are not all alike in age or condition. Once formed, a crater is subject to degradation.

The chief erosional process on the moon seems to be micrometeorite bombardment. Without a protective atmosphere even particles of small size are able to reach and strike the surface, causing abrasion (sandblasting effect) or small-scale cratering. This process is sometimes referred to by lunar investigators as "gardening." It is a slow process: Some estimate the erosion rate on the moon to be no more than .04 in. (1 mm) every million years. This rate is nonetheless sufficient to subdue the sharpness and relief of craters' rims and other features (22–12). In addition to micrometeorite bombardment, ejecta and other fine particulate matter tend to drift about and settle in low areas, such as the floors of craters, contributing to their ultimate obliteration (Figure 22–13).

Lunar rocks and soils. In general we can say that rocks found in mare areas are basaltic and rich in iron and titanium (Figure 22–14). Highland rocks consist of anorthosite (an igneous rock containing more than 65% Ca-plagioclase) and breccias of varying mineralogic character (Figure 22–15). This leads to the tentative conclusion that older highlands represent lighter differentiated lunar crust, which was broken and disrupted by later megameteorite bombardment that formed the younger mare areas. Highland breccias were formed probably by later impact events.

Lunar soils contain an array of minerals, including ilmenite (a titanium-rich mineral), pyroxene, olivine, plagioclase, and the less familiar *cristobalite* and *troilite*. Cristobalite is a high-temperature form of quartz found in some volcanic rocks on earth. Troilite is an iron sulfide mineral present in most meteorites. It is probable that its origin is extralunar, because meteorite material (carbonaceous chondrites) has been found mixed in with lunar soils.

The soils also contain fragments of basalt, breccia, anorthosite, and glass particles. Apollo 11 soils contained 4% anorthosite, which may have drifted in from the lunar highlands. The glass fragments are interesting for their abundance and shape. Dark brown and black glass assume such shapes as spheres, teardrops, and dumbbells (Figure 22–16). They appear to have formed when hot fluid ejecta struck the lunar surface. The amount of glass present in a lunar soil may be an index to its maturity. John Lindsay, who studied lunar soils from Apollo 11 and 12, suggests that continual meteoric impact progressively vitrifies the material, and hence, older lunar soils will contain more glass. On the basis of this assumption, Apollo 11 soils are more mature than Apollo 12 soils. Absolute age dating confirmed this.

Apollo 17 astronauts discovered what they described as an orange soil and took samples of it (Figure 22–17). This excited scientists, because it suggested possible venting of steam during volcanic activity, or at least a water-bearing mineral such as limonite. The orange soil turned out to consist of glass spheres and may or may not have been volcanic. The cause of the "rust" on the glass spheres was goethite, a hydrated mineral. Minerals containing water, as might be expected, are extremely rare on the moon.

Summary of lunar history. The moon probably formed about 4.6 billion years ago along with the earth and the rest of the solar system. A heat buildup because of accretion

Figure 22–12. Craters in various stages of degradation. (Courtesy of National Space Science Data Center.)

430
ENVIRONMENTS
HERE
AND
BEYOND
THE
EARTH

Figure 22–13. Crater floored with mare material. (Courtesy of National Space Science Data Center.)

Figure 22–14. Lunar basalt. (Courtesy of National Space Science Data Center.)

Figure 22–15. Lunar breccia. (Courtesy of National Space Science Data Center.)

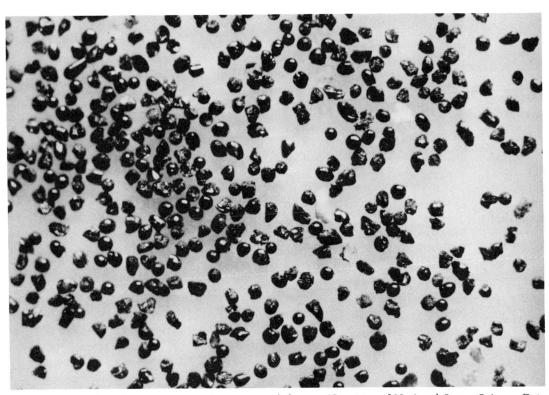

Figure 22–16. Glass fragments assume a variety of shapes. (Courtesy of National Space Science Data Center.)

432

ENVIRONMENTS
HERE
AND
BEYOND
THE
EARTH

Figure 22–17. Sample site during Apollo 17 mission. (Courtesy of National Space Science Data Center. NASA photo AS17-137-21009.)

took place during initial formation or shortly thereafter. This accretionary phase was the same one that involved other members of the solar system during which planets swept out areas of debris-clogged space and contributed substantially to their own growth.

The heat buildup caused at least partial fusion of the early moon, and differentiation took place. Seismic studies show that the moon now has a crust 15–40 mi (25–70 km) thick. Below that is a denser mantle (pyroxene?). Remanent magnetism found in igneous rocks 3.5 billion years old suggests

that the moon had a molten metallic core. If this is true, then the earth and moon are similar in having a gross threefold arrangement of crust, mantle, and core, although the moon's lithosphere appears to be much thicker and more rigid than that of the earth.

The next major event in the moon's history was the cataclysmic bombardment that created the extensive mare areas. This appears to have happened about 4 billion years ago. The gaping craters opened to subcrustal molten material and ushered in an era of vulcanism enduring until about 3.2 billion years

Table 22–3. Geologic time scale for the moon.

Copernican	Fresh craters show rays and secondary craters (ejecta)
Erathosthenian	Rims of craters still sharp, but ejecta field not clear
Imbrian	Time of formation of mare basins
Pre-Imbrian	Events prior to mare formation

ago. During this period several upwellings of basaltic lavas filled in the scars left by the bombardment. When vulcanism subsided, the moon became essentially quiescent. For the past 3 billion years, it has suffered only surface modification because of meteoroid impact. Why the moon suddenly became "dead" 3 billion years ago is still not understood. Geologists have devised a tentative geologic time scale for the moon (Table 22–3).

MARS

Mariner 4 reached the vicinity of Mars in July 1965 after a 228-day journey from earth. At a distance of 6118 mi (9241 km) Mariner 4 began clicking off a series of 19 usable photos at a rate of 1 per minute. These photos covered a strip 200 mi (322 km) wide and 3000 mi (4838 km) long, representing about 1% of the surface area of Mars. Up until this time, observations of Mars by telescope were difficult at best because of its distance from earth and the fact that it so often lies in line of sight with the sun. Still it was known that Mars had an atmosphere, polar caps, and surface markings that changed seasonally. It offered the best hope of harboring some extraterrestrial form of life in the solar system. The results were disappointing. The long awaited pictures revealed a desolate, cratered landscape resembling that of the moon. There was no evidence of the famous "canals."

The conclusion that Mars was just another moonlike body was, however, premature. Mariner 9 was launched May 30, 1971, on a historic 167-day voyage of 287 million mi (462 million km). It became the first spacecraft to be placed into orbit around another planet. During a useful lifetime of 349 days in orbit, Mariner 9 mapped the entire surface of Mars and transmitted 7329 television pictures back to earth. Pictures of the two small Martian satellites, Deimos and Phobos, were also obtained. Results from Mariner 9 showed dramatic differences from the results of the Mariner 4 expedition.

The highlights revealed by Mariner 9 include: (1) violent dust storms, (2) four gigantic volcanoes, (3) numerous craters, and (4) channels and gullies that appear to be water eroded.

Volcanoes

At the time of Mariner 9's arrival, a dust storm of incredible proportions was raging in the thin Martian atmosphere, obscuring virtually the entire surface of the planet. A topographic feature was observed, however, as a dark albedo area within the dust storm. When the storm finally subsided after several days, this feature turned out to be an extremely large volcano. It was named Olympus Mons. Three other aligned volcanoes, smaller than Olympus Mons, were located nearby (Tharsis Ridge) (Figure 22–18). These volcanoes were not detected by Mariner 4 nor by Mariners 6 and 7, which passed by Mars in 1969. The size of Olympus Mons is awesome: 310 mi (500 km) in diameter at the base and rising to a height of 15 mi (25 km) or three times the height of Mount Everest. At its summit, the crater, or caldera, is 40 mi (65 km) across.

These volcanoes show that at least part of the interior of Mars is hot. In fact, the planet may be in the initial stages of a prolonged period of vulcanism, which could lead to an enriched atmosphere containing more water than at present. Thus, instead of being an old and dying planet, Mars may be very youthful in its evolution. The general absence of volcanoes elsewhere on the surface supports this belief. But this assumes that, in some

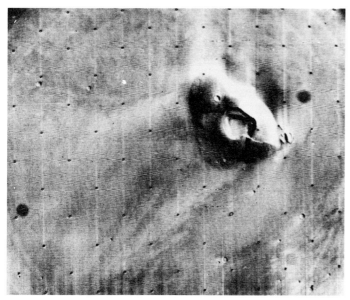

Figure 22–18. Evidence of volcanic activity on Mars. (Courtesy of NASA.)

earlier period of Martian history, no extensive vulcanism (and subsequent destruction of its evidence) took place. The eruptions of Olympus Mons and its companions may have contributed significantly to the present atmosphere, which contains abundant carbon dioxide and very small amounts of water. Volcanoes on earth discharge considerable carbon dioxide, water, and nitrogen. By analogy we might assume the Martian atmosphere contains nitrogen also, which may be attributable to solar wind, but only traces have been found.

Channels and gullies

Pictures from Mariner 9 show channels with tributaries or distributaries not unlike drainage networks on earth produced by running water (Figure 22–19). These sinuous channels do not appear to have formed as a result of wind activity or flowing lavas.

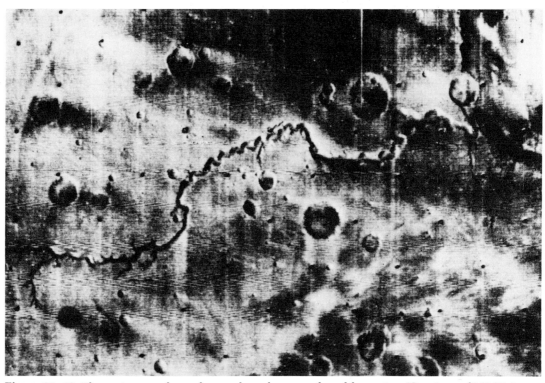

Figure 22–19. These sinuous channels may have been produced by water. (Courtesy of NASA.)

Some scientists, aware of the lack of water on Mars, are searching for alternative models to explain these drainage features. Others insist they could have formed only from aqueous activity. The problem of water on the surface is admittedly difficult. Martian atmospheric pressure is less than 1% that of earth. This means that any surface water would tend to evaporate or "boil away" very quickly. The white polar caps, once thought to be water, turn out to be frozen carbon dioxide (similar to dry ice) less than 10 ft (3 m) thick.

One suggestion from Daniel Milton of the U.S. Geological Survey may explain the channels. He observes that the braided and meandering channels in the Martian equatorial zone are extremely wide—up to 24 mi (40 km). He believes these channels could have been formed by water under catastrophic flood conditions that lasted only a short time. The source of water would not be water as we know it but carbon dioxide hydrate ($CO_2 \cdot 6HOH$). The melting of such *ground ice* during the Martian summer (with temperatures rising briefly to 77°F) and recession of the ice caps would be the most likely cause of temporary flooding. This hypothesis would also explain the so-called "chaotic terrain" on Mars, caused by decay of ground ice and resultant fracturing and slumping. The validity of this hypothesis will probably have to await manned landings on the planet.

Other features

Although the dust storm that obscured Mars at the time of the arrival of Mariner 9 was unusual—none of that magnitude had occurred in more than a century—dust storms are nonetheless common, with wind velocities of 100–200 mi/hr (160–320 km/hr) or more. These storms seem to be a chief cause of changes in surface markings, which were noted by astronomers. Dust storms in an arid world, such as Mars, would likely produce sand dunes; indeed they do. Hellas,

a conspicuous circular area with a diameter of more than 900 mi (1600 km), has been known for 200 years. Mariner 9 photos show it to be a huge dune field with dunes far larger than any on earth (Figure 22–20).

Another Martian enigma is the great canyon, Margaritifer Sinus, trending east–west along the equator. It is 3000 mi (4800 km) long, averages 62 mi (100 km) wide, and is almost 4 mi (6.4 km) deep (Figure 22–21). Its origin is unknown but may be related to tectonic activity. Fractures or cracks, some up to 1 mi (1.6 km) in width, also cut across the Martian surface. These are perhaps tensional features produced by expansion of the crust.

It is known that Mars has no magnetic field. This may agree with the concept that Mars is a young planet still undergoing internal differentiation to produce a metallic core. On the other hand, if Martian rocks, when they are obtained, display remanent magnetism, it would suggest formation of a metallic core much earlier in the history of the planet and reverse ideas on the present stage of evolution of Mars.

During its orbiting of Mars, Mariner 9 obtained photos of both Deimos and Phobos, the two satellites. Phobos, the inner moon of Mars, is cratered and has the shape (as some have expressed it) of an Idaho potato (Figure 22–22). Phobos measures are 13 by 16 mi (20 by 25 km). Mars now has a third satellite— Mariner 9. Its instruments are shut down and it is silent. It will continue to orbit Mars every 12 hr for the next 50 years or so. As a final gesture, Mariner 9 will then crash into the Martian surface.

OTHER PLANETS

We know considerably less about the other planets in the solar system than we now know about the moon and Mars, even though unmanned probes have reached some of them. Mariner 5 was launched toward Venus on June 14, 1967, and on arrival sent back information about its atmosphere

436

ENVIRONMENTS
HERE
AND
BEYOND
THE
EARTH

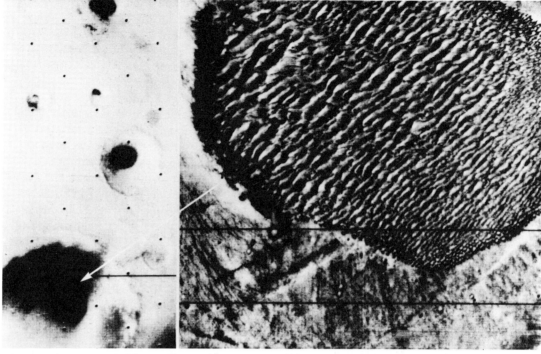

Figure 22–20. Dune field of Hellas. (Courtesy of NASA.)

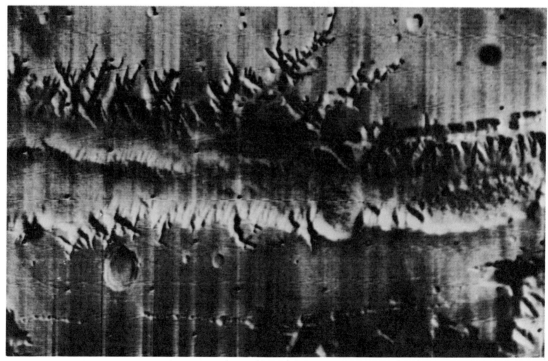

Figure 22–21. Gigantic canyons on Mars are still an enigma. (Courtesy of NASA.)

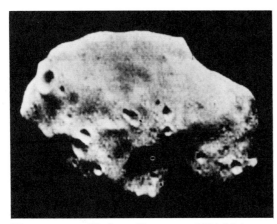

Figure 22–22. Phobos. (Courtesy of NASA.)

and temperatures. The Soviets were able to land a space vehicle on the planet, but it ceased to function soon after and presumably was not able to send back much information.

Even with the most powerful telescopes Venus was a frustrating object to observe. The planet is shrouded with a dense atmosphere that conceals surface details. Mariner 5 shows, however, the swirling patterns of atmospheric circulation (Figure 22–23). This atmosphere was once thought to be mostly water vapor, and a hot, steamy planet was visualized. We know now that at least the upper part of the atmosphere consists chiefly of carbon dioxide, with scanty amounts of water, carbon monoxide, hydrochloric acid, and hydrofluoric acid. Surface temperatures are greater than 300°C, far too hot for terrestrial-type life. In June 1972, scientists at Cal Tech's Jet Propulsion Laboratory penetrated the thick atmosphere using radar signals and, with the aid of a digital computer, were able to determine that the Venusian surface is pock-marked with large, shallow impact craters similar to those on the moon. This is strange, because the dense atmosphere should protect against extensive meteoric bombardment.

Mariner 10 came to within 400 mi (700 km) and photographed Mercury between March 23, 1974, and April 3, 1974. About 2000 pictures were taken (Figure 22–24). Major landforms include numerous craters, basins (some flooded with lava), ridges, and plains. Mercury is similar to the moon in its surface features. Conspicuous scarps suggest compressive forces that set up as Mercury cooled and shrank. Mercury has a magnetic field and probably differentiated into a silicate crust and an iron-rich core while undergoing meteoroid bombardment. More recent craters show ray systems comparable with those of the moon. Its tenuous atmosphere consists of helium. There is no evidence of wind activity as we see on Mars.

Finally the largest planet in the solar

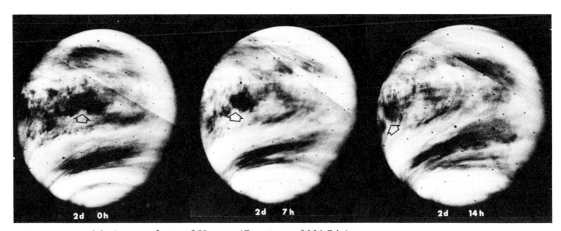

Figure 22–23. Mariner 5 photo of Venus. (Courtesy of NASA.)

438
ENVIRONMENTS
HERE
AND
BEYOND
THE
EARTH

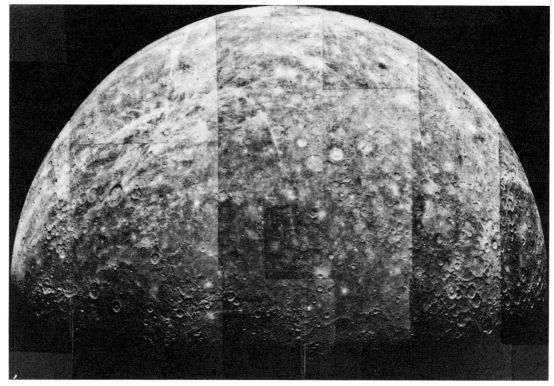

Figure 22–24. Mercury. (Courtesy of NASA.)

system—Jupiter—is a swirling ball of hydrogen that fell short of achieving nuclear reaction. Otherwise it may have become a luminous body, such as the sun, and would have changed the solar system into a binary star system.

SUMMARY

Planetology deals with the comparative nature and origin of planetary bodies. Current knowledge of the moon and Mars has resulted from highly successful space exploration programs including six manned landings on the moon and the photographic mapping of the entire Martian surface.

The moon's structure resembles that of the earth, possessing crust, mantle, and core. Its surface can be broadly divided into mare regions of basaltic lava flows and highlands composed of lighter anorthosite and brec-

cias. Mare regions resulted from large meteorite bombardment and upwelling of basalt. Highlands may contain vestiges of original differentiated lunar crust. The heavily cratered surface is subject to the erosional process of micrometeorite bombardment, resulting in progressive degradation of craters. The oldest age of rocks on the moon is 4.6 billion years. Mare areas are younger than highlands areas.

Mars displays (1) violent dust storms, (2) enormous but few volcanoes, (3) numerous craters, and (4) sinuous channels that may have resulted from erosion by liquid carbon dioxide hydrate. Mars may be youthful in its evolution rather than old and dying.

Questions for thought and review
1. Define planetology.
2. If an astronaut weighs 180 lb on earth,

what would he weigh on the moon? On Mars?

3. What is meant by albedo?

4. How would you differentiate between a lunar crater formed by volcanic action and one formed by meteor impact?

5. Scientists speak of "lunar soil." What is the difference between lunar soil and soil as we know it on earth?

6. If the Gulf of Mexico and Hudson Bay actually resulted from impact by asteroids, what field evidence would you seek in these areas to substantiate this?

7. What is the chief erosional process on the moon?

8. What is remanent magnetism?

9. What hazards would an astronaut face in landing on Mars?

10. How would you account for the craters on Venus in view of its thick atmosphere?

Selected readings

Allen, J. P., 1972, Apollo 15: Scientific Journey to Hadley-Apennine, *American Scientist*, vol. 60, no. 2, pp. 162–174.

Dunne, J. A., et al., 1974, Mariner 10 Mercury Encounter (and other reports), *Science*, vol. 185, no. 4146, pp. 141 ff.

Goles, G. G., 1971, A Review of the Apollo Project, *American Scientist*, vol. 59, no. 3, pp. 326–331.

Lewis, J. S., 1971, The Atmosphere, Clouds, and Surface of Venus, *American Scientist*, vol. 59, no. 5, pp. 557–566.

Muehlberger, W. R., and Wolfe, E. W., 1973, The Challenge of Apollo 17, *American Scientist*, vol. 61, no. 6, pp. 660–669.

Murray, B. C., 1973, Mars from Mariner 9, *Scientific American*, vol. 228, no. 1, pp. 48–69.

Mutch, T. A., 1970, *Geology of the Moon*, Princeton Univ. Press, Princeton, N.J.

23 Earth resources

STUDENT OBJECTIVES

At the conclusion of this chapter the
student should:

1. understand the concepts of reserves and
   different types of resources

2. be able to differentiate between primary
   and secondary deposits

3. know something about the geology and
   occurrence of several important metallic
   and nonmetallic mineral deposits

4. understand the relationships between
   fossil fuels; their origin, distribution,
   and limitations; and the need to develop
   other energy resources for the future

5. be aware of the close ties between
   limited resources, increasing world
   population, and the need to limit growth
   to prevent the depletion of resources

*Beans and corn you can plant and harvest
again and again with the seasons. Unlike
beans and corn, iron, copper and zinc you
harvest only once. Then they're gone,
maybe forever.*

Anonymous

Figure 23–1. A scene from suburban America in the 1970s. How many things in this photograph are directly or indirectly derived from Earth resources? Will this scene be significantly different in the 1980s?

Look about for a minute or so. How many things surrounding you are themselves earth-derived materials or have been shaped by tools made from earth resources driven by fuel or electricity made from other earth resources (Figure 23–1)?

Homo sapiens has been dependent on earth materials for many centuries. Hundreds of thousands of years ago early peoples made the first tools and weapons from stone. Next they learned to refine a few metals and to produce tools and weapons from bronze. Later the Greeks and Romans added iron, copper, lead, and certain pre-

cious ores to the list of metals. Our civilization today fabricates many metals in large quantities and in ever increasing amounts. Iron for making steel, copper, lead, zinc, and (since World War II) aluminum are the principal metals used in our industrial society.

Energy consumption has followed similar routes. Early societies used wood. Olive oil served the Romans as a major source of energy. During the Middle Ages animal tallow was used considerably. Widespread consumption of coal began in the nineteenth century. Although the first oil well in this country was drilled in 1859 by E. L. Drake in

Pennsylvania, most petroleum was utilized in the form of kerosene until near the turn of the twentieth century. The internal combustion engine was invented in the late 1800s, and gasoline was found to be the best fuel for this engine. Today we are highly dependent on petroleum and coal as fuels. Most electricity today is generated with either coal- or petroleum-fired generators. However, we are presently entering a time during which nuclear energy will become much more important. We may also see other forms of energy tapped.

The United States and most of the other of the industrialized nations felt the effects of the major petroleum shortage during early 1974. This shortage was brought about by a combination of circumstances, including increased demand for petroleum products in the United States and elsewhere and decreased production in the United States and curtailment of supplies to western Europe and the United States by oil-exporting countries to increase prices. This shortage was therefore not strictly attributable to the lack of recoverable petroleum in th earth, and the dearth was practically nonexistent by mid-1974. However, governments of nations which either did not produce large amounts of petroleum for export or which had large domestic demands warned that the nonavailability of petroleum in 1974 should serve as a warning of things to come.

Petroleum is not the only raw material

444

ENVIRONMENTS
HERE
AND
BEYOND
THE
EARTH

likely to be in short supply. Metals will be in greater demand and in even shorter supply than petroleum by the turn of the next century. The U.S. Bureau of Mines estimates that our demand for petroleum will more than double by the year 2000, but that for metals will more than triple. Each year it may be noted that lower grades of metals are mined (Figure 23–2). As lower grades are mined, recovery costs increase. In addition to costs, the amount of energy needed to extract the metal from the ore increases.

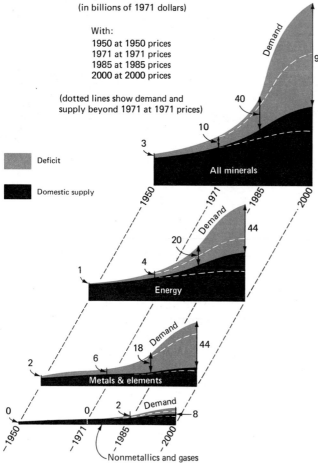

Figure 23–2. Past, present, and projected future resource needs of the United States to the year 2000 showing portions that can be supplied domestically and that which must be imported. (From U.S. Bureau of Mines, Mining and Minerals Policy 1973, Part 1.)

FUNDAMENTAL CONCEPTS AND DEFINITIONS

The term *ore* is probably one of the most misunderstood terms in geology and mining technology. An ore is one or more minerals from which a valuable material may be extracted at a profit. Most commonly this is applied to metals, but the term may also be used to describe a deposit of a valuable nonmetal, such as diamond or mica. However, for something to be extracted at a profit a sizable accumulation of the material must be present. A few pounds (kilograms) of a material containing 10% copper would be interesting but not an ore. Yet several million tons of this material would be a very rich ore deposit. The waste material that must be separated from the ore is called *gangue.*

It is important to (1) know the size of a deposit, (2) know the total amount of recoverable material of some type, such as iron, (3) estimate, for future needs, amounts of unproven quantities, and (4) speculate on amounts of underdiscovered materials. These computations yield values called *reserves.* Reserves may be measured and estimated for a single deposit to determine initially if the deposit is worth mining and, if so, the projected lifetime of mining activities at that site. For example, calculations of reserves for all iron ore deposits of the United States may combine with rates of consumption to project for our needs for several years into the future.

Resources may be classified as *renewable* or *nonrenewable.* Most mineral deposits are nonrenewable, because they are not replenished as they are mined. Agricultural and forest products are renewable resources; another crop may be grown after one is harvested. Groundwater is also a renewable earth resource.

Resources may also be thought of in terms of their overall value. *High-value resources* are mined regardless of where they occur. After mining, high-value resources may be transported halfway around the world to be used. They have a high unit value or price per unit of weight. *Low-value resources* are generally common enough not to be mined and transported great distances. They have a low unit value or price per unit of weight. For example, a sizable copper deposit would be mined wherever it was found, but a limestone suitable for highway aggregate would be mined only if a highway were being built nearby.

TYPES OF DEPOSITS

Mineral deposits may be classified into two major groups: primary and secondary. *Primary deposits* include those formed as a result of the processes of igneous activities, including the formation of veins, metamorphic processes, and processes involved with creating sedimentary rocks. Hydrothermal deposits, which are formed from hot-water solutions usually released during the late stages of cooling of igneous bodies at depth, are also considered to be primary deposits.

Secondary deposits may result in the concentration of valuable minerals. Weathering of a vein or body of material exposed at the surface may yield minerals in quantities sufficient to mine (see Chapter 11). The weathering process may also serve to release certain resistant minerals, which may then be eroded, transported, and deposited elsewhere. Deposits of this type are called *placers*. However, for a mineral to accumulate in a placer deposit, it must possess one or more of the following characteristics: (1) It must be chemically inert, or nearly so, so that it will not be attacked by the weathering processes; (2) it should either be soft and coherent or hard to be resistant to mechanical abrasion; (3) it should have a higher-than-average specific gravity so that it will tend to accumulate and not to be dispersed in the sedimentary environment. Table 23–1 lists several valuable minerals that accumulate as placers.

Table 23–1. Some placer minerals.

Mineral	Composition
Gold	Au
Platinum	Pt
Diamond	C
Rutile	TiO_2
Monazite	$(Y, Ce, Th, La)PO_4$
Zircon	$ZrSiO_4$
Corundum	Al_2O_3
Garnet	$Fe_3Al_2(SiO_4)_3$

METALLIC MINERAL DEPOSITS

Mineral deposits that contain one or more extractable metals are called *metallic mineral deposits*. The metal may be combined in various ways; or it may occur in the native state, as is the case with most of the gold mined in the world. Combinations as sulfides and oxides are probably the most common minerals from which metals are extracted. Table 23–2 contains a summary of

Table 23–2. Important ore minerals.

Metal	Minerals	Chemical composition
Iron	Hematite	Fe_2O_3
	Magnetite	$FeFe_2O_4$
	Limonite	$HFeO_2$
	Siderite	$FeCO_3$
Copper	Chalcopyrite	$CuFeS_2$
	Chalcocite	CuS
	Cuprite	Cu_2O
	Malachite	Copper carbonate
	Native copper	Cu
Aluminum	Bauxite[1]	Complex oxides and hydroxides of aluminum
Gold	Native gold	Au
	Pyrite	FeS_2 (with traces of gold)
Silver	Argentite	Ag_2S
	Native silver	Ag
Lead	Galena	PbS
Zinc	Sphalerite	ZnS

[1] Bauxite is actually a rock composed of several aluminum-rich minerals, such as diaspore, $AlO(OH)$, and gibbsite, $Al(OH)_3$.

446

ENVIRONMENTS
HERE
AND
BEYOND
THE
EARTH

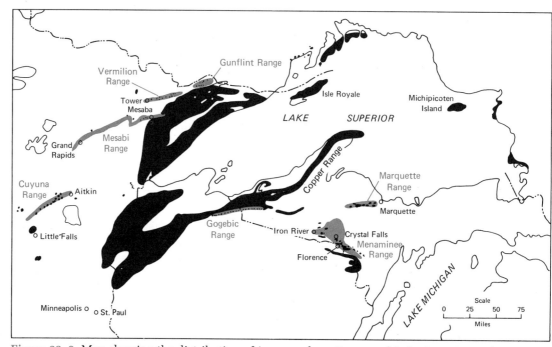

Figure 23–3. Map showing the distribution of iron ore deposits around Lake Superior. (After A. M. Bateman, *Economic Mineral Deposits,* New York, John Wiley and Sons, 1950.)

some of the more important ore minerals and the metals that are extracted from them.

Iron

Iron is probably the most important metal in our civilization today; it has been for several centuries. The most important ore mineral of iron is hematite. The large iron ore deposits in the Lake Superior region of the United States and Canada are of hematite (Figure 23–3). These deposits have been worked for many years and much of the high-quality ore has been removed. However, large quantities of lower-grade (about 30% iron) taconite either have not been mined in the past or have been mined and discarded. This material is presently being mined, but it requires additional treatment so that it can be concentrated before being sent to the smelter. Large deposits of taconite also are found in Brazil, Venezuela, India, Australia, and the Soviet Union.

Large sedimentary iron ore deposits of rich hematite and lower-grade taconite were all formed in the marine environment at surface conditions during the Precambrian, when atmospheric conditions may have been different from those of today. Later in the Precambrian these deposits were deeply buried and metamorphosed. So, to understand why and how Precambrian sedimentary iron ore deposits of the world formed, it is necessary to study the nature of the Precambrian atmosphere and also to determine the composition of the oceans at that time. Such an understanding would enable us better to predict where other deposits of this type may be located.

Magnetite is also an important source of iron. Some magnetite occurs in the Lake Superior region with hematite and in taconite. In many deposits the principal or only ore mineral is magnetite. These are generally thought to have formed in association with intrusive igneous activity. The large magnetite deposit at Kiruna, Sweden, has been mined for years. Other large magnetite deposits occur in Wyoming and the Adirondacks of New York.

Copper

Before the turn of this century the major source of copper in the United States was the native copper deposit in the Keweenawan Peninsula of northern Michigan. This was the largest such deposit in the world. Soon after, the copper sulfide deposits of the western United States surpassed those in Michigan in production. Their lead has not been challenged since.

The native copper deposit in northern Michigan resulted from deposition of copper in cavities in Late Precambrian basalt flows. A few of the individual masses of copper were the size of railroad boxcars and presented their own special problems of mining. (How would you attempt to break up a mass of material that, when a hole is drilled and a charge set, makes the drill hole serve as the barrel of a cannon when the charge is exploded?) Most, however, was in small lumps and granules that could be removed by crushing the rock.

The large copper deposits in Utah, Nevada, Arizona, and New Mexico are predominantly sulfide deposits in which the sulfides were introduced into igneous plutons by hydrothermal (hot-water) solutions. These *porphyry deposits* are of relatively low grade, but they contain huge volumes of material from which ore minerals may be recovered with ease. Generally only one or two valuable minerals are present, and separation processes remain simple. Most of these mines are open-pit mines (Figure 23–4). Copper concentrations as low as 1% or less are being recovered profitably in some areas.

The largest known copper deposit in the world is the Chuquicamata deposit in Chile. It, too, is a porphyry deposit. Other large copper deposits are present in northern Mexico, Rhodesia, Zambia, Zaïre, and Canada.

Lead and zinc

Lead and zinc are found together. There are several types of lead–zinc deposits. Mississippi Valley type deposits were named for their occurrence in Missouri, Illinois, and Wisconsin. Others of this type are present in Virginia and Tennessee (Figure 23–5). Some of these deposits contain minable lead and zinc; others contain only one of these minerals. For example, the deposits in southeastern Missouri make up the so-called Missouri lead belt. Those in eastern and central Tennessee and Virginia are zinc deposits.

Mississippi Valley type deposits are located almost totally within carbonate rocks of Early Paleozoic age. They have a simple mineralogy and have formed either from very low-temperature hydrothermal solutions, ancient groundwater solutions, or a combination of both.

A large lead–zinc deposit is located at Coeur d'Alene, Idaho. This deposit is thought to have formed at higher temperature because of the greater numbers of minerals present and the presence of high-temperature gangue minerals. The Coeur d'Alene deposit is localized within Precambrian carbonate rocks.

Gold

Deposits of gold have been responsible for the movement of sizable numbers of people into areas in quest of riches. The United States experienced several gold rushes during the nineteenth and early twentieth centuries — to Georgia in 1829, to California in 1849, to Colorado in 1891, and to Alaska in 1896. The gold production of the United States was surpassed by South Africa in 1905. The United States remained second until 1931, when the Soviet Union and Canada pushed the United States into fourth place. Gold deposits have been mined for more than 1000 years and have been the symbol of riches and the world monetary standard for centuries.

Gold accumulates either as a primary or a

Figure 23–4. The large open pit mine at Bingham, Utah. (Courtesy of Kennecott Copper Company.)

secondary mineral. It most commonly develops in economic deposits as native gold but can also be found as a minor element in pyrite and in a few other minerals.

The most common primary occurrence of gold is in quartz veins. These veins are thought to have formed by hydrothermal introduction of gold and quartz as some igneous body was cooling at depth. This is the major explanation for gold in California.

Placer gold is mined in many places. Weathering of rocks containing gold as either native gold or gold in pyrite releases this metal into the surface environment, where it accumulates with stream gravels and sands

as a placer deposit. Here the image of the prospector panning for gold is properly focused (Figure 23–6). Large-scale mining of placers by dredges or hydraulic techniques makes recovery of much lower concentrations profitable (Figure 23–7). Moreover, the increase in the price of gold in recent years may reopen many formerly uneconomic deposits. At present most of the world's gold comes from South Africa. Most of the gold produced in the United States comes from South Dakota, and some is mined in Nevada and Alaska.

Deposits in South Dakota and South Africa are in conglomerates thought by some

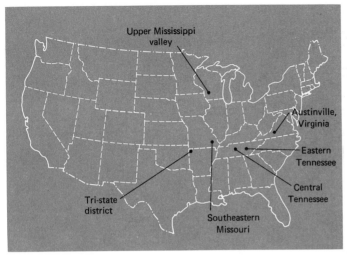

Figure 23-5. Map showing the distribution of Mississippi Valley type lead-zinc deposits in the eastern United States.

Figure 23-6. Prospector panning for gold in Alaska around 1900. (A. H. Brooks, U.S. Geological Survey.)

geologists to be "fossil placers." There is no genetic relationship between the two deposits except, perhaps, that similar processes acted at different times to cause the accumulation of a valuable metal.

NONMETALLIC MINERAL DEPOSITS

Nonmetallic mineral deposits refer to those deposits of minable materials that are generally utilized without recovery of a specific element. Exceptions to this are mining and recovery of sulfur, phosphates, and fluorite (for fluorine). However, nonmetallics also include such things as sand and gravel, gemstones, abrasives, limestone (agricultural), limestone (for cement), building stone, clays (for making bricks), and inorganic fertilizer materials (potash and phosphates). Some of these materials are used after being concentrated; others are mined and used directly.

Nonmetallic deposits are present in many places and under a variety of conditions. Some, such as the gemstones, are generally concentrated in small deposits; others, such as sand or stone deposits, may occur over wide areas. Sand, gravel, and crushed stone comprise the largest economic value among nonmetallic mineral deposits. We might think that gemstones would be more valuable until we stop to think that each building, dam, and highway in the world requires many tons of sand and gravel for construction.

High-value nonmetallics

When the subject of high-value nonmetallic minerals arises, we usually think first of gemstones. These are certainly high-value resources, because they are ordinarily mined wherever they may be found in sufficient

Figure 23–7. Dredge mining gold near Fairbanks, Alaska. (U.S. Geological Survey.)

concentrations, regardless of their location. Some precious gems include diamond, emerald, ruby, sapphire, and precious opal.

Diamonds have been valued for many centuries. The first gems were discovered in India about 800 B.C. and were used for the eyes of a Greek statue dating about 400 B.C. The Romans apparently valued diamonds, but the art of diamond cutting was not developed until some time during the Middle Ages.

Diamonds have a unique primary occurrence all over the world in a variety of dark, peridotite that is called *kimberlite*. Kimberlite is found in near-vertical cylindrical pipes and dikes. Diamonds occur scattered throughout some kimberlite bodies, but most are barren of diamonds. Kimberlite is generally highly altered and contains minerals in addition to diamond, which indicate a very high-temperature–high-pressure origin, such as might exist in the mantle. Diamonds are not really abundant in these rocks; a yield of a few carats of diamonds per 100 tons of ore is considered profitable.

The secondary occurrence of diamonds is as placer deposits in stream gravels and alluvium and in beach sands. The Orange River draining the diamond region of South Africa has along it many diamond-bearing gravels. Also, beaches around the mouth of the river are being mined.

Principal diamond-producing countries are Zaïre, South Africa, Ghana, and Sierra Leone. Several other African countries produce diamonds in marketable quantities. They are also mined in Brazil, India, and the Soviet Union. Most gem quality diamonds are mined in South Africa; the other countries produce the majority of industrial diamonds used primarily for abrasives and cutting tools.

Other minerals generally thought of as gemstones frequently occur as materials not suitable for use as gems. Many of these, such as corundum (ruby and sapphire), are very hard and are suitable for use as abrasives.

A number of other high-value nonmetallic minerals are used in electronics industries, making fertilizers, ceramics, and other essential materials. Table 23–3 lists several high-value minerals, their use, and their sources.

Low-value nonmetallics

The most important low-value nonmetallic mineral deposit is sand and gravel. This material is the basis of construction industries everywhere.

Sand may be produced from stream deposits, recent marine beach deposits, and weathered bedrock deposits. It may also be produced from poorly cemented sandstones by mining and crushing. Gravel for concrete and road aggregate may be derived from bedrock, glacial, surface, marine, and stream deposits. Any rock type—whether igneous, sedimentary, or metamorphic—may be used for road or concrete aggregate provided it meets the desired characteristics. In areas that have no surficial gravel or consolidated bedrock material available for aggregate, shells of oysters and clams or some other abundant substitute have been used. When neither bedrock nor surface sources are available, aggregate may be transported, making it a more costly resource.

Limestone is another low-value nonmetallic deposit. It is used primarily for aggregate, fertilizer, and cement. It may be shipped for moderate distances into areas

Table 23–3. Some high-value nonmetallic minerals

Mineral(s)	Use(s)	Source(s)
Phosphates	Fertilizers, detergents	United States (Florida, Montana, Wyoming, Tennessee, North Carolina) Morocco Soviet Union Tunisia
Sulfur	Sulfuric acid	United States (Texas, Louisiana, other states as a by-product of sulfide mining) Italy Japan Soviet Union
Potash minerals	Fertilizers, dyes, soaps	United States (New Mexico, California, Utah) Canada France Soviet Union East Germany West Germany
Kyanite, sillimanite, andalusite (Al_2SiO_5)	Refractory ceramics	United States (North Carolina, California, Georgia) India Kenya South Africa
Graphite	Refractories, lubricants, pigments	Korea Austria Soviet Union China Mexico
Asbestos	Textiles, insulating materials	Canada Soviet Union South Africa Rhodesia

452

ENVIRONMENTS
HERE
AND
BEYOND
THE
EARTH

lacking limestone accumulations. Chemical-grade limestone contains 98% (or more) calcium carbonate. It is quite valuable for making lime for industrial, agricultural, and municipal uses.

ENERGY RESOURCES

Today's industrialized world is powered by coal and petroleum (Figure 23–8). Nuclear power and hydroelectric power as yet make up a relatively small portion of our energy needs. While the *fossil fuels*, coal and petroleum, remain as our principal sources of energy, it is important to understand the reasons for their occurrence, to examine their modes of origin, and to see how much of these resources remain for future use.

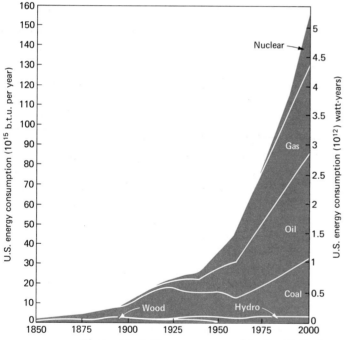

Figure 23–8. Energy consumed in the United States from 1850 projected to the year 2000. Note that fossil fuels will probably remain as our principal energy source at least until the turn of the century. (From Chauncey Starr, "Energy and Power." Copyright © 1971 by Scientific American. All rights reserved.)

Coal

Coal consists of an accumulation of plant debris that formed in great swamps during the geologic past (see Chapter 9). This plant material was removed from contact with the atmosphere as it sank beneath the waters of the swamp. It was later covered by sediment. As the weight of overlying sediment increased, the mass of vegetation became compressed and began to lose some of the water and other volatiles contained in it. Throughout time this plant material was changed into coal. The material first becomes low-rank coal (lignite or subbituminous); if pressures remain or increase, it becomes higher-rank coal, such as bituminous coal (Figure 23–9). Anthracite is produced under still higher pressures and somewhat higher temperatures.

Coal is one of the most abundant resources that the United States has at present (Figure 23–10). Deposits of bituminous coal in the eastern part of the country have been most extensively mined because of their quality and ease of access to industry. Much of the industrial growth in the East during the nineteenth and twentieth centuries may be attributed to the availability of abundant coal resources.

Western coal is generally lower-rank bituminous, subbituminous, and lignite. How-

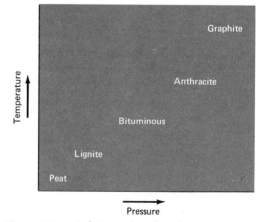

Figure 23–9. Relation of coal rank to temperature and pressure.

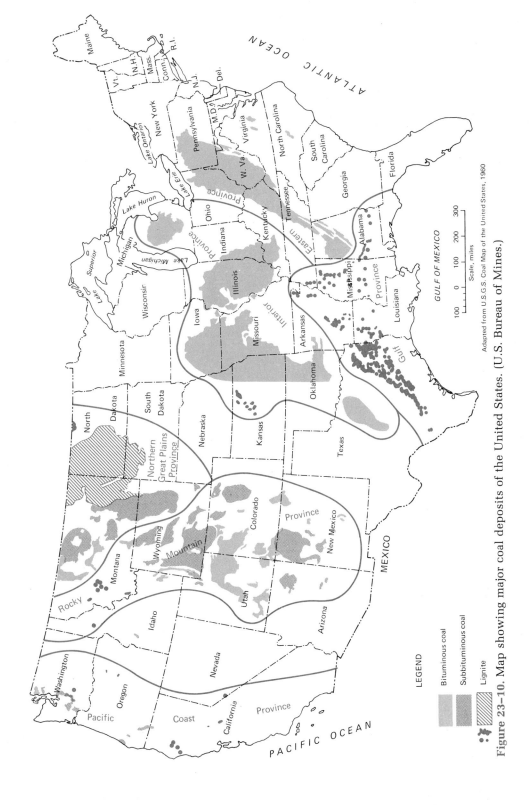

Figure 23–10. Map showing major coal deposits of the United States. (U.S. Bureau of Mines.)

Adapted from U.S.G.S. Coal Map of the United States, 1960

LEGEND

Bituminous coal

Subbituminous coal

Lignite

454

ENVIRONMENTS
HERE
AND
BEYOND
THE
EARTH

ever, it is high-quality coal that contains a sulfur content lower than most of the coal in the East. It also occurs in thicker beds over a larger area. Therefore coal reserves for the future lie in the West, principally in Montana, Wyoming, and the Dakotas. The western coal was formed primarily during the Mesozoic and Cenozoic eras; coal in the East is predominantly Paleozoic (see Table 4–2).

The United States is well endowed with coal reserves. This country will likely be self-sufficient in this commodity for several hundred years. Yet one-half the known coal reserves in the world are found in Asia, principally in the Soviet Union and China. Japan has some coal deposits. The United States and Canada share about one-fourth of the known world coal reserves. The coal deposits of England and Wales, Germany, France, Belgium, and Poland are partially responsible for the industrialization of these countries. Africa has little coal. Most of this is in Rhodesia and South Africa. Australia has the largest coal deposits of the Southern Hemisphere, but these are not large when compared with the Asian or North American deposits. Deposits in South America and Antarctica are relatively limited in size.

Petroleum and natural gas

Petroleum and *natural gas* occur primarily in marine sedimentary rocks. They accumulate after migration from their place of origin to some site where they are trapped. So, in studying the geology of petroleum, it is necessary to understand its mode of origin, migration, and entrapment.

Origin of petroleum. Whether the source of petroleum is terrestrial or marine has been debated for many years. Yet lack of agreement on this point may have helped rather than hindered the search for petroleum because of all that we have learned about it. Petroleum has formed mostly in marine sedimentary rocks or is generally found close to them. Therefore one possible place that it

may originate is the marine environment.

Abundant organic material is deposited in the marine environment along the coasts of continents. It is also being produced in large quantities in the ocean. There is disagreement over whether vegetable or animal organic matter is the primary source for petroleum. But the dominant occurrence in marine sediments has led to the conclusion that petroleum likely forms in the marine environment.

There is minor support for an inorganic origin of petroleum. However, the presence of complex compounds that can be directly related to living things and the association of petroleum with sedimentary rocks lead most to support the organic origin of petroleum. Also, the *source rock*, a single rock type in which the origin occurs, is probably a shale or siltstone. But petroleum may originate in large quantities in other sedimentary rock types.

Petroleum migration. Once organic matter (or inorganic material) is transformed into petroleum, it will migrate from its site of origin to a point of entrapment. Migration is governed partially by the fact that water has a higher density and that petroleum will float on water, because the two liquids will not mix. Its movement is also governed by the laws of fluid dynamics and by the "plumbing" of the geologic system in which it occurs. That is, interconnecting openings must exist between grains in the rocks in order for migration to proceed.

Petroleum migration is thought to take place in two stages: *primary migration* from the source to the reservoir rocks and *secondary migration* within the reservoir rocks to the trap. Petroleum may be forced from its source rocks by the weight of compacting sediment, by water forcing it upward, by pressure of expanding gases, or by other means that may operate on a smaller scale.

Once petroleum is forced from the source rocks into the more permeable reservoir, sec-

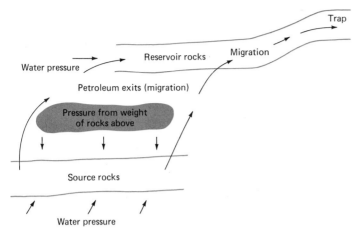

Figure 23–11. Movement of petroleum from the source rocks into the reservoir rocks and a trap.

ondary migration, governed in part by the same factors listed above, has begun (Figure 23–11). Considerable discussion has been vented on the subject of how far petroleum can migrate. Some believe that petroleum can migrate more than 90 mi (150 km). Others feel certain that petroleum cannot migrate more than a few miles (kilometers). The argument for short-distance migration is: No matter how porous a reservoir may be, at some point it very likely will change over any appreciable distance. Another factor that would hinder long-distance migration is changes in structure. Faults, folds, and other breaks may prevent movement.

Reservoir rocks and traps. Any rock type may serve as a petroleum reservoir as long as it is porous, possesses a seal or a cap and, is able to give up its petroleum once it is tapped. This actually restricts reservoir rocks to relatively few types. It particularly excludes most igneous and metamorphic rocks, along with many sedimentary rocks. The most common reservoir rocks are sandstones and porous or fractured carbonate rocks, but many other rock types have yielded petroleum under proper conditions.

Porosity in reservoir rocks may be reduced by cementation, compaction, recrys-

tallization, and granulation. For permeability there must exist an interconnection of pores as well as pores of sufficient size to allow the liquid to pass through. All of the factors that reduce porosity may also reduce permeability.

Reservoir traps may be divided into three major groups: structural, stratigraphic, and combination traps. The earliest oil fields were located in *structural traps*, principally anticlinal folds (see Chapter 18). So dominant was the thinking concerning the occurrence of petroleum in anticlines that most exploration during the latter part of the nine-

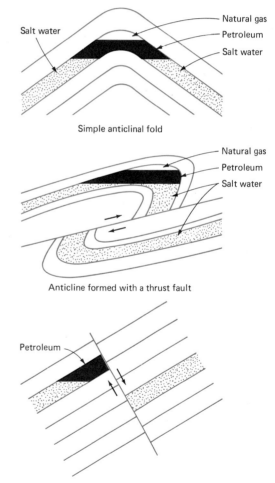

Simple anticlinal fold

Anticline formed with a thrust fault

Trap formed by a seal along a normal fault

Figure 23–12. Traps formed by anticlinal folds and different kinds of faults.

455

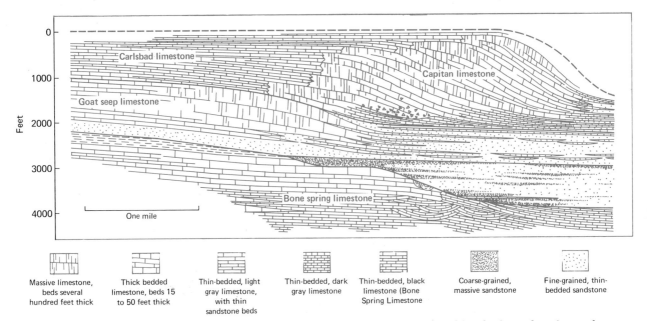

Figure 23–13. The facies relationships and stratigraphic trap formed by the buried reef complexes in West Texas and eastern New Mexico. (From *Geology of Petroleum*, Second Edition, by A. I. Levorsen. W. H. Freeman and Company. Copyright © 1967.)

teenth and early part of the twentieth century was dominated by the *anticlinal theory* of petroleum accumulation. Anticlinal folds of many types, including domes, tight and open anticlines, faulted anticlines, and anticlines formed over buried hills and ridges, have been important producers of petroleum.

Faults may also serve as traps for petroleum and natural gas. Normal faults, thrust faults, and strike-slip faults (see Chapter 18) may form a seal that prevents the upward migration of petroleum (Figure 23–12). They may also fracture brittle rocks near faults so that porosity is increased, thereby increasing the potential volume of oil in a reservoir.

Stratigraphic traps usually result from a lateral change in sedimentary rock type, called a facies change (see Chapter 9). Not until well into the twentieth century was the importance of stratigraphic traps caused by facies changes recognized. As a result several important discoveries were made. An excellent example of this type of trap lies in the Permian Basin of West Texas and eastern New Mexico where several buried reef, reef–

flank, and basin facies successions serve as traps in one of the largest oil-producing areas in the United States (Figure 23–13).

Unconformities may also trap petroleum (Figure 23–14). Angular unconformities are the most obvious candidates for traps, but disconformities may also trap petroleum.

Combination traps may involve folding and facies changes, facies boundaries cut by faults, and even regional tilting of sedimentary rock units containing rock types that thin and wedge (Figure 23–15). The growth faults (see Chapter 18) on the Gulf Coast of Texas and Louisiana serve as combination traps because their location and movement

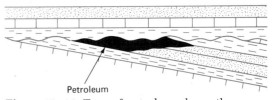

Petroleum

Figure 23–14. Trap of petroleum beneath an unconformity.

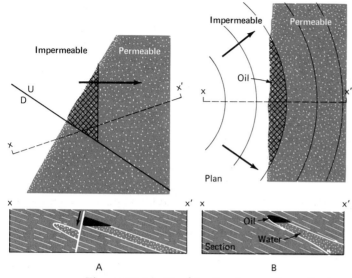

Figure 23–15. Combination traps. (From *Geology of Petroleum*, by A. I. Levorsen. W. H. Freeman & Co. 1967.)

may control sedimentation and facies changes.

Petroleum-producing areas. The United States was the world's main producer of petroleum during the 1940s (Figure 23–16). By 1972 this country produced less than one-fourth of the world's refined petroleum products, largely because of intensified development of petroleum resources in other parts of the world.

The first petroleum in the United States came from wells drilled in Pennsylvania and Kentucky. Large fields were discovered in Oklahoma, Wyoming, California, and Texas in the early 1900s. The large fields on the Gulf Coast of Texas and Louisiana, some located on the continental shelf, the West Texas fields, and the fields in and near Los Angeles produced most of the petroleum during the 1960s and early 1970s. All these, except that in West Texas, involve combination traps. The West Texas fields are developed in stratigraphic traps.

The North Slope field along the northern coast of Alaska is a large field that will eventually yield as much as 9 billion barrels of petroleum. It is developed in a large gentle anticlinal structure containing stratigraphic pinchouts. Other large fields probably exist along the coast and continental shelf of other parts of North America.

One field located in the area of the Mackenzie River delta along the northern coast of Canada will add significantly to the energy resources of that country. However, this is primarily a natural gas field. Stratigraphic and combination traps in rocks underlying the Great Plains and foothills of the Canadian Rockies of Alberta presently serve as the main source of petroleum for Canada.

The largest known petroleum reserves in the world, about 62%, are in the Middle East, mainly in Saudi Arabia and Kuwait. Iran and Iraq each have resources about as great as those of the United States. Most of the petroleum in the Middle East occurs in combination traps in limestones of Mesozoic and Tertiary age.

Most of South America lacks known petroleum resources, except for Venezuela, a major exporter to the United States. The east side of the Andes in Peru and Bolivia also contains some fields, but exploration has proceeded slowly. The few small fields in Brazil are not of sufficient size to supply that country's needs.

Africa has some large fields north of the Sahara Desert, principally in Libya, but the southern part of the continent contains very few. Some petroleum discoveries have been

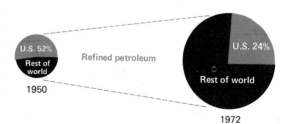

Figure 23–16. Comparison of U.S. petroleum production to the rest of the world in 1950 and 1972. The relative size of the circles indicates the magnitude of total production. (From U.S. Bureau of Mines, Mining and Minerals Policy, 1973.)

457

458

ENVIRONMENTS
HERE
AND
BEYOND
THE
EARTH

made along the western coast, but thus far they are not of the magnitude of the large fields in other parts of the world.

Western Europe has few petroleum resources. The fields in the North Sea yield petroleum and natural gas but will supply only a small proportion of the needs of the continent. The Soviet fields along the Caspian Sea have produced significant amounts of petroleum for many years. Other fields in the Soviet Union make that country practically self-sufficient.

Most of Asia contains little in known petroleum resources, except for those in the Soviet Union and China, somewhat unknown quantities in themselves. Japan imports almost all of its petroleum from the Middle East. The petroleum production in the East Indies and Australia is not enough for the industrialized nations of this area to be self-sufficient.

Most petroleum resources, like other resources, are concentrated in some areas and impoverished in others. Figure 23–17 summarizes petroleum reserves for major oil-producing nations of the world.

Other petroleum sources. Oil shale and tar sands are two promising future sources of petroleum. Both must be mined and require heat to release or distill the liquid to be recovered. *Oil shale* does not contain petroleum; organic matter is in the form of a residue called *kerogen*. Distillation breaks down the kerogen and releases a liquid similar to petroleum. Oil shale deposits in Wyoming and Colorado are estimated to contain more than 1000 billion barrels of petroleum. More than 400 billion barrels of petroleum may eventually be recovered from this resource. Sizable deposits of oil shale also occur in Brazil, the Soviet Union, central Africa, and western Europe.

Tar sands consist of sand or sandstone cemented by an asphaltic residue of a former liquid petroleum accumulation. The largest tar sands in the world are in western Canada.

The most valuable deposits are in Alberta; some exist in Saskatchewan. The tar sands occur in rocks of Cretaceous age. They have to be mined and processed by heating in order to release hydrocarbons.

Both oil shale and tar sands will undoubtedly serve as major sources of petroleum in the future. However, for this to come about, environmental problems associated with surface mining and disposal of waste from processing of these materials must be solved, particularly in the case of the oil shale. Other factors that prevent oil shale and to some extent tar sands from being economic at present are the number of steps they must be taken to produce the same raw material that comes directly from an oil well in liquid form. Oil shale must be mined, transported to the processing plant, and then returned to break down the kerogen to produce a refinable liquid.

Future energy sources

Coal and petroleum are present on the earth in finite quantities. As our demands for rates of consumption of these sources of energy continue to increase, the limited aspect of their occurrence will become more apparent. We shall look briefly at some of the alternatives to these traditional energy sources.

Nuclear energy. Nuclear energy promises to be an abundant energy source for the future. Our use of nuclear energy is now limited to fission reactors fueled with U-235–enriched uranium fuel. Reactors of this type burn fuel in a nonrecyclable fashion much like a conventional coal- or petroleum-fired power plant. Uranium resources can therefore be depleted in the same way that coal or petroleum are. The breeder reactor, once developed on a commercial scale, will likely prevent complete depletion of uranium resources, because it produces more fuel, in the form of fissionable plutonium-239, than it uses. Therefore there will be an artificial

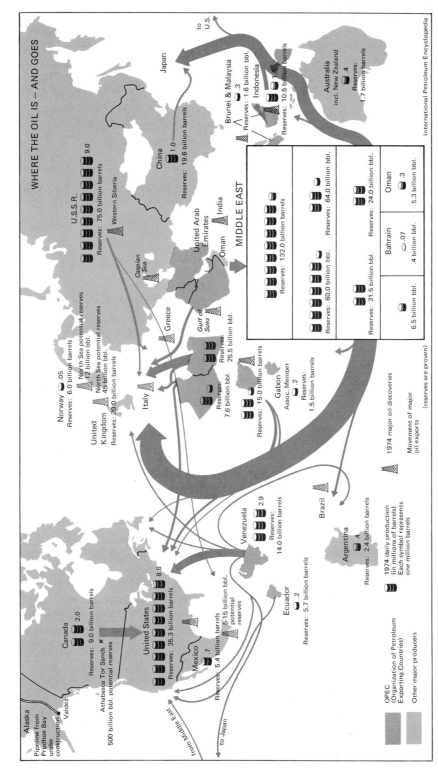

Figure 23–17. Distribution of petroleum reserves and where it is used throughout the world. (Reprinted by permission from TIME, The Weekly Newsmagazine; Copyright Time Inc.)

460

ENVIRONMENTS
HERE
AND
BEYOND
THE
EARTH

source of reactor fuel in the future, provided that the breeder reactor is employed before uranium resources are completely depleted.

The fusion reactor may be the major energy source of the next century. Fusion involves hydrogen (H^2 and H^3 isotopes) fuel that reacts at very high temperatures to produce helium and energy. This is the same process that keeps stars, such as the sun, burning for billions of years. Problems involving control of the fusion process, so that energy cn be released slowly, must still be resolved.

Uranium for fission reactor fuel occurs in minable quantities either as primary uranium minerals, such as uraninite (UO_2) in igneous or metamorphic rocks, or as secondary minerals, such as carnotite (a complex potassium uranyl vanadate) in sedimentary rocks. Principal deposits in the free world are in the United States (mainly in Colorado, Utah, New Mexico, and Wyoming), Canada, South Africa, France, and Australia.

One geologic problem arising from the use of nuclear energy is the disposal of nuclear waste products into the physical environment. Nuclear wastes present problems that normal chemical wastes do not: They cannot be broken down by the addition of other chemicals and must be stored for hundreds or thousands of years until radioactive elements decay to safe levels. Low-level wastes, because of their low heat productivity, may be stored in impermeable underground cavities excavated in silicate rocks. High-level heat-generating wastes must be stored in a heat-conducting material. One of the few natural materials that needs this requirement is salt.

Geothermal energy. Heat is released continuously from the interior of the earth, with some areas releasing heat at faster rates than others. The rate at which temperature increases with depth is called the *geothermal gradient*. Areas with a high geothermal gradient are potentially areas that could be tapped for geothermal energy. However, other factors restrict the number of suitable areas to a relative few (Figure 23–18). Major geothermal heat sources are found in Iceland, New Zealand, northern Italy, Japan, Mexico, the Soviet Union, and California.

The most obvious requirement for a potential geothermal area is heat. Recent volcanic areas that are either still active or have cooling igneous bodies at depth are the most common geothermal areas. Most geyser regions could serve as geothermal energy sources, but tapping them would probably destroy the geysers.

Tapping geothermal energy for use is commonly accomplished by drilling wells to depths that will likely yield superheated steam. However, if water is not present in the first place, there will be no steam. So there must be an adequate source of water in the form of surface water (or in some cases trapped seawater) to be heated and maintained at a high temperature. Attempts to pump water down to be heated and brought back to the surface have not been successful in most areas because of the small area from which heat is withdrawn and the slow rate of heat recharge (this is attributed to the insulating quality of most common rocks).

Disposal of waste water is a problem in some areas because of the presence of dissolved salts. Hot water has a greater solvent capability than cooler water; thus over a long period of time large quantities of materials may be dissolved. Such a problem exists in the Cerro Prieto district near the Salton Sea along the California–Mexico border. Sulfur (SO_2) pollution of the air is also a problem in some geothermal areas. According to some estimates, the amount of sulfur pollution may be as much as that associated with fossil fuels.

Geothermal energy will not supply more than a small percentage of our total energy needs in the future, largely because of its restricted occurrence. However, there is a possibility of utilizing geothermal heat from

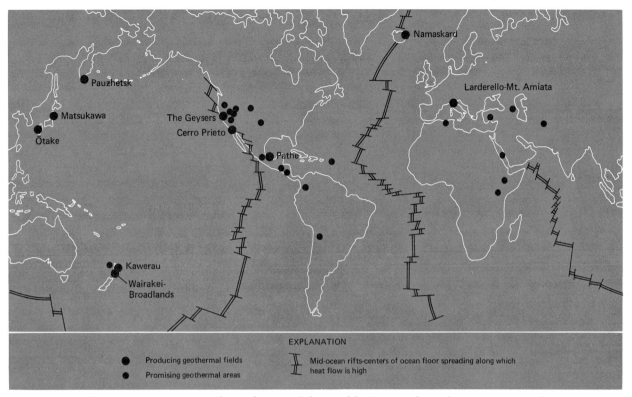

Namaskard

Larderello-Mt. Amiata

Pauzhetsk

Matsukawa

Ōtake

The Geysers

Cerro Prieto

Pathe

Kawerau

Wairakei-
Broadlands

EXPLANATION

● Producing geothermal fields

● Promising geothermal areas

▮ Mid-ocean rifts-centers of ocean floor spreading along which
heat flow is high

Figure 23–18. Major geothermal areas of the world. (U.S. Geological Survey News Release, January 27, 1971.)

more normal areas. If this is developed, geothermal energy could become a significant energy source.

Tides. Tidal energy has been considered as a potential energy source. This could possibly be tapped in coastal estuaries and other restricted channelways possessing large tidal ranges, such as in the Bay of Fundy between Nova Scotia and New Brunswick. In such areas gates would be left open so that water would flow in during high tide. Then the gates would be closed, trapping water and directing it through conventional hydroelectric turbines. This energy source will not supply a large amount of our total needs but could cut consumption from other sources of energy on a local scale.

Solar energy. A large amount of energy from the sun reaches the earth each day,

equivalent to that produced by burning about 5000 billion tons of coal. Attempts to tap this energy on a large scale have been unsuccessful. However, development of home-sized solar generators or solar furnaces may be practical in the near future.

Wind. In some areas of the earth the wind blows almost continuously. Windmills have been used for many years to pump and even to generate electricity. Small-scale use of this energy source will likely continue, but large-scale development will probably not take place for some time.

RESOURCE NEEDS: PAST, PRESENT, AND FUTURE

We have observed that there is an unequal distribution of resources on our planet and that this distribution is related to the

462

ENVIRONMENTS
HERE
AND
BEYOND
THE
EARTH

geologic history of the earth. We can also see the trend toward mining lower and lower concentrations of resources (Figure 23–19). Paralleling these relationships are (1) the increasing rate of energy consumption, (2) the immediately apparent slowdown in use of fossil fuels because of the incapability of reserves to meet long-range demands, and (3) the concern about environmental quality. Will these trends continue? If so, toward what end? Will we have enough resources of all kinds to meet projected demands?

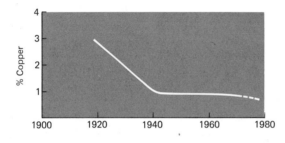

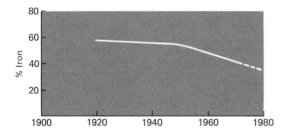

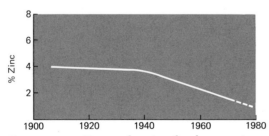

Figure 23–19. Curves showing the decrease in concentration of metals in minable ores. This is due to increased efficiency of mining, increase in value of metals, and depletion of higher grade ores.

Many things are involved in answering these questions, not the least of which is political-socioeconomic structures that obtain in this country and in the rest of the world and how they will change in the next century. Another is that practically all judgments related to resources and other commodities are based on economics. Moreover, the dominant economic philosophy is that of growth: growth in production, standards of living, and availability of consumer goods. These are readily equated with a better quality of life and the "American dream." Yet all these things require increased utilization of resources and energy. As more nations become industrialized, will they achieve the standard of living we have experienced in the United States? A few may achieve it, but most will not. The future will be clouded by uncertainties of a population not brought under control, by dwindling resources, and by increased use of energy. A great deal of energy will have to be expended to recover resources from waste. Already we are recycling large amounts of aluminum, copper, lead, iron, and even glass bottles that we use today.

Increases in food production will be made difficult by problems arising from increased resource needs for support of agriculture, as in production of chemical fertilizers (mined as potash and phosphates) and energy resources for producing, running, and maintaining agricultural machinery.

What about our continued abilities to convert to the use of more abundant substitutes for resources that become depleted? Are we not already substituting plastics and ceramics for many metals? And what about the promise of almost unlimited amounts of energy from nuclear fusion? Also, are we not able to mine profitably smaller and smaller concentrations of resources from the earth? All of these fit into several of what geologist Preston Cloud calls "cornucopian premises."

Cloud calls one of these premises the

"technological fix." It states that technology will provide us with continued high standards of living and inexhaustible energy sources, abundant food, and a clean environment. But will it? Technology has given us many things, but there is a limit to what it can do, particularly in the light of the finite nature of the earth's resources.

The premise that economics is the sole governing factor in determining the minable concentration of any metal implies that a continuum exists, from whatever we mined in the past to the average crustal abundance that could become minable under the proper economic conditions. It ignores the fact that mineral deposits are anomalous features resulting from processes operating within the earth, which concentrate these elements many times above their average crustal abundances (Figure 23–20). It is unlikely that we will be mining lead at its average abundance of 0.00013%, considering its economically minable concentration today of about 2%. Realize also that many rocks contain almost no lead. The same is true for most valuable minerals.

So where will this lead us? We should first stabilize world population as a limit to growth. Our rates of energy and other resource consumption should then be lowered or at least leveled off. If we do not attempt to stabilize the system of which we are a part, nature will do it for us, perhaps drastically. Because we are a part of the earth system, we are subject to the laws that attempt to maintain it in a state of equilibrium. The solutions may not be so much to technological as to legal, socioeconomic, and educational problems.

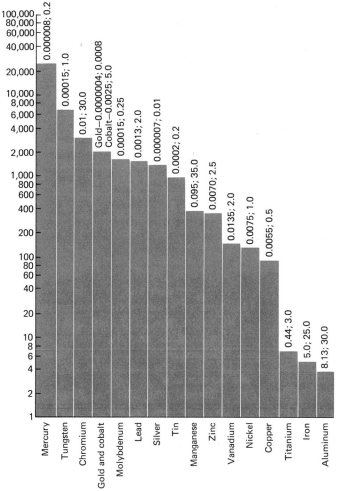

Figure 23–20. The relationship between average crustal abundance, minable concentration, and the concentration factor above crustal abundance. (From Cargo and Mallory, *Man and His Geologic Environment*, Reading, Mass., Addison-Wesley, 1973.)

SUMMARY

Human beings have been dependent on mineral resources for many centuries. Today we use a wide variety of metallic and nonmetallic minerals as well as fuels derived from earth resources.

Resources may be divided into renewable and nonrenewable resources and into high- and low-value resources. Estimates of reserves are made to plan for mining on a local scale or imports and exports on a national or an international scale.

Mineral deposits may be grouped into primary and secondary deposits. Primary deposits most commonly occur in igneous or metamorphic rocks, but many important pri-

464

ENVIRONMENTS
HERE
AND
BEYOND
THE
EARTH

mary deposits are present in sedimentary rocks. Secondary deposits usually occur on the surface as a result of weathering. Placer deposits may contain important concentrations of durable, chemically inert, dense minerals.

Mineral deposits containing one or more metals that may be extracted economically are metallic mineral deposits. Some metals may concentrate in the native state; others combine as sulfides and oxides.

Nonmetallic mineral deposits are those from which a nonmetallic element is recovered or which are mined and some compound is extracted. Examples of high-value nonmetallics include gemstones, certain fertilizer components, and materials used in ceramics and electronics industries. Low-value nonmetallics include sand and gravel, brick clays, and material for cement.

The fossil fuels—coal and petroleum—remain as our principal sources of energy. Coal was formed as an accumulation of land plants in swamps that were buried and much of the volatile material forced out under pressure. Different ranks of coal result from different amounts of volatiles being forced out. Present coal resources of the United States are sufficient for many decades.

Petroleum probably forms in the marine environment from an accumulation of either land- or marine-derived plant or animal organic matter. Once petroleum forms, it migrates from the source beds into reservoir beds, where it accumulates in structural, stratigraphic, or combination traps. The Middle East contains most of the known petroleum reserves in the world. The United States, the Soviet Union, and Venezuela are also major producers. Oil shale and tar sands may provide additional petroleum resources in the future.

The most likely large energy source in the future will be nuclear energy. Geothermal energy, tides, solar energy, and wind energy will account for a small percentage of future energy needs.

The immediate need to control world population and to cut back on the use of resources may result in a continued availability of resources. Limitations in quantity will force substitutions, but this is also finite.

Questions for thought and review

1. What makes a mineral deposit a minable ore?
2. How can limestone be both a high-value and a low-value resource?
3. What are the characteristics of a mineral that would accumulate in a placer deposit?
4. Why has taconite not been widely sought in the past?
5. Why is sand and gravel the most valuable of all nonmetallic mineral deposits?
6. How can a low-rank coal, as subbituminous, still be a high-quality coal?
7. What determines the manner in which petroleum will migrate?
8. Why does most petroleum accumulate in combination traps?
9. Why has oil shale not been economic in the past?
10. Why will geothermal, tidal, solar, and wind energy not supply much of our future needs on a large scale?
11. What solutions exist concerning directions to take in problems of increased demands for metals, energy, and other resources?

Selected readings

Ayers, Eugene, 1956, The Fuel Situation, *Scientific American*, vol. 195, no. 4, pp. 43–49.

Bachmann, H. G., 1960, The Origin of Ores, *Scientific American*, vol. 202, no. 6, pp. 146–156.

Cloud, Preston E., Jr., 1972, Realities of Mineral Distribution. In McKenzie, G. D., and Utgaard, R. O., *Man and His Physical Environment*, Burgess, Minneapolis, pp. 194–207.

de Nevers, Noel, 1966, Tar Sands and Oil Shales, *Scientific American*, vol. 214, no. 2, pp. 21–29.

Ehrlich, P. R., 1968, *The Population Bomb*, Ballantine, New York.

Flawn, P. T., 1966, *Mineral Resources*, Rand-McNally, Chicago.

Hubbert, M. K., 1971, The Energy Resources of the Earth, *Scientific American*, vol. 225, no. 3, pp. 61–70.

Marsden, S. S., Jr., 1958, Drilling for Petroleum, *Scientific American*, vol. 199, no. 5, pp. 99–111.

Mero, J. L., 1960, Minerals on the Ocean Floor, *Scientific American*, vol. 203, no. 6, pp. 64–72.

Rona, P. A., 1973, Plate Tectonics and Mineral Resources, *Scientific American*, vol. 229, no. 1, pp. 86–95.

Star, Chauncey, 1971, Energy and Power, *Scientific American*, vol. 225, no. 3, pp. 36–49.

24 Control of the environment

STUDENT OBJECTIVES

Upon completing this chapter the student should:

1. be able to compare and contrast environmental problems in today's world with those of the past

2. be aware of environmental problems introduced by irrigation in arid and semiarid regions

3. recognize the nature of industrial pollution of air and water

4. understand changes caused by water pollution in oceans

5. recognize the role of municipal sewage in the pollution of streams and lakes

6. be able to identify alternatives to the internal-combustion engine as a means of transportation

7. suggest ways in which geologists, in cooperation with other scientists and urban planners, can improve our environment

Speak to the earth, and it shall teach thee.

Job XII.8

468
ENVIRONMENTS
HERE
AND
BEYOND
THE
EARTH

The setting was autumn, 1948, in the small industrial town of Donora, Pennsylvania. The Friday night Halloween parade was being conducted in a rather thick fog. But this was no ordinary fog. It was caused by a temperature inversion—cold air trapped beneath a layer of warm air. The fog (or smog) was composed of an assortment of exhausts from steel and zinc mills, trains, cars, trucks, and other sources trapped and concentrated by the inversion. Later analysis showed that the fog mixture included, among other ingredients, sulfur, arsenic, lead, cadmium, iron, zinc, chlorides, and acid gases. The fog, described by one as "so thick you could taste it," persisted for five days. Over the weekend 20 people in Donora died as a result of breathing this air, and thousands were seriously ill. A state of emergency was in effect until wind finally dispersed the fog.

Another scene: An average water sample is drawn from the Mississippi River just south of St. Louis, Missouri. A fish is placed in this water. It dies. The water is diluted 10 times with pure water (not from the Mississippi). A fish is placed in the diluted water. It dies within 1 min. Even when the river water is diluted 100 times with pure water, the fish dies overnight.

Human beings have invaded, adjusted to, and prospered in a variety of earthly environments from deserts to tropical jungles to frozen tundra. In the past, primitive groups adjusting to these environments accepted the dictates of nature. There was little choice. With advancing civilization and technology, however, *Homo sapiens* acquired increasing power and opportunity to affect the natural environment and to alter it to his needs. In so doing, equilibrium systems in nature have become upset. The incidents at Donora and in the Mississippi River illustrate this point. These examples also suggest that perhaps we do not have so much control over our environment as we think, that in our ignorance we have upset long-established balances in nature without

due reflection or analysis of the consequences.

We have alluded to the fact in this book that water and atmosphere make our planet unique. In this case, one would think that intelligent inhabitants of this planet would treat these two substances with great reverence. In general, however, this has not been so. The history of humankind is one contrasting great achievements with great follies. This chapter proposes to review some of the problems related to imprudent use of our water and atmosphere and to consider possible remedial measures that might be introduced to protect them.

LESSONS FROM THE PAST

Lost civilizations are a microcosm of world civilization. Archaeologists assure us that ancient civilizations arose, flourished, and then slipped into oblivion. The Aztecs, Mayas, and Incas were in the Western Hemisphere; on the other side of the world were the Babylonians, Persians, and Minoans. Historians explain their disappearances as great civilizations in terms of lost leadership, complacency, decadence, or aggressive neighbors. It would appear, however, that archaeological findings have uncovered another, perhaps more important, cause for their disappearance. The cause may have been their inability to live in harmony with nature.

The disappearance of the Babylonian civilization from the once rich Tigris–Euphrates basin is well documented. The downfall began innocently enough by the efforts of the Babylonians to improve their standards of living. A program for the irrigation of the rich but dry lands of the Tigris–Euphrates basins was the first step toward disaster. While irrigation advanced in the Babylonian Empire, the quality of water dropped drastically as more and more dissolved salts that washed out of the farmlands found their way into principal streams.

Archaeologists have traced cycles of salinization and land abandonment in the Tigris–Euphrates basin of the former Babylonian Empire that date back 6000 years. Evidence supporting the cyclic theory of salinity and land abandonment has been revealed in records of ancient temple surveyors, who noted the presence of saline ground. These records are further supported by parallel studies of grain crops raised in this region. When cultivation of this arid region began through irrigation, the proportion of wheat and barley crops was equal. Within 1000 years the less salt-tolerant wheat crop dropped below 2% of the total grain crop. The shift to barley as a principal grain crop was associated with a decline in the fertility of the soil, attributable to salinization of the soil. In one area studied, the crop yield declined nearly 50% in a period of 400 years.

The government of Iraq is attempting to reclaim this land, which has been a desert since the twelfth century. The concern for water by Iraq is underscored by the fact that anyone caught deliberately damaging irrigation works in that country is punished with death. One may wonder, however, whether the costs will balance the gains. At best, irrigation is but a temporary measure in attempts to feed the hungry of the world.

Human impact on the chemical content of water is profound. Farming developments have turned semiarid regions into lush gardens of Eden with the use of chemical fertilizers and irrigation but not without paying a price in terms of damage to the biosphere (Figure 24–1).

In the southwestern United States we are faced with a critical salinization problem, produced by increased use of the Colorado River for irrigation. The problem has become aggravated during the past several years by demand for the fresh fruits and vegetables that grow so abundantly in warm regions. Salts washed from farms in the upper basin of the Colorado River have increased the salinity of the water in the lower part of the river by 30% to 1.3 tons of salt per acre-foot (1000 ppm). The agricultural empire of the Imperial Valley of California depends almost entirely on irrigation water from the Colorado. Salt concentrations have resulted in once-fertile farms being abandoned and are responsible for increasing costs of operation at the remaining farms.

Are we traveling the same route as the Babylonian Empire? It would seem so. Not only are areas in the United States threatened, but the same pattern can be observed in irrigated lands in Egypt, the Soviet Union, China, Australia, and Africa. Irrigation is not a permanent solution to feeding a populace, because once-fertile land is eventually withdrawn from the tillable land bank.

WATER LOSSES AND DAMS

Although we sometimes associate the construction of dams with irrigation, dams are erected frequently for purpose of water control and prevention of floods. Regardless of the reasons, dams may actually result in a loss of water for useful purposes.

Behind each dam a lake is formed. Water that moved swiftly through a valley is now impounded to form a broad expanse of open water. This broad water surface increases loss of water to the atmosphere through evaporation. Much of this water vapor will eventually return to earth in some form of precipitation but usually at some distance from where it was taken.

The rising water level behind a dam may introduce additional water losses. The process of stream valley evolution considers balances in nature. The stream bed is usually a hard-rock surface, and valley walls provide a balance between water absorption and maintenance of stream flow. Impounded water behind a dam must find a new balance. This adjustment may result in heavy water losses.

Sandstones in the southwest soak up water quickly. Such sandstones are found in

470

ENVIRONMENTS
HERE
AND
BEYOND
THE
EARTH

Figure 24-1. The typical American desert landscape *(left)* can be transformed into a lush Garden of Eden *(right)* through irrigation. Where once grew cacti and mesquite now grow cotton, lettuce, and broccoli. The blossoming landscape hides an ugly problem. With irrigation, water is being pumped from the soil faster than nature can return it. (Courtesy of American Airlines.)

the Imperial Valley water transportation system. This system includes Lake Mead, Davis Dam, Parker Dam, and Imperial Dam (Figures 24–2 and 24–3). Here, as water levels in impounded lakes reach permeable sandstone formations, water from the lakes passes through the sandstones as does water through a sieve. As much as 10–15% of the water of the Colorado River is lost in this way. Adding this loss to water lost by evaporation means a sizable loss of water vitally needed in an arid region.

INDUSTRIAL AND URBAN WATER POLLUTION

Industrial growth is closely associated with population growth. The greater the population, the greater is the need for goods and services. In an industrial society, growth rate exceeds ability to solve problems created by growth. Before we explore pollution problems produced by industry and urban communities, we should review our water resources.

About one-third of the water flowing in streams of the United States is used at least once. At the present rate of increase of water usage, by the year 2000 we will use the total stream flow. Use of water does not necessarily mean that water has been consumed. When we take a bath or a shower, we do not consume water. The water was used by us, but it can be restored either by natural means or by water treatment. There is, however, a significant amount of water used. Of the water applied to irrigation, only 40% is returned to streams. By contrast, it is estimated that 90% of water used by municipalities or by industry is returned to streams. A close

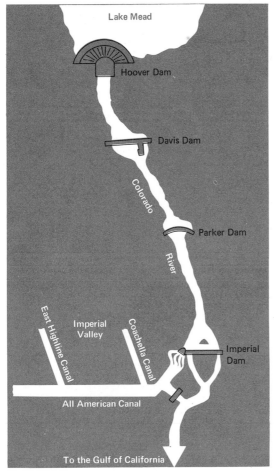

Figure 24–2. Today a vast complex of dams and reservoirs along the Colorado River and the Rio Grande put much of the water flowing through these rivers to work before the water moves down channel. As water flows from irrigation ditches through irrigated fields, it leaches out salts in the soil. After this bitter water returns to the river, it increases the water's salinity, limiting its use for human consumption.

appraisal of urban and industrial use of water will show that we are moving slowly but surely in the direction of chronic water shortage.

Industrial water pollution

When water is used to rotate turbines in hydroelectric stations, only the energy of the water is used. It is neither lost nor polluted. If water is used in industry as a coolant, it is changed only in its temperature. Usually this is a temporary change, although it can have profound effects on water organisms which are found in the vicinity of the industry. Far more serious effects result, for example, when water used in chemical treatments in industry becomes contaminated with acids, phenols, polyvinyl chloride, and chromium. To recover this water would be a major and costly undertaking. We are rapidly approaching the point at which such major and costly undertakings must be carried out. We might say that a philosophy of "unmanufacturing" our products must be assumed to restore the integrity of natural systems.

There are many well-documented cases of discharges of mercury into waterways. Mercury is many times heavier than water, and it was assumed that waste mercury could be disposed of by dumping it safely into nearby lakes and streams. It was not fully realized that mercury would undergo changes when introduced into water to form the compound methyl mercury. Methyl mercury enters the food chain in the lower orders of plants and then advances into the marine food chain step-by-step until it may end up on our table.

The effect of methyl mercury on the human body is not fully understood. Some extreme cases of mercury poisoning have been reported. The Minimata incident in Japan is perhaps the best known. The bay into which an industrial plant was discharging mercury was also a source of fish for people living in the area. As time passed, strange things began to happen. At first it appeared that the cat population was undergoing nervous disorders. Shortly after that, similar disorders appeared among the people. More than 100 people died or suffered serious damage to their nervous systems. After that the world learned of the danger from this source. How does one remove methyl mercury from a stream? Obviously the answer is

472

ENVIRONMENTS
HERE
AND
BEYOND
THE
EARTH

Figure 24–3. Behind each dam an artificial lake is formed. The broad lake surface increases loss of water through evaporation. Higher water levels behind each dam may reach permeable rock formations. Penetration of impounded water into rock increases water losses. (Courtesy of U.S. Geological Survey.)

by not permitting mercury wastes to enter waterways. Although greater control is now exercised, the problem is still with us. Awareness of the problem is, in itself, an important step forward.

Mercury is only one of a number of trace elements; we also have copper, molybdenum, cadmium, beryllium, lead, selenium, arsenic, and strontium. The effects of these elements, singly or in combination, when introduced into the human system have not been fully explored. Some trace elements are beneficial, even necessary, for human health. Others such as lead and mercury adversely affect the nervous system, and beryllium, selenium, arsenic, and cadmium are known to be carcinogenic in animals in laboratory studies. High concentrations of cadmium have been found in oysters. Cadmium poisoning is the cause of a disease called *itai-itai* (ouch-ouch) in Japan, because it causes severe distress to the victim. The U.S. Geological Survey has detected concentrations of cadmium above U.S. Public Health Service standards in raw waters used for drinking in 20 U.S. cities. Long-term as well as short-term effects of trace elements require more investigation.

Municipalities and water pollution

Industry has often been singled out as the prime cause of water pollution. This is not true. Although it does indeed contribute to degradation of waterways, the prime contributor of pollution are the many metropolitan areas that have developed in the last 30 years or so. In 1900, 1 in 20 Americans lived in urban areas. By 1970, 15 in 20 Americans lived in urban areas. During that period of time, while urban population was increasing by 2800%, rural population was decreasing by 30%. Demographers predict that, by 1985, 4 in 5 people in the United States will live in a metropolitan area.

The rapid growth of urban areas outstripped the capacity of most sewage treatment plants to treat waste waters adequately.

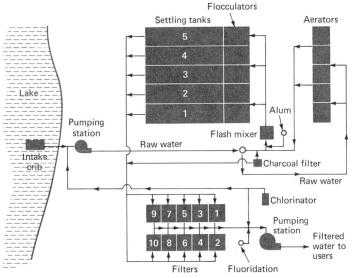

Figure 24–4. In this water treatment system the charcoal filter was used primarily during the summer high-algae season to remove the unpleasant taste from drinking water. The charcoal filter in the future may be used continuously to remove suspected carcinogens from chlorinated drinking water.

Sewage treatment plants designed for a population of 500,000 people are frequently servicing cities of 1,500,000 people. The same can be said for many suburbs adjoining these cities. Sewage treatment plants can be classified into primary and secondary (many large cities have only primary sewage plants). Primary sewage treatment plants remove only solids. This means that chemicals such as phosphates, which are in solution, pass through these plants unaltered. Secondary treatment plants can remove some amounts of phosphate and nitrogenous wastes, but the process requires careful control of activated sludge involved in the removal. Secondary plants are relatively few in number and are usually overtaxed in their capacity. Thus millions of gallons of untreated or partially treated sewage water are dumped into lakes, streams, and rivers daily.

The problem is compounded by a city's increased needs for tap water in homes, offices, schools, and industries located within its boundaries. One city's sewage water may be another city's drinking water. Washroom graffiti at a small bus stop in Missouri reads, "Flush the toilet. They need the water in St. Louis." A recent survey shows that only 350,000 Americans live in urban areas that adequately treat their waste waters. The remainder live in areas where sewage waste waters receive less than adequate treatment. Probably 1 in 4 American cities and towns has no sewage treatment facilities at all.

Providing safe and adequate drinking water for large communities is a monumental task. The problem has been made exceedingly more complex as a result of research findings associating chlorine with cancer. Most public health service standards in the United States and elsewhere are concerned primarily with prevention or spread of communicable diseases, such as cholera and typhoid. Municipalities rely heavily on chlorine treatment of water (Figure 24–4). Chlorine destroys the bacteria that produce these diseases. But recent studies indicate that

474

ENVIRONMENTS
HERE
AND
BEYOND
THE
EARTH

chlorine may react with hydrocarbons in water to form chlorinated derivatives. Chlorinated hydrocarbons, known carcinogens, are suspected cancer producers.

Research conducted at the University of New Orleans by John Laseter and his associates shows that tap water in New Orleans contains more chlorinated hydrocarbons than does untreated Mississippi River water.

The Environmental Protection Agency (EPA) has also confirmed the presence of a number of organic compounds suspected as carcinogens in the New Orleans water supply. Among the compounds found were chloroform, carbon tetrachloride, and the pesticide dieldrin. It would thus appear that untreated water from the Mississippi River is safer for human consumption than the Mississippi water treated with chlorine.

The statistical study *The Implications of Cancer-Causing Substances in Mississippi River Waters* (Environmental Defense Fund, Washington, D.C., November 1974) indicates that cancer mortality rate is 15% higher among white males who drank water from the Mississippi River than among those who obtained their drinking water from wells. Groundwater supplying the wells, in contrast to the Mississippi River water, is not contaminated by the chemicals suspected of being carcinogens.

Natural cycles for the purification of water can no longer function. If a few people living along a river dump their sewage into the stream, organisms in the water can easily convert wastes into harmless substances. As the population along the river increases, the river's ability to reduce wastes becomes overstrained. It is then necessary that sewage or intake water be treated if the river water is to be safe to drink. If population continues to increase, more elaborate treatment is required to make the water safe for human consumption. We can conclude that, as population increases, the per capita cost of water purification increases. The problem that we

must now face as a result of the New Orleans studies is, "What can we use to purify water if chlorine is potentially more hazardous than the bacteria it kills?"

The ability of the carbon molecule to filter out impurities has been recognized by scientists. It was used during World War I to protect soldiers from poisonous gas attacks. It is used in goldfish bowls today to protect our piscatorial friends. Perhaps it can help us in our efforts to provide safe drinking water.

An average household can get safe drinking water if the water treatment plant introduces charcoal filter beds into its purification system. Many municipalities are using charcoal filters during the summer high-algae period. Charcoal removes the algal taste in water. Using charcoal filter beds, many cancer-causing chlorinated hydrocarbons may be filtered from drinking water.

THE VULNERABILITY OF THE OCEANS

There is a close association between oceans and atmosphere. We have seen this relationship in the hydrologic cycle (see Chapter 4). It does not end there. Phytoplankton living in oceans are the primary replenishers of oxygen in the atmosphere. A forest during daylight produces a considerable amount of oxygen by photosynthesis, but during darkness much of it is removed. The net gain of atmospheric oxygen from land plants is very small. Thus phytoplankton in oceans replenish the oxygen that we remove each day as we breathe, drive, or burn anything in open air.

Perhaps the most ominous ecological news of the 1960s was found in a journal article by Charles F. Wurster entitled "DDT Reduces Photosynthesis by Marine Phytoplankton." The report noted that reduction in photosynthesis appears even when DDT concentrations were as small as a few parts per billion (ppb). Should photosynthesis be significantly reduced, not only would our own oxygen supply be jeopardized but also life in oceans as well.

Drastic shifts in populations of phytoplankton species have been recognized. Huge blooms of one or a few species have produced what are commonly called *red tides*. Red tides have appeared frequently in the past decade along the coast of Florida, adversely affecting marine life in that region (Figure 24–5). Red tides have now moved up to the coast of New England, causing concern in the fishing industry. It has been suggested that DDT is a principal cause of

Figure 24–5. Dead fish, like grains of sand, cover a Florida beach after a red tide bloom. (U.S. Geological Survey.)

red tide blooms, but other factors, such as changes in climate and currents, could be involved as well.

Oil spills and oceans

Another threat to oceans is oil. Oil reaches the sea in a variety of ways. Accidental spills from oil tankers and offshore oil well blowouts receive much publicity, but these account for less than 0.1% of the total problem. Oil spillage associated with refueling of ocean-going vessels is a more serious source of oil pollution (Figure 24–6). In the *Torrey Canyon* tanker disaster in 1967, about 100,000 tons of crude oil spilled into the sea. By comparison, routine discharges from tankers and commercial vessels add about 3.5 million tons of oil to the ocean every year. Another source of oil is accidents related to extraction of offshore oil, although the number of accidents is very small compared with the number of offshore wells safely drilled each year.

Subsea oil pipelines are another potential hazard. Continental shelves carry 20,000 mi (32,250 km) of these conduits. Pipes are slowly rusting, and it is only a matter of time before they pose the threat of leakage. Older pipelines should be watched closely.

An oil spill of catastrophic proportions took place in Japan on December 24, 1974, under extremely puzzling circumstances. The biggest oil spill in Japan's history (18,750 barrels of heavy oil) occurred at a refinery in Mizushima, located midway between the cities of Osaka and Hiroshima. The spill spread through the Inland Sea, polluting an area 80 mi (129 km) long and 20 mi (32 km) wide. The ooze blighted numerous beaches and destroyed fish and cultivated seaweed beds. Damage to beaches, fisheries, and seaweed was estimated in excess of $20 million during the first week of the spill.

Officials of the Mitsubishi Oil Company, operators of the refinery, offered no explanation for the reason of the spill. Geologists

Figure 24–6. *Left*-Oil spillage connected with the refueling of ocean-going vessels is responsible for nearly 99% of oil pollution. (Ward's Natural Science Establishment, Inc.) *Right*-Accidental oil spills related to offshore drilling are, at present, minor contributors to the oil pollution problem. The demand for oil has produced a rash of requests for offshore drilling rights on the continental shelves on both the west and east coasts of United States. The danger of oil pollution in our coastal waters will quickly multiply as more and more oil drilling rigs are erected.

theorize that the oil spill resulted from the collapse of land beneath oil storage tanks. The refinery was built on land reclaimed from the sea. It is believed that the limitations of this land could have been predicted by geologists had they been consulted prior to the erection of the storage tanks.

Another type of oil spill is actually nature's doing. Natural seeps from subsurface oil traps are well known and have served as a vivid clue in the search for petroleum. Such seeps occur on the ocean floor as well as on the continents.

Effects of oil pollution are not fully understood. We are aware of the diminishment of populations of sea birds, fish, and shellfish in areas of oil spills. Effects vary with type of oil. Some components of oil are toxic;

some are carcinogenic. Although weathering may reduce the toxicity of oil, carcinogenic components are extremely long lived. Oil that reaches beaches may linger for a few months, but marine life may need many years to recover after an oil spill.

SOME ASPECTS OF AIR POLLUTION

The Industrial Revolution is hailed by historians as a great turning point in the history of civilization, and so it is in terms of goods and services generated. Yet it remains a mixed blessing when we weigh the advantages of industrialization against the resultant damage to ecosystems (Figure 24–7).

In consuming fuels to provide power for industrial plants and automobiles, harmful

Figure 24–7. Is the atmosphere a topless sewer? Industrial nations have learned that this is not so. The amount of pollutants introduced into our atmosphere must be greatly reduced if we hope to maintain an environment compatible with man. (Ward's Natural Science Establishment, Inc.)

waste products are introduced into the atmosphere. During the decades of increasing industrial development it was thought that the atmosphere had an endless capacity to absorb waste products without adverse effects. We have learned especially in the past one or two decades that pollutants introduced into the atmosphere do not simply disappear harmlessly but bring about changes in the natural environment that are detrimental to our existence.

Research supported by both government and industry has been directed toward determining how the atmosphere dilutes wastes. Pollution control codes in most cities have reduced the amount of industrial pollution. In 1970 the U.S. Congress passed the Clean Air Act, setting standards for air pollution from automobiles and factories. This was an important piece of legislation, because it approached the problem on a national scale.

In the meantime, however, geologists and engineers must devise and urge the initiation of programs designed to develop alternatives to the present dirty fuels and inefficient combustion machines.

A prime contributor to air pollution is sulfur, found in fossil fuels. In the absence of new sources of clean fuels, new ways should be found to remove sulfur and other impurities from present fuels. The automobile industry has introduced the catalytic converter (Figure 24–8). Cars so equipped burn lead-free gasoline. Alternatives to the internal-combustion engine in automobiles are being developed; the rotary Wankel engine shows promise (Figure 24–9).

Some scientists feel that the ultimate fuel for the automobile may be not oil but hydrogen. Hydrogen can be produced from water; when it unites with oxygen in the process of combustion, it again forms water.

478

ENVIRONMENTS
HERE
AND
BEYOND
THE
EARTH

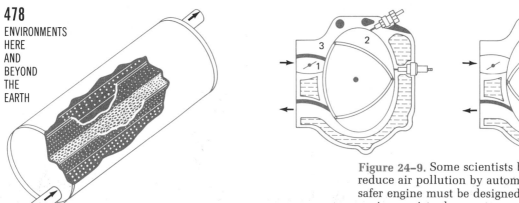

Figure 24–8. The greatest single contributor to the air pollution problem is the automobile. As of 1975, American-built automobiles must meet pollution emission standards. Many 1975 models use a catalytic converter to reduce harmful emissions passing through the exhaust system. Catalytic converters neutralize unburned hydrocarbons and carbon monoxide, while exhaust gas recirculation reduces the emission of nitrogen oxide.

Figure 24–9. Some scientists believe that to reduce air pollution by automobiles, a new and safer engine must be designed. The Wankel engine, a pistonless motor combining simplicity and a superior power-to-weight ratio, is being tested in America. The engine produces twice the power of a piston engine of equal weight but is only one-third as large. It produces less nitrogen oxide than a conventional engine.

No harmful by-products are produced in the process. But some believe that hydrogen is too dangerous for use in automobiles. (The same opinion was voiced by many at the end of the nineteenth century regarding the use of gasoline.)

GEOLOGY AND URBAN PLANNING

Two phenomena are characteristic of the second half of the twentieth century. The first is the population explosion and the second a trend toward concentration of populations in a few metropolitan areas. In most instances rapid growth of metropolitan regions was not preceded by adequate planning for expansion. As a result of lack of planning, many highly populated areas are now confronted with seemingly insurmountable problems.

Problems include inadequate sewage treatment, scarcity of fresh water, noise pollution, air pollution, inadequate traffic arte-

ries, housing blight, power shortages, and an endless string of similar frustrations. We no longer can permit spontaneous growth of urban areas. The time has come for cooperative action on the part of city officials, urban planners, and geologists to transform city jungles to livable space for humans.

Land developers often begin housing projects without prior investigation of subsurface conditions. Homes on the coast of California and along the inland canyons of that state have, after heavy rains, been found precariously perched or, even worse, have fallen into the waters below (Figure 24–10). Knowledge about the behavior of earth materials is available, but it will be of little use unless it is applied in planning the environment.

A number of unfortunate cases of total disregard for known dangers has been documented. In 1966 the California Division of Mines and Geology developed an urban geologic map of the San Clemente area in the southern part of the state. The community was rapidly expanding. The geologist working on the map project discovered that a new civic center was being planned to occupy a landsite that was, what he considered, a

Figure 24–10. A picturesque setting for a home may prove to be a hazardous one. These homes along a canyon wall near Los Angeles, California, are in serious trouble. Subsurface conditions were not investigated by the land developer. Here coastal terrace deposits overlying deformed sedimentary rocks, when lubricated by heavy rains, slip from under the home. (U.S. Geological Survey.)

landslide zone. He brought this to the attention of the city engineer, who promptly disposed of the report in his file cabinet. Shortly after, newspapers reported large cracks in the new city hall located near the proposed site of the civic center.

In 1959 U.S. Geological Survey Bulletin 1093 reported that the Bootlegger Clay, which underlies much of Anchorage, Alaska, was an unstable material; when wet it could easily be dislodged by a shock. The report suggested that, because of the low cohesiveness of this clay, buildings erected in Anchorage must have foundations that could counteract slippage. It not only stated that Anchorage was located in an earthquake zone, but it listed a number of previous slumps and landslides in the region. Five years later, the Good Friday earthquake of March 27, 1964, destroyed much of the city (Figure 24–11).

What has happened in California and Alaska is happening in every other state. Although the need is apparent, there is still little effective teamwork among geologists, city planners, land developers, and engineers. In California extensive property damage brought about by heavy rains ended in a costly lawsuit against Los Angeles. Irate property owners sued the city for $7.5 million for property damage resulting from improper planning.

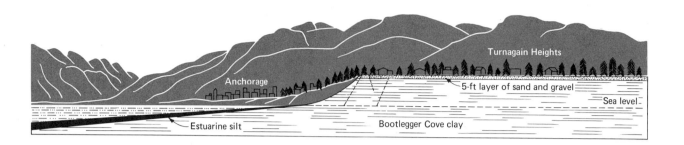

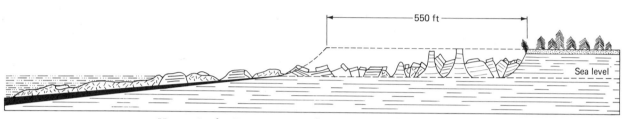

Figure 24–11. Homes in the Turnagain Heights area of Anchorage, Alaska, were built on layers of sand and gravel that overlie the Bootlegger Cove clay, an unstable material which, when wet, can easily be dislodged by a shock. The earthquake of 1964 produced an extremely large earth slide here. Two minutes after the earthquake began, cracks appeared near the edge of the bluff. *(top).* Three minutes later, 550 ft (168 m) of the bluff had moved 700 ft (213 m) into the sea *(bottom).*

Building codes and ordinances in major cities that were adequate 50 years ago will not meet demands of today. Zoning ordinances should protect future home buyers from catastrophic losses. Ordinances and building codes must be adhered to, or they will have little value. The policy of granting exceptions or deviations from existing codes is a common practice. Land developers frequently circumvent restrictions imposed by the codes. Deviations that violate geological and engineering reports should be considered criminal acts. Failure to apply scientific knowledge to environmental development problems will result in a costly and unnecessary drain on a community's resources. Lack of proper planning may produce expensive remedial charges for public works as well as individual homes of the taxpayers in a community.

DISPOSING OF SOLID WASTES

Every man, woman, and child in the United States contributes to our environmental problem 5.3 pounds per day of solid waste in the form of garbage, paper packaging, bottles, cans, plastic containers, junked cars, worn-out appliances, and so on (Figure 24–12). Agriculture generates 2 billion tons of solid wastes each year. This incredible amount of waste material must be put somewhere. We do not appreciate the immensity of the problem until we are confronted with an event such as a garbage strike in New York City (Figure 24–13).

Several techniques for disposal of solid wastes have been developed. None, however, is the answer to the staggering problem. Incineration of solid wastes adds to the problem of air pollution. Sanitary landfill disposal presents the problem of where to go

Figure 24–12. Disposal of solid waste is an ever-increasing environmental problem in United States. On pickup days, trash lines city streets. This incredible pile of discarded packaging, bottles, cans, and plastic containers, must be disposed of somewhere. What to do with it has become an increasingly perplexing problem. (Brown Brothers, 220 West 42nd Street, New York, N.Y. 10036.)

when landfill sites are exhausted. It also presents the problem of groundwater pollution. The geologist must become involved in this problem. Knowledge of groundwater pollution will help communities to solve this problem.

Ultimately different patterns of life style must be developed in America if we are to maintain acceptable standards of environmental quality in the face of mounting piles of refuse (Figure 24–14).

SUMMARY

The hydrosphere and atmosphere have become dumps for wastes from our industrialized and urbanized society. The environment, with its delicately balanced natural cycles of cleansing and replacement, cannot cope with the demands of a rapidly growing world population.

Food demands have multiplied in the past several decades. The problem has been met temporarily through irrigation of arid and semiarid lands. Past experiences indicate that reliance on irrigation to increase the food supply eventually may have an adverse effect on soil fertility. The Babylonian civilization disappeared because of the salinization of the once-fertile Tigris–Euphrates valley. What happened there is now happening in the Imperial valley of California and in other places.

Dams erected to prevent periodic flooding or as reservoirs for the irrigation of farmlands produce artificial lakes. The broad water surface of these lakes increases loss of

Figure 24–13. The enormity of the solid-waste disposal problem cannot be fully appreciated unless periodic removal of trash is interrupted for a period of time. New York City experienced this calamity during a recent strike of garbagemen. Within a short time, streets were separated from sidewalks by hills of trash piled so high that a pedestrian could not see cars. (U.S. Geological Survey.)

482

ENVIRONMENTS
HERE
AND
BEYOND
THE
EARTH

Figure 24–14. We deserve something better. The life style in America has caused our environment to become rapidly degraded. Oil-polluted water, discarded auto tire and muffler, decaying building materials, paper cups, vegetable cans, beer cans, sludge along the shore, and foaming streams are symptoms of a disease that must be cured if we are to maintain acceptable standards of environmental quality. (Ward's Natural Science Establishment, Inc and U.S. Geological Survey.)

water through evaporation. Frequently newly established water levels behind dams reach permeable rock strata. Water from the reservoir will penetrate these rock zones. The net effect of evaporation and seepage losses reduce by as much as 15% the amount of water available for useful purposes.

Although industry is not the prime source of water pollution, industrial wastes that enter streams present a serious survival problem. Wastes, such as acids, phenols, polyvinyl chloride, chromium, and mercury, that are entering our hydrosphere, persist in the environment for long periods of time.

Water pollution associated with urban sewage treatment presents the greatest challenge. Most of the sewage dumped into streams is improperly treated. As world population increases, the immensity of the sewage treatment problem becomes staggering. Most urban communities discharge their sewage into the same water from which they draw their supply of drinking water.

The ultimate sink of polluted waters is the oceans. Humankind depends to a large degree on phytoplankton living in oceans as primary replenishers of oxygen in the atmosphere. Contamination of ocean water with DDT and other wastes has adversely affected the process of photosynthesis so essential to the production of oxygen.

A conscious effort is being made at present to reverse these threats to water and air. The federal government has begun a long-range program of assistance to local communities to improve sewage treatment facilities. The Clean Air Act has as its goal the reduction of harmful discharges into the atmosphere. Hopefully these efforts may keep in check the degradation of our environment. The problem, however, is worldwide in scope.

The rapid growth of large urban centers has introduced social, psychological, and environmental problems. Urban planning that recognizes the geological nature of a region will in the long run result in more functional and less stress-producing urban societies.

The role of the geologist in environmental planning has not been adequately utilized by municipalities, land developers, and industry. There is great need for close cooperation between city and industrial planners on the one hand, and geologists and other environmental scientists on the other, in developing a fresh new approach to urban living, transportation, and industrial development.

Questions for thought and review

1. Irrigation is only a temporary measure in our attempts to feed the hungry of the world. Describe the problems introduced into the environment by irrigation.
2. Why does the salinity of the Colorado River increase as it flows through southern California?
3. Explain how the construction of dams in semiarid regions may increase water loss.
4. What is the difference between water usage and water consumption?
5. The problems of water pollution produced by municipalities are different from those produced by industry. Discuss the differences in terms of pollutants introduced into streams and lakes, and the degree of difficulty in the reclamation of waters so polluted.
6. The oceans not only provide humanity with food but also help replenish oxygen

in the atmosphere. Explain the source of a recently discovered threat to our survival.
7. How do oil spills affect marine life?
8. What alternatives to the internal-combustion engine are now used in automobiles?
9. Suggest ways in which geologists can assist in planning a better environment.

Selected readings

Campbell, Ian, 1966, *California*, The State Geologists Journal, vol. 18, no. 2, pp. 6–8.

Cooke, R. F., 1969, *Oil Transportation by Sea*. In Hoult, D. P., ed., *Oil on the Sea*, Plenum, New York, pp. 93–102.

Elsaesser, H. W., 1971, Air Pollution: Our Ecological Alarm and Blessing in Disguise, *Transactions of the American Geophysical Union*, vol. 52, no. 3, pp. 92–100.

Findley, Rowe, 1973, The Bittersweet Waters of the Lower Colorado, *National Geographic*, vol. 144, no. 4, pp. 540–569.

Jacobsen, Thorkild, and Adams, R. M., 1958, Salt and Silt in Ancient Mesopotamian Agriculture, *Science*, vol. 128, no. 3334, pp. 1251–1258.

Lindsay, Sally, 1970, How Safe Is the Nation's Drinking Water? *Saturday Review*, May 2, pp. 54–55.

Marx, Wesley, 1967, *The Frail Ocean*, Ballantine, New York.

Miller, R. D., and Dobrovolny, E., 1959, *Surficial Geology of Anchorage, Alaska, and Vicinity*, U.S. Geological Survey Bulletin 1093, p. 128.

Slosson, James E., 1959, *Engineering Geology — Its Importance in Land Development*, The Urban Land Institute Technical Bulletin 63, p. 20.

UNESCO, 1961, *Salinity Problems in the Arid Zones*, Proceedings of the Tehran Symposium, Paris, UNESCO, Arid Zone Research, vol. 14, p. 395.

Wurster, Charles F., 1968, DDT Reduces Photosynthesis by Marine Phytoplankton, *Science*, vol. 159, no. 3822, pp. 1474–1475.

————, 1971, *Persistent Insecticides and Their Regulation by the Federal Government*, Testimony before U.S. Senate Committee on Agriculture and Forestry, March 25.

Appendix A Physical properties of minerals

Name	Chemical composition	Hardness	Specific gravity	Color
Quartz	SiO_2	7	2.7	Colorless, white, yellow, green, gray, purple
Opal	$SiO_2 \cdot nH_2O$	5–6	2.1–2.3	White, gray, tan, yellow, green
Orthoclase	$KAlSi_3O_8$	6	2.6	White, pink, gray
Plagioclase	$NaAlSi_3O_8$ to $CaAl_2O_8$	6	2.6	White, pink, gray
Muscovite	Hydrous potassium aluminum silicate	$2-2\frac{1}{2}$	2.8–3.1	Colorless, light yellow, green, brown
Biotite	Hydrous potassium iron magnesium aluminum silicate	$2\frac{1}{2}-3$	2.8–3.2	Black, brown, dark green
Chlorite	Hydrous iron magnesium aluminum silicate	$2-2\frac{1}{2}$	2.6–2.9	Dark grayish green
Talc	Hydrous magnesium silicate	1	2.7–2.8	White, gray, greenish white
Serpentine	Hydrous magnesium silicate	2–5	2.2–2.7	Green, yellowish brown, brown
Kaolinite	Hydrous aluminum silicate	$2-2\frac{1}{2}$	2.6	White, tan
Montmorillonite	Hydrous aluminum silicate	$1-1\frac{1}{2}$	2.5	White, gray
Amphibole, variety Hornblende	Hydrous calcium, iron magnesium aluminum silicate	5–6	3.2	Dark green to black
Pyroxene, variety Augite	Calcium iron magnesium silicate	5–6	3.2–3.4	Dark green to black
Tourmaline	Boron, aluminum, iron, magnesium, silicate	$7\frac{1}{2}$	3–3.25	Black, pink, brown, green
Epidote	Hydrous calcium aluminum iron silicate	6–7	3.4–3.5	Yellowish to dark green
Staurolite	Hydrous iron aluminum silicate	$7-7\frac{1}{2}$	3.7–3.8	Brown to brownish black
Kyanite	Al_2SiO_5	5 parallel to length of crystal, 7 across	3.6–3.7	Blue to gray with blue or green centers
Sillimanite	Al_2SiO_5	6–7	3.2	White, tan, gray

Streak	Luster	Cleavage or fracture	Form	Other properties
White	Vitreous	Conchoidal fracture	Massive or 6-sided prisms and dipyramids	
White	Resinous	Conchoidal fracture	Massive	
White	Vitreous	2 cleavages at 90°	Massive or prismatic crystals	
White	Vitreous to pearly	2 cleavages at 90°	Prismatic crystals or massive grains	Striations on one cleavage face
White	Vitreous to silky	1 perfect cleavage	Cleavage fragments, 6-sided crystals	
Dark green	Vitreous	1 perfect cleavage	Cleavage fragments	
White	Vitreous to pearly	1 perfect cleavage	Massive or small crystal aggregates	
White	Pearly to greasy	1 perfect cleavage	Massive or in foliated aggregates	
White	Greasy, waxy, silky	Conchoidal (massive variety)	Massive or fibrous (crysotile asbestos)	
White	Dull, earthy	Uneven fracture	Massive	Earthy odor when damp, plastic when wet
White	Dull, earthy	Uneven fracture	Massive	Plastic when wet
Light green to white	Vitreous	2 perfect cleavages intersecting at 56 and 124°	Long prismatic crystals and cleavage fragments	
Greenish gray	Vitreous	2 perfect cleavages intersecting at 90°	Short prisms and cleavage fragments	
White	Vitreous to resinous	Indistinct clearage	Long 3-sided crystals	
White	Vitreous	1 good cleavage	Usually granular to massive; may be in prismatic crystals	
White to light tan	Resinous to vitreous	Irregular fracture	Prismatic crystals, frequently twinned as a cross or an "X"	
White	Vitreous to pearly	1 perfect cleavage perpendicular to length of crystals	Long prismatic crystals and bladed aggregates	
White	Vitreous	1 perfect cleavage parallel to length of crystals	Long prismatic or fibrous crystals	

Name	Chemical composition	Hardness	Specific gravity	Color
Garnet	$Fe_3Al_2(SiO_4)_3$	$6\frac{1}{2}$–$7\frac{1}{2}$	3.5–4.3	Red, brown, green, black
Olivine	$(Mg, Fe)(Mg, Fe)SiO_4$	$6\frac{1}{2}$–7	3.3–4.4	Olive green to grayish green to brown
Zircon	$ZrSiO_4$	$7\frac{1}{2}$	4.7	Brown, colorless, green, reddish brown
Nonsilicates				
Calcite	$CaCO_3$	3	2.7	White, colorless, gray, blue, yellow, black, pink
Dolomite	$CaMg(CO_3)_2$	$3\frac{1}{2}$–4	2.85	White, gray, pink, brown
Fluorite	CaF_2	4	3.2	Light green, blue, purple, white, colorless, yellow, pink
Barite	$BaSO_4$	3–$3\frac{1}{2}$	4.5	White, gray, bluish white, yellow
Corundum	Al_2O_3	9	4.0	Brown, gray, white, red, blue, pink
Apatite	$Ca_5(PO_4)_3(F, OH, Cl)$	5	3.2	Green to brown to blue
Halite	$NaCl$	$2\frac{1}{2}$	2.2	Colorless, white
Graphite	C	1–2	2.2	Black
Gypsum	$CaSO_4 \cdot 2H_2O$	2	2.3	Colorless, white, pink
Economic minerals — metallic				
Pyrite	FeS_2	6–$6\frac{1}{2}$	5.0	Pale brass yellow
Chalcopyrite	$CuFeS_2$	$3\frac{1}{2}$–4	4.1–4.3	Brass to golden yellow
Pyrrhotite	FeS	4	4.6–4.7	Brownish bronze
Chalcocite	Cu_2S	$2\frac{1}{2}$–3	5.5–5.8	Lead gray
Magnetite	Fe_3O_4	6	5.2	Black

Streak	Luster	Cleavage or fracture	Form	Other properties
White	Vitreous to resinous	Irregular fracture	12-sided crystals or fragments	
White	Vitreous	Conchoidal fracture	Granular crystal masses	
White	Adamantine	Irregular fracture	Elongate 4-sided prisms and dipyramid	
White	Vitreous to earthy	Rhombohedral cleavage (3 directions not at 90°)	Crystals or cleavage fragments	
White	Vitreous	Rhombohedral cleavage (3 directions not at 90°)	Small rhombohedral crystals or cleavage fragments	
White	Vitreous	Octahedral cleavage (4 directions)	Cubic crystals or cleavage fragments	
White	Vitreous	3 cleavage directions	Massive, granular, or radiating flowerlike crystals	
White	Adamantine to vitreous	Rhombohedral and basal cleavage (actually parting)	Massive, granular, barrel-shaped 6-sided crystals	
White	Vitreous to subresinous	1 poor cleavage uneven fracture	Massive 6-sided crystals	
White	Vitreous	3 directions at 90°	Cubic crystals, massive cleavage fragments	Salty taste
Black	Metallic to dull	1 perfect cleavage	Foliated or scaly masses	Greasy feel
White	Vitreous, silky pearly	4 cleavage directions (usually see 3)	Massive, prismatic, or fibrous crystals	
Greenish to brownish black	Metallic	Conchoidal fracture	Massive cubic crystals, 12-sided crystals with pentagonal faces	
Greenish black	Metallic	Irregular fracture	Massive also as 4-sided crystals	Iridescent tarnish
Black	Metallic	Irregular fracture	Massive	Slightly to strongly magnetic
Grayish black	Metallic	Conchoidal fracture	Massive	Imperfectly sectile, tarnishes black
Black	Metallic	Irregular fracture	Massive or octahedral (8-sided) crystals	Strongly magnetic

Name	Chemical composition	Hardness	Specific gravity	Color
Chromite	$FeCr_2O_4$	$5\frac{1}{2}$	4.6	Black to brownish black
Ilmenite	$FeTiO_3$	$5\frac{1}{2}$–6	4.7	Black
Galena	PbS	$2\frac{1}{2}$	7.4–7.6	Lead gray
Hematite	Fe_2O_3	$5\frac{1}{2}$–$6\frac{1}{2}$	5.26	Reddish brown to silver to black
Copper	Cu	$2\frac{1}{2}$–3	8.9	Copper red

Economic minerals—nonmetallic				
Limonite	Uncertain	3 or less	4 or less	Yellowish to reddish brown
Goethite	$FeOOH$	5–$5\frac{1}{2}$	4.4	Yellowish brown to dark brown
Cassiterite	SnO_2	6–7	6.8–7.1	Brown to black
Sulfur	S	$1\frac{1}{2}$–$2\frac{1}{2}$	2.1–2.2	Yellow
Bauxite	Mixture of several minerals	1–3	2–2.6	White, gray, yellow, pink
Uraninite	UO_2	$5\frac{1}{2}$	7.5–9.7	Black
Carnotite	Hydrous potassium uranyl vanadate	1–2	4.7–5	Bright yellow to greenish yellow
Sphalerite	ZnS	$3\frac{1}{2}$–4	3.9–4.1	Yellow, brown, reddish brown, black
Azurite	$Cu_3(CO_3)_2(OH)_2$	$3\frac{1}{2}$–4	3.8	Bright azure blue
Malachite	$Cu_2CO_3(OH)_2$	$3\frac{1}{2}$–4	3.9–4	Bright green

SOURCE: Berry, L. G., and Mason, Brian, 1959, *Mineralogy: Concepts, Descriptions, Determinations:* San Francisco, W. H. Freeman, 630 p.
Frye, Keith, 1974, *Modern Mineralogy:* Englewood Cliffs, N.J., Prentice-Hall, 325 p.
Hurlbut, C. S., Jr., 1971, *Dana's Manual of Mineralogy,* 18th ed.: New York, John Wiley and Sons, 579 p.
Mason, Brian and Berry, L. G., 1968, *Elements of Mineralogy:* San Francisco, W. H. Freeman, 550 p.

Streak	Luster	Cleavage or fracture	Form	Other properties
Dark brown	Metallic to submetallic	Irregular fracture	Massive	
Black to brownish red	Metallic to submetallic	Irregular fracture	Massive	
Lead gray	Bright metallic	3 directions at 90° (cubic)	Cubic crystals and cleavage fragments	
Reddish brown	Metallic to dull	Rhombohedral and basal cleavage (actually partial)	Massive, foliated, or in leaflike crystals, ellipsoidal grains (oolites)	
Copper red	Metallic	Hackly (needlelike fracture)	Irregular masses, plates, wirelike forms	Ductile and malleable
Yellowish brown	Dull to earthy	Irregular fracture	Massive botryoidal	
Yellowish brown	Adamantine to dull silky	1 direction (usually not seen)	Massive or reniform	
White	Submetallic to dull	1 cleavage (usually not seen)	Massive or reniform	
White	Resinous	Conchoidal to uneven fracture	Massive, earthy incrustations	
White	Dull to earthy	Irregular fracture	Massive	
Brownish black	Submetallic to dull	Irregular fracture	Massive or botryoidal	Radioactive
Light yellow	Dull to earthy	1 direction (usually not seen)	Granular powdery aggregates	Radioactive
White to light yellowish brown	Resinous to submetallic	6 directions (dodecahedral)	Massive with cleavage faces	
Blue	Vitreous	3 directions (usually not seen)	Small crystals and granular aggregates	
Green	Adamantine to dull	Irregular fracture	Massive, granular botryoidal	

Appendix B Matter and energy

Matter is anything that possesses mass (weight on earth) and occupies space. Matter is composed of atoms, which are made up principally of protons, neutrons, and electrons (see Chapter 5). Matter may exist in three principal states: solid, liquid, or gaseous. How do they differ and why do they exist?

Solids, liquids, and gases may be thought of as matter in different states of order or disorder. Solids, particularly crystalline solids, have the greatest degree of order with a definite arrangement of their constituent atoms, ions, or molecules. Liquids possess less order and are not rigidly bound together. Glasses have open structures similar to liquids but are more rigidly bound. Gases have practically no bonds, and their atoms or molecules have greater distances between them than those of solids or liquids, hence the least order.

Any solid may be transformed into a liquid and any liquid into a gas. Some solids exist in more than one form. For example, at least five solid forms of SiO_2 exist in addition to liquid and gaseous forms. All solid forms of anything, as well as its liquid and gaseous phases, are stable at different temperatures and pressures. This is the reason for their existence.

Yet some materials persist at temperatures and pressures above or below levels where they should change into something else. Why, for example, do we still see igneous and metamorphic rocks at or near the surface containing minerals that are well outside their respective realms of stability? Minerals such as muscovite, biotite, orthoclase, kyanite, and pyroxene do not form on the earth's surface. Still we can go to many places on the surface and find them. The answer is that these materials are *metastable*. They are not in equilibrium with their surroundings but will not change into something else unless energy is added to them. What, then, is energy? *Energy* is the capacity to do work. The kind of work that must be done to transform metastable materials into something more stable in the environment is that commonly supplied by *thermal energy*.

Each of the several kinds of energy may function in the operation of geologic processes. Thermal energy is involved in the crystallization of magmas, metamorphism, and even the surficial processes of evaporation and freeze–thaw. This thermal energy is derived from several sources; surface processes are driven by heat from the sun, whereas internal processes are driven by heat generated during radioactive decay. Energy is given off as *radiation energy* during radioactive decay and partially converted to heat.

Mechanical energy as *kinetic energy* (energy of things in motion) and *potential energy* (energy attributable to position) function in many geologic processes. Kinetic energy may be calculated with the following equation:

$$KE = \frac{1}{2}mv^2$$

where m is mass and v is velocity. The kinetic energy of a stream increases many times when it changes from a slow-moving stream at low flow to a fast-moving stream at flood stage. Both the mass of the stream and its velocity increase dramatically. Thus the ability of the stream to erode and transport material, to accomplish *work*, is enhanced.

Potential energy may be calculated from:

$$PE = mgh$$

where m is mass, g is acceleration due to

gravity (32 ft/sec² or 9.80 m/sec²), and h is height of the material above a place where it would come to rest if set in motion. For example, an unstable mass of rock attached to a mountainside 125 ft (400 m) above a valley would have a certain potential energy. If this mass became detached and developed into a landslide, the potential energy at rest would become transformed into kinetic energy as long as it were in motion. Friction and impacts with objects would eventually slow the slide through transfer of kinetic energy to other objects by doing work on them. The slide would then come to rest at the eleva-

tion where it no longer had the ability to move. Having lost its kinetic energy, it would probably not move again, if it came to rest at the bottom of the valley. Therefore it would have lost its potential energy as well.

Mechanical energy may be converted to thermal energy, and vice versa. Some of the kinetic energy of the slide discussed above would probably be lost as heat generated through friction. Some of the heat from melting of rocks at depth may function in the uplift of a mountain system and thus in the mechanical processes of folding and faulting.

Appendix C   Units of measurement

Throughout this book a dual system of measurement has been used. When units of length, distance, weight, or volume are mentioned, such familiar units as inches, feet, miles, pounds, or cubic feet are given with approximate equivalents (in parentheses) in the metric system, such as centimeters, meters, kilometers, kilograms, or cubic meters.

All through our lives we measure many things. How did ways of measurement start? Probably thousands of years ago primitive men and women used their own arms, fingers, or feet as units of length. Weight may have been indicated in terms of an animal killed for food. However, some people have longer arms or bigger feet than others. What was needed were *standards:* units of measurement that everybody agreed to use.

Kings in earlier times attempted to solve this problem by proclaiming that their own personal measurements would be the standard that everyone in the kingdom would use. This system may have worked for a while, at least until a new king came along. Then of course the "standards" would change. By such hit-or-miss methods has the system of measurement used in the United States, called the English system, evolved. It is cumbersome, antiquated, and difficult to use.

More than 150 years ago Thomas Jefferson recommended that the United States convert to the metric system. Unfortunately this was not done. Today the United States is the only major country not on the metric system. This impedes us in matters of commerce and other communication. Two exceptions are the facts that the metric system has been used for years in this country by scientists and by physicians.

The metric system was developed in France. It is logical and easy to use. Conversion from one unit to another is accomplished quickly and accurately. For example, how many feet are there in 3 mi? You have to stop and think. Now, how many meters are in 3 km? The answer, quickly determined, is 3000 m because there are exactly 1000 m in 1 km. The metric system is a simple decimal system based on 10.

Four measures include: (1) length or distance, (2) volume, (3) weight, and (4) temperature. The most common units in the metric system are:

LENGTH OR DISTANCE
Basic unit = the meter (slightly longer than a yard)
1 meter can be subdivided into:
100 centimeters
1000 millimeters
1000 meters = 1 kilometer (0.62 miles)

VOLUME
Basic unit = the liter (slightly less than 1 quart)
1 liter = 1000 milliliters
4.5 liters = 1 gallon

WEIGHT
Basic unit = the gram (1 gram weighs about the same as a paper clip)
1 gram = 1 milliliter of water
1000 grams = 1 kilogram (2.2 pounds)
45 kilograms = 100 pounds

TEMPERATURE

Measured by the Celsius thermometer where

0°C = freezing point of water (32°F)

100°C = boiling point of water (212°F)

Here are some familiar equivalents.

F°		C°
70°	comfortable	21°
98.6°	body temperature	37°
250°	warm oven	120°
450°	hot oven	230°

Appendix D Field and laboratory identification of rocks*

PROCEDURE

Practice identifying large, known specimens first. When the properties of these known rocks have been found in the Table and verified in the specimens, the student is ready to try identifying some unknown chips. Sets of labelled hand specimens may be purchased from several laboratory supply houses.

Whenever possible, examine freshly fractured surfaces on the rock. Specimens typically supplied in teaching and museum collections have already been fractured, and no further preparation is necessary. With experience, even weathered rocks and rounded pebbles become recognizable, but the beginner should not start with these.

The dilute hydrochloric acid used as an effervescence test for calcium carbonates (calcite, aragonite, dolomite) should not be applied indiscriminately to unknown specimens, for it will eventually alter many kinds of rock without giving any immediate information; the acid residue on the acid-insoluble specimens is inconvenient at best, and could build up to somewhat hazardous quantities. When used properly on carbonate-bearing rocks, the acid is soon neutralized.

Experience has shown that very good accuracy can be expected in classroom use of this Table. The limiting factors are (a) the individual's ability to discover and diagnose the important properties of an unknown rock; (b) the natural lack of a stereotyped appearance of many rocks; (c) lack of usable macroscopic characteristics in a few rocks. Systematic use of the Table will guide even the beginner to a remarkable degree of success in rock identification.

The only equipment needed for determination of rocks using this Table is a magnifying glass (8x to 12x is about the best magnification range); a steel knife blade or a steel needle in a convenient handle; a small hammer for breaking and

trimming rock chips; and dilute hydrochloric acid (1 part of concentrated acid in 5 of water) in a container from which it can be withdrawn conveniently a drop at a time.

Before attempting to use the Rock Identification Table, the student should review and practice identification of the major properties upon which it is based, and should also review and practice identification of the main minerals that constitute the common rocks.

PROPERTIES FOR RECOGNIZING ROCKS

Major Categories of Rocks

Three major groups of rocks are fairly easily distinguished. These are IGNEOUS ROCKS, formed by the solidification of cooling molten rock, or magma, on or beneath the surface of the earth; SEDIMENTARY ROCKS, formed from particles of rock washed downhill by rain water, transported in some cases long distances by rivers to lakes or to the sea, and there deposited and generally cemented together by materials deposited from solutions; METAMORPHIC ROCKS, formed by the transformation of earlier formed rocks (whether igneous or sedimentary or even previously metamorphosed), by the action of heat, pressure, and in many cases hot, watery solutions.

Igneous rocks are formed, as stated, by the cooling and crystallization of hot, molten rock materials. The igneous rocks may have cooled at some depth in the earth's crust, usually after being squeezed or intruded into crevices or other kinds of openings: such rocks are called intrusive, and they are distinguished by coarse grain-size, with recognizable minerals in a coarse, equigranular texture (see below). Magma that flows out (extrudes) on the surface of the earth at volcanoes is called lava. It cools and solidifies much faster than intrusive rock and forms extrusive igneous rock. The fast cooling of extrusive rocks gives rise to a fine-grained texture (see below) which may make identification of individual mineral grains difficult.

Sedimentary rocks are generally those formed by agencies having to do with weathering of earlier formed rocks. Such agents include rain, chemical constituents such as carbon-dioxide and oxygen in the air or dissolved in surface waters, and other chemicals in the rivers, lakes, and seas. The principal agents of transportation are gravity (sliding down hillsides), water, and in some areas, wind. Sedimentary rocks form not only from mechanical sediments such as calcium carbonate, and less commonly sediments composed of sand and clay particles, but also from chemical salts deposited from evaporating sea-water, as rock salt (halite), gypsum, etc. When these deposits are buried by further materials, they tend to become cemented together, and then they are called rocks. A significant component of sedimentary rocks is the content of fossilized plants and animals (especially shells and skeletal materials).

Sedimentary, igneous, and sometimes previously metamorphosed rocks buried deeply within the earth's crust become *metamorphic rocks* through the transformations that result from the heat and high pressure of burial, and the alterations of texture and composition that are caused by recrystallization of the constituents under the influence of these agents. Water and, in many cases, other chemically active components contribute also to metamorphism. The most important characteristics of metamorphic rocks are foliation (see below), and the presence of minerals not found in the source rocks.

Regional metamorphism alters rocks over large areas. Contact metamorphism affects the rock locally surrounding an intruding igneous magma, in an "aureole" or alteration-zone extending usually no more than a few miles (in some cases, only a few feet) from the contact of the intrusive body with the surrounding rock, called country rock.

Texture

The size or sizes of mineral particles or grains, and their shapes and arrangement in the rock, determine the texture of the rock. The types of texture common in most rocks are as follows:

a) *Coarse Equigranular* or *Coarse-Grained* (also called phaneritic, granitoid, or granitic). Most of the mineral grains making up the rock are of the same general size, larger than 2 mm across, and although some may be platy, others acicular (needle-like), the majority seen on a freshly broken surface are about as long as they are broad. (Examples are gabbro, granite.)

b) *Fine Equigranular, Fine-Grained.* Most of the grains are about the same size, and grains are smaller than 2 mm. (E.g., basalt, marble.)

c) *Inequigranular or Porphyritic.* Grains may be of any size, but some grains are decidedly larger than others. (E.g., porphyries, schist, porphyroblastic gneiss.)

d) *Aphanitic or Dense.* The rock is so finely crystalline or finely fragmental that the grains are indistinguishable even using a magnifying lens. (E.g., lithographic limestone, claystone, some felsites, some rhyolites, obsidian.)

Porous texture can be useful in recognizing certain extrusive rocks and some sedimentary rocks.

Fabric

Fabric is defined as the pattern (or lack of pattern) of orientations assumed by grains or by groups of grains in the rock. This includes patterns of crystallographic orientation that are detectable only in the laboratory, but it is more commonly associated with patterns that are brought out by parallelism of fibrous or of platy grains, and it is also commonly associated with selectively grouped grains such as alternating layers of larger and smaller grains, or layers of light-colored and dark-colored minerals. Such an arrangement of grains produces a *directed fabric*.

Some layered structures are nearly planar, as with bedding in sedimentary rocks when they are originally deposited; other layered structures are bent or folded in various degrees, including the curved and folded fabrics of some schists and gneisses.

In some rocks, the parallelism of platy grains, as in grains of mica, may be so excellent that the rock will split more or less readily along these surfaces. The surfaces in such rocks are called planes of foliation. Foliation is developed to a very high degree in schists, to a lesser degree in gneisses; and there is no foliation at all in certain granites. It is possible to find rocks of all three of these types that have the same chemical composition, and the degree of foliation is really a continuously variable property.

One uniformly fine-grained rock, called slate, that splits easily into smooth, lustrous plates, is said to have *slaty cleavage*.

Rocks having no discernible preferred orientation and no selective grouping of any of the grains are said to have a *random fabric*. Such rocks look the same when viewed from any angle: they are without structure.

Joints are widely spaced parallel or intersecting structural cracks. They are not considered as indications of a directed fabric because they may be present in any rock, and indeed individual joints may cut across boundaries between two kinds of rock. Joints have nothing to do with the naming of a rock.

Fragmental (Clastic) and Crystalline (Non-clastic) Rocks

Fragmental rocks are composed of mineral grains that are usually cemented together. Even if only a few of the grains in the rock are composed of pieces of earlier-formed, broken rock, the rock should be considered fragmental. If the fragments are rounded the rock is called a conglomerate; if they are mostly angular, it is a breccia. Almost all of the fragmental rocks listed in the Table are sedimentary.

Crystalline or non-clastic rocks are formed by the precipitation of mineral matter from solution, by crystallization from a magma, or by recrystallization of solid materials under the influence of heat, pressure, and/or warm solutions. This group includes igneous, metamorphic, and some sedimentary rocks. These rocks are *coherent*, in many cases very hard to break, without any cement between the grains because the grains, forming in contact with each other, tend to stick together both by chemical bonding and by physical, interlocking shapes. Grain abrasion (so commonly seen in clastic rocks) almost never can be detected in such rocks because the grains have grown in place without rubbing against each other.

Very fine-grained rocks that cannot be studied effectively with simple tools such as hand-lens and knife-blade are called *dense*.

Mineral Content

Most rocks are named according to their mineral content. A simplified list of minerals is used in this Table as the basis of identifying the coarse-grained igneous and metamorphic rocks. This requires a knowledge of the physical properties of minerals as described in standard texts on mineralogy, but the number of minerals essential for this is not large—perhaps 20 to 30.

Acid Test

The rocks containing calcium carbonate (calcite, aragonite) and calcium-magnesium carbonate (dolomite) can be distinguished by applying a tiny drop of dilute hydrochloric acid, noting whether bubbles are formed (the "fizz" can sometimes be heard when it is too weak to be seen with the hand-lens). The bubbles consist of carbon dioxide. They form very slowly or not at all on dolomite, unless the grains of dolomite are crushed to a fine powder. Fragments of carbonate shells or mineral grains in a sandstone or conglomerate should not be confused with the main mass of the material of the rocks, most of which may not react at all. Similarly, calcium carbonate that may be present as cementing agent between grains of sandstone should not be confused with similar material making up the bulk of a limestone rock.

DESIGN AND USE OF THE TABLE

The Table gives a simplified classification of rocks to be used in sight identification of hand specimens.

Igneous rocks are classified on the basis of textures and minerals present.

Texture is indicated on the left side of the Table. The rock names are located in the vertical columns.

The position of a rock along a horizontal row indicates its chemical or mineral composition (its position in the sequence commonly referred to as acidic [left] basic [right]).

Acidic rocks (also called silicic or persilicic rocks) have a high content of silica (SiO_2); quartz and feldspar predominate in them, and they are typically light in color and low in specific gravity (e.g., granite, rhyolite).

Basic rocks (subsilicic) have a lower content of silica and more iron and magnesium. Iron and magnesium are contained in minerals such as pyrozene, amphibole, biotite, and olivine. These minerals make basic rocks darker and heavier, even though feldspar is frequently present (e.g., gabbro, basalt). Basic rocks with almost no feldspar are known as *ultra-basic*. An example is peridotite.

The distinction between acidic and basic rocks is arbitrary, for there is a complete gradation from one extreme to the other. It is therefore convenient to consider a group of intermediate rocks, whose compositions lie between those of acidic and basic rocks.

The quantity of the important rock-forming minerals present in igneous rocks is indicated in the Table by shading ranging from dark to white. The darker the box referring to a particular rock composition, the more important is that particular mineral constituent. When the space for a mineral in a particular rock is white, the mineral may be present but is not essential for naming the rock.

Sedimentary rocks

Some sedimentary rocks are composed of particles of minerals from a previous source. These constitute the group known as the clastic or detrital sediments. They are distinguished by their content of such particles, mechanically derived from other rocks. They are often classified by reference to the size of the particles. They may also be classified by geologic history or by agent of transportation (wind or water), but these details may be unknown, and are in many cases only the result of geological interpretation.

The name of a clastic sedimentary rock is based on the size of the fragments, the *texture*, and the compositions of grains and of the cementing material between them.

Mineral matter that is dissolved in water will precipitate, or crystallize and settle out, if the water evaporates. The rocks thus formed are called evaporites. Evaporites are generally classed with the sedimentary rocks. Among the constituents of evaporites are silica (quartz), carbonates of calcium and magnesium (calcite, dolomite), and a number of salts such as sodium chloride (halite), calcium sulfate (gypsum and anhydrite), and less abundant salts containing phosphorus, manganese, iron, and barium.

Although deposits of metallic ores do sometimes accumulate as sediments, these are rare enough to be of only local interest. The commonest of these is iron ore such as that found in the Clinton Formation in southeastern United States, and perhaps some of the source beds of the metamorphosed iron formations of the Lake Superior area.

Metamorphic rocks

Para-metamorphic rocks were formed from sedimentary rocks; ortho-metamorphic rocks, from igneous rocks.

In the Table, the grade (or degree) of metamorphism is indicated roughly by the figures 1, 2, 3, 4, in which 1 represents the lowest grade, as defined by an arbitrary combination of the temperature and the pressure to which a rock must be subjected to produce the particular mineral or minerals observed. The precise representation is usually done by naming the critical minerals that are found present, or "co-existing" in the rock at the metamorphic conditions. The subdivisions, "Essential" and "Accessory," are approximate and partly arbitrary because the *amount* of a mineral is generally determined by the bulk composition of the rock and by the metamorphic conditions. This information is important for detailed classification, but not for the preliminary purposes of the Table.

A dot in front of the mineral name indicates that this mineral may be a principal component of the rock mentioned in the row in which a mineral is shown.

The basis of the complete classification of metamorphic rocks is easily understood. Every mineral may be characterized in terms of the ranges of temperature and pressure within which it can form stably in any specified medium (the magma, the rock undergoing metamorphism, etc.) in which crystallization is taking place. Thus, in principle, if we know the assemblage of all the minerals present, and the composition of the whole rock, we can determine the limits of temperature, pressure, etc., at which the metamorphism took place. This has great importance in prospecting for ore deposits, and in understanding the geologic history of a region.

Glossary[1]

AA Type of lava flow with a rough, fragmental surface.

ABLATION Process by which glacier ice below the snowline is wasted by evaporation and melting.

ABRASION Erosion of rock through friction as solid particles moved by water, ice, or wind strike the rock surface.

ABSOLUTE TIME Time measured in years since an event occurred.

ABYSSAL PLAIN Flat region of the sea floor, usually at the base of the continental rise, formed by deposition of sediments that obscure pre-existing topography.

ACCESSORY MINERAL Mineral that occurs in small quantities in a rock.

ALLUVIAL SOIL Soil on floodplains and deltas that is actively in the process of being formed.

ALPINE GLACIER Glacier confined to a stream valley. It is usually fed from a cirque. Also referred to as *valley glacier* or *mountain glacier*.

AMPHIBOLE Member of a group of dark, rock-forming, hydrous silicate minerals rich in iron and magnesium. Hornblende is the most common amphibole.

ANDESITE Fine-grained, extrusive, intermediate igneous rock composed predominantly of plagioclase and hornblende.

ANGULAR UNCONFORMITY Unconformity in which the bedding of rocks on either side of the unconformity is not parallel.

ANTECEDENT STREAM Stream that maintains, after uplift, the same course it followed prior to uplift.

ANTHRACITE Commonly called hard coal. The highest metamorphic rank of coal, it is black and burns with a blue, smokeless flame.

ANTICLINE Upfold in which oldest rocks are found in the core, and rocks dip away from the core.

APHANITIC Texture of igneous rocks in which grains are too small to be seen without the aid of magnification.

AQUIFER Permeable rock unit through which groundwater moves.

ARÊTE Narrow and rugged crest of a mountain range formed by cirques developing from opposite sides into the mountain ridge.

ARKOSE Feldspar-rich, clastic sedimentary rock, usually reddish to buff, and often derived from rapid disintegration of granitic rocks.

ARROYO Flat-floored channel of an intermittent stream typical of arid and semiarid climates. Synonymous with wadi and wash.

ASH Fine pyroclastic material less than 4 mm in diameter; usually relatively unconsolidated.

ATMOSPHERE Mixture of gases that surround the earth; chiefly oxygen and nitrogen, with some argon and carbon dioxide and minute quantities of helium, krypton, neon, and xenon.

ATOM Smallest part of an element possessing the properties of the element.

AUTHIGENIC Formed or generated in place; usually refers to the formation of a mineral in a body of sediment after deposition.

AZONAL SOIL Soil lacking well-developed horizons; resembles parent material.

BARCHAN Crescent-shaped dune with horns pointing downwind.

BASALT Dark- to medium-dark, commonly extrusive, mafic igneous rock composed chiefly of calcic plagioclase and clinopyroxene in a glassy or fine-grained groundmass; extrusive equivalent of gabbro.

BASE LEVEL For a stream, a level below which

[1] These definitions are consistent with those in the *Glossary of Geology*, published by the American Geological Institute (1972). This authority should be sought for definitions not in this glossary.

it cannot erode. For a region, a plane extending inland from sea level sloping gently upward from the sea.

BASIN Symmetrical structure in which all beds dip toward a common point; younger units are found in the center.

BATHOLITH Discordant pluton with more than 40 mi² (100 km²) of surface exposure.

BEDDING PLANE Nearly planar surface that separates rock layers.

BENIOFF ZONE Inclined surface along which earthquakes occur as oceanic crust subsides into the mantle.

BENTHONIC Forms of marine life that are bottom dwelling.

BENTONITE Soft, light-colored rock composed of clay minerals produced by weathering (chemical alteration) of glassy igneous material.

BIOFACIES Subdivision of a stratigraphic unit distinguished by its faunal or floral assemblage.

BIOSPHERE All the area occupied or favorable for occupation by living organisms.

BITUMINOUS Commonly called soft coal. It is dark brown to black and burns with a smoky flame.

BLOCKS Lava with a surface of angular blocks.

BLOWOUT Basin scooped out of unconsolidated deposits by deflation.

BOULDER TRAIN Series of glacial erratics from the same bedrock source. It is arranged across the landscape in the shape of a fan with the apex at the source and widens in the direction of glacier movement.

BRECCIA Clastic rock composed of angular fragments greater than 2 mm in diameter and cemented with a finer grained matrix.

BUTTE Isolated, flat-topped hill or mountain with relatively steep slopes or cliffs; smaller than a mesa.

CALCITE Carbonate mineral, CaCO₃; the major constituent of limestone.

CATACLASTIC METAMORPHISM Kind of metamorphism that occurs within or near a deep-seated fault zone involving crushing and granulation of rocks.

CAVITATION Process of erosion in a stream channel caused by the sudden collapse of vapor bubbles against the channel wall.

CEMENTATION Process by which chemically precipitated mineral matter in pore spaces of sediment binds it into a coherent mass.

CHEMICAL WEATHERING Weathering produced by decomposition (chemical breakdown) of the minerals of a rock into minerals that are more stable under surface conditions.

CINDER CONE Relatively steep-sided volcanic cone formed by cinders; usually associated with more explosive volcanoes.

CIRQUE Steep-walled hollow in a mountainside formed by ice plucking and frost action; shaped like an amphitheater.

CLASTIC Type of rock composed of broken fragments derived from preexisting rocks.

CLAYSTONE Clastic sedimentary rock composed of clay-sized particles; differs from shale in that it lacks *fissility*.

CLEAVAGE Breaking of a mineral along planes of weakness within the crystal structure. The internal structure of a mineral controls the kinds and numbers of cleavages.

COAL A combustible sedimentary rock composed chiefly (more than 50% by weight) of carbonaceous material; formed from plant remains. It is classified in one of four categories (rank) depending on degree of metamorphism.

COBBLE Rock fragment with a diameter in the range of 64–256 mm.

COELACANTH Fish that has evolved since Middle Devonian time and still exists off the coast of South Africa.

COLLISIONAL MOUNTAINS Mountains that result from folding, faulting, and metamorphism in the area of collision of continents or of a continent with an island arc, for example, the Himalayas.

COLLUVIUM Mixture of soil, loose rock, and other debris that has accumulated on a slope as a result of mass movement.

COMPACTION Reduction in bulk volume or thickness within a body of sediment because of increasing weight of overlying material.

COMPETENCE Measure of the maximum size of particles that a transporting agent, such as a stream, glacier, or wind, can move.

COMPOSITE VOLCANO See *Stratovolcano*.

COMPRESSIONAL FORCES Forces that push things together, occasionally resulting in a decrease in volume.

CONCORDANT PLUTON Intrusive igneous body that was emplaced parallel to dominant layering (bedding or foliation) in intruded rocks.

CONGLOMERATE Sedimentary rock composed of subangular to rounded grains larger than 2 mm in diameter.

CONNATE WATER Water trapped in a sedimentary deposit at the time the deposit was laid down.

CONSEQUENT STREAM Stream that follows a course that is a direct consequence of the original slope of the surface on which it developed.

CONTACT AUREOLE Zone of contact metamorphism surrounding an igneous body.

CONTACT METAMORPHISM Process of recrystallization produced predominantly by the heat of an intrusive igneous body.

CONTINENTAL DRIFT Idea that continents were once in contact and have since drifted to their present positions.

CONTINENTAL GLACIER Ice sheet that covers large sections of a continent's mountains and plains.

CONTINENTAL RISE That portion of the continental margin that lies between the continental slope and the abyssal plain.

CONTINENTAL SHELF That portion of the continental margin that lies between the shore and the continental slope. It is characterized by a very gentle slope.

CONTINENTAL SLOPE That part of the continental margin that lies between the continental shelf and a continental rise or an oceanic trench.

CONTINUOUS REACTION SERIES Portion of Bowen's reaction series that involves a continuous reaction without change in the mineral involved (e.g., the change from calcium to sodium plagioclase).

CORDILLERAN MOUNTAINS Mountains formed by a continental mass colliding with a subsiding mass of oceanic crust (e.g., the Andes).

CORE Central or innermost zone of the earth. It lies below a depth of 2900 km and is believed to be composed of nickel and iron.

CORIOLIS FORCE The tendency of particles of matter in motion on the earth's surface to be deflected to the right in the northern hemisphere and to the left in the southern hemisphere because of the earth's rotation.

CORRELATION Definite correspondence in character and in stratigraphic position between geologic formations or fossil faunas of two or more separate areas.

COUNTRY ROCKS Rocks intruded by some igneous body; also called wall rocks.

CRATON Portion of the earth's crust that has undergone no significant tectonic activity over a long period of time; it is stable.

CREEP Slow, gradual downslope of soil and rock under the influence of gravity.

CROSS-BEDDING Series of thin beds of sediment deposited at an angle to the original surface of deposition; exhibits truncation of beds at the top and tangential orientation to underlying rocks.

CROSS-STRATIFICATION See *Cross-bedding*.

CREVASSE Fissure in glacier ice; a break in a natural level.

CRYSTAL Homogeneous solid of uniform composition with an orderly internal structure; bounded by smooth planar faces.

DEBRIS AVALANCHE Rapid type of mass movement that involves a mixture of unsorted soil and rock material.

DEBRIS FLOW Rapid type of mass movement of mixed coarse debris and water.

DEBRIS SLIDE Landslide that involves a slow to rapid downslope movement of relatively dry and unconsolidated material.

DEEP-SEA FAN Sediment fan on the ocean floor, commonly seaward from a large river or other source of sediment.

DEFLATION Erosion process in which wind carries off unconsolidated material from the land surface.

DENSITY CURRENT Gravity-motivated flow of more dense water beneath surrounding water. Temperature, salinity, or suspended sediment differences may produce differences in density.

DESICCATION Loss of water from pore spaces of a sediment; results from compaction.

DETRITAL Refers to material, generally sedimentary, formed by mechanical means.

DIAGENESIS All changes or modifications undergone by a sediment after deposition, including such things as compaction, cementation, and desiccation.

DIFFERENTIAL WEATHERING Weathering that occurs at irregular or different rates because of variation in composition of a rock body or differences in weathering intensity.

DIKE Discordant tabular pluton.

DIP Angle between some planar feature and the horizontal, measured at right angles to the strike.

DISCONFORMITY Unconformity in which beds above and below are parallel but may have considerable topographic relief on the erosion surface.

DISCONTINUITY Boundary within the earth implying change in rock type. Seismic wave velocities change abruptly here.

DISCONTINUOUS REACTION SERIES Reaction series that involves a change from one mineral to another with each reaction (e.g., olivine reacts to form pyroxene, then amphibole).

DISCORDANT PLUTON Intrusive body that cuts across the layering in country rocks.

DOLOMITE Rock-forming mineral with composition $CaMg(CO_3)_2$; reacts slowly with cold dilute hydrochloric acid.

DOLOMITIZATION Process by which limestone is converted to dolostone by the replacement of original calcium carbonate (calcite) by magnesium-calcium carbonate (dolomite).

DOLOSTONE Carbonate sedimentary rock composed chiefly (more than 50% by weight) of the mineral dolomite.

DOME Circular or elliptical anticlinal structure in which the layering (bedding or foliation) dips gently away from a central point.

DRIFT Any material laid down directly by ice or deposited in streams, lakes, or oceans as a result of glacial activity. Unstratified glacial drift is called *till*.

DRUMLIN Streamlined hill composed of glacial till. The long axis is parallel to movement of glacier ice. The blunt end points in the direction from which the ice came. The gently sloping sides points in the direction that the ice was flowing.

DUCTILE BEHAVIOR Continuous or plastic deformation.

DUNE Mound or ridge of sand piled up by wind.

DYNAMO THEORY Belief that the earth's magnetic field is attributable to dynamo action in the fluid core. The conducting liquid is supposed to flow in such a pattern that the electric current indicated by its motion through the magnetic field sustains that field.

EARTHFLOW Nonrotational type of mass movement underlain by a well-defined movement surface parallel to the ground surface.

EJECTA Rock material thrown out during crater impact.

ELASTIC REBOUND Movement of rocks along a fault involving a sudden release of energy, which causes an earthquake.

END MORAINE Ridge of till marking the farthest advance of a glacier. Also called *terminal moraine*.

ENTRENCHED MEANDER Meandering course of a stream that flows in a deep valley following rejuvenation; also known as an incised meander.

EPEIROGENY Vertical movements of crustal segments (e.g., mountains, continents).

EPICENTER Point on the earth's surface directly above the focus. Most earthquake damage occurs here.

EQUILIBRIUM State of rest, a condition of balance in any system.

EROSION Surface process of breakdown and removal of rock and soil material by mechanical and chemical means.

ERRATIC In glaciation, a rock or boulder that is carried by ice to a place where it rests on or near bedrock of different composition.

ESKER Sinuous ridgelike deposit of stratified glacial drift.

ESTUARY River valley that has been flooded by the sea.

EUSTATIC SEA LEVEL CHANGE Worldwide raising or lowering of sea level.

EVAPORITE Member of a group of sedimentary rocks formed by precipitation of salts from an aqueous solution and concentrated by evaporation. Some examples are rock salt, anhydrite, and gypsum.

EVOLUTION Development of an organism toward complete adaptation to environmental conditions to which it has been exposed with the passage of time.

EXFOLIATION Removal of concentric sheets or slabs of rock from exposed rock masses by predominantly mechanical means.

EXTRUSIVE ROCKS Igneous rocks that crystallize on the earth's surface; volcanic rocks.

FACIES Sum of all primary lithologic and paleontologic characteristics exhibited by a sedimentary rock.

FAULT Fracture along which appreciable movement has taken place.

FAUNAL SUCCESSION Observed chronologic sequence of life forms through geologic time.

FELSIC ROCKS Igneous rocks that are rich in feldspar and silica.

FIRN Granular ice formed by the recrystallization of snow. It is sometimes called *névé*.

FISSILITY Tendency of a rock to break easily into thin sheets, usually shale.

FISSURE Surface of fracture or a crack in a rock along which there is distinct separation.

FJORD Glacially deepened valley that is flooded by the sea to form a long, narrow inlet.

FLOODPLAIN Area that borders a stream covered by water in time of flood.

FOCUS Point within the earth where rock movement takes place, causing earthquakes.

FOLD Bend or flexure in the layering in a rock.

FOLD BELT Zone of intensive development of faults, metamorphism, and intrusive bodies; an orogenic belt.

FOLIATION Planar arrangement of micas or other minerals, slaty cleavage, or other textural or structural features in a rock.

FOSSIL Direct or indirect remains of a plant or animal that has been preserved in rock.

FOSSIL FUEL Hydrocarbon deposit that may be used as a source of energy.

FRACTIONAL CRYSTALLIZATION Process whereby certain minerals crystallize at progressively lower temperatures from a melt.

FREEZE-THAW Process of water's freezing and then melting as it applies to the mechanical weathering process; frost action.

FROST HEAVING Uneven lifting or upward movement of surface material by a growing subsurface ice mass.

FROST WEDGING Breaking and prying of rock or soil as water expands on freezing in cracks.

GABBRO Mafic phaneritic igneous rock composed predominantly of plagioclase and pyroxene.

GALAXY Large systems of stars, nebulae, star clusters, and interstellar matter that make up the universe.

GANGUE Valueless rock or mineral material that is separated from an ore.

GEOLOGIC CYCLE Repetitive internal or external geologic processes.

GEOLOGIC MAP Map on which is recorded geologic information, such as the distribution and nature of rock units and the occurrence of structural features.

GEOLOGY Science of the earth.

GEOSYNCLINE Elongate portion of the earth's crust that receives enormous thicknesses of sediment and volcanic rocks and is eventually uplifted into a mountain chain.

GEOTHERMAL GRADIENT Rule of increase in temperatures in the earth with increasing depth; the rate varies from place to place.

GEYSER Type of thermal spring that intermittently ejects steam and water with considerable force.

GLACIER Mass of ice formed by recrystallization of snow.

GNEISS Foliated, regional metamorphic rock in which bands of light material alternate with bands of darker material or bands of different texture.

GONDWANALAND Ancient continent composed of today's continents of South America, Africa, India, Australia, and Antarctica.

GOSSAN The iron-rich, weathered surface zone of a sulfide deposit.

GRADED BEDDING Type of sedimentary bedding that exhibits a gradual change in grain size from bottom to top (usually coarse on bottom and fine on top).

GRADIENT Slope of a stream bed.

GRANITE Plutonic igneous rock composed mainly of orthoclase feldspar and quartz with smaller amounts of mica, hornblende, and plagioclase feldspar.

GRANITIZATION Metamorphic process whereby initially sedimentary (or volcanic) rocks are transformed into igneous-looking granitic rocks.

GRAVIMETER Device that measures differences in gravitational attraction between an initial station of known gravity and other points of interest.

GRAVITY Resultant force on any body of matter at or near the earth's surface because of the attraction by the earth.

GRAVITY FAULT Fault that moves under the influence of gravity; generally a normal fault.

GRAYWACKE Dark gray or greenish black, dense, coarse-grained sandstone composed of poorly sorted, usually angular quartz and feldspar with a high percentage of small, dark rock fragments.

GROUND MORAINE Material that is deposited from a glacier on the ground surface over which the glacier has moved. It forms a gently rolling surface.

GROUNDWATER Underground water within the zone of saturation.

GUYOT A flat-topped *seamount*.

HALF-LIFE Amount of time necessary for radioactive elements to lose half their mass by decay into stable products.

HANGING VALLEY Valley that has a greater elevation than the main valley at the point of their junction. It is often, but not always, formed by a deepening of the main valley by a glacier.

HISTORICAL GEOLOGY Major branch of geology that is concerned with evolution of the earth, its atmosphere, and its environment from its origins to its present-day forms. It includes such studies as stratigraphy, paleontology, and geochronology.

HOMOCLINE Rocks over a large area that dip gently in the same direction.

HORN Spire of rock left when cirques have intersected from three or more directions. Also called a matterhorn.

HORNFELS Fine-grained, dark, nonfoliated, contact metamorphic rock.

HOT SPRING Spring that brings hot water to the surface. Water temperature is usually 15°F or more above mean air temperature. Also called thermal spring.

HYDRAULIC GRADIENT Head of underground water that is divided by the distance of travel between two points. If the head is 10 ft for two points 100 ft apart, the hydraulic gradient is 10%.

HYDROLOGIC CYCLE Constant circulation of water from the sea, through the atmosphere, to the land, and its eventual return to the sea.

HYDROSPHERE Waters of the earth, including waters of ocean, rivers, lakes, groundwater and other bodies of water as liquid or ice.

IGNEOUS ROCK Rock formed by cooling and solidification of molten magma.

INTRUSIVE Magma that is emplaced in preexisting rock.

IONOSPHERE Region of the upper mesosphere. Absorption of solar rays by atmospheric gases produce ions responsible for auroras.

ISLAND ARC Chain of islands (e.g., the Aleutians), that rise from the sea floor and near a continent.

ISOGRAD Line on a map that separates a zone of higher metamorphic grade from one of lower grade. The line is then a line of equal metamorphic grade and marks the first appearance of an index mineral.

ISOMORPHISM Substitution of an atom, ion, or molecule for another in a mineral without changing its structure.

ISOSTASY Condition of equilibrium, comparable with floating of units of the crust above the plastic mantle.

ISOTOPE One of two or more species of the same chemical element. An isotope contains the same number of protons but has a different number of neutrons with a corresponding difference in atomic mass.

JOINT Fracture along which there has been no appreciable movement.

JUVENILE WATER Water reaching the surface for the first time or added to groundwater by magma.

KAME Steep-sided hill of stratified glacial drift.

KARST TOPOGRAPHY Topography characterized by sinkholes, streamless valleys, and disappearing streams. Features are associated with regions underlain by limestone that have been eroded extensively.

KETTLE Depression in the ground surface formed by melting of a block of ice buried by glacial drift.

LACCOLITH Concordant, mushroom-shaped igneous pluton that arches layers above but not those below the pluton.

LAMINAR FLOW Type of flow in which stream lines remain distinct from one another; a flow of current without turbulence.

LAPILLI Volcanic gravel; pyroclastic fragments that range from 1–64 mm in diameter.

LATERAL MORAINE Ridge of till along the side of a valley glacier.

LAVA Molten rock material that is extruded onto the earth's surface.

LIMESTONE A sedimentary rock composed essentially of calcium carbonate.

LITHIFICATION Conversion of an unconsolidated sediment into solid rock.

LITHOFACIES Mappable subdivision of a stratigraphic unit distinguished by its lithologic character.

LITHOSPHERE Rigid outer part of the earth; the solid part of the earth.

LONGSHORE CURRENT Localized ocean current that is parallel and close to the shore generated by wave refraction.

LOPOLITH Large concordant pluton that has a lenticular shape with the central part sunken below the edges.

MAFIC ROCK Igneous rock rich in iron and magnesium, expressed as prominence of minerals such as olivine, pyroxene, and amphiboles.

MAGMA Molten rock material.

MAGNETIC ANOMALY Deviation from the calculated (or expected) magnetic field in an area. A positive anomaly indicates that the magnetic field is more intense; a negative anomaly indicates that the field is less intense than expected.

MANTLE Zone of the earth below the crust and above the core.

MARBLE Metamorphosed limestone.

MARE (MARIA) Dark area of the moon.

MASSIVE Describes a fine-grained mineral aggregate that does not contain easily visible grains; rock that is of uniform texture.

MASS MOVEMENT Downslope movement under the influence of gravity.

MATURE SOIL Soil containing a well-developed B horizon.

MEAN Arithmetic average of a set of values.

MEANDER Turn or loop in a stream's course.

MECHANICAL WEATHERING Weathering process that involves disintegration of material by purely mechanical means.

MEDIAL MORAINE Ridge of till that develops when two adjacent lateral moraines form a junction.

MESA Isolated flat-topped landmass standing above its surroundings and bounded by very steep slopes on all sides. Mesas are larger than buttes.

METALLIC MINERAL DEPOSIT Mineral deposit from which a mineral that contains a valuable metal may be extracted.

METAMORPHIC FACIES Assemblage of minerals or rocks of any composition that has reached equilibrium during metamorphism under a set of physical conditions.

METAMORPHIC ROCK Rock formed by the effects of heat, pressure, and/or chemical action on preexisting rocks.

METAMORPHISM Alteration of rocks through heat, pressure, and/or chemical action.

METASOMATISM Solution-assisted recrystallization or replacement under the range of temperatures and pressures of metamorphism.

METEORIC WATER Groundwater that is derived primarily from precipitation.

MIGMATITE Mixed rock that looks much like igneous rock but has originated from metamorphic processes.

MINERAL Naturally occurring, inorganic crystalline solid with a definite chemical composition; possesses distinctive physical and chemical properties.

MONOCLINE Local steepening of an otherwise uniform dip in some direction.

MORAINE General term applied to a landform composed of glacial till.

MOUNTAIN GLACIER See *Alpine Glacier*.

MUDFLOW Rapid-flowing mass of predominantly fine-grained material with a high degree of fluidity.

MUDSTONE Clastic sedimentary rock composed of silt and clay; differs from shale in that it lacks *fissility*.

MYLONITE Very fine-grained flinty rock that results from extreme crushing or granulation. Mylonite may be faintly banded.

NAPPE Large thrust, sheet, or fold that has moved horizontally for several miles (kilometers).

NÉVÉ See *Firn*.

NONCONFORMITY Unconformity in which rocks beneath the erosion surface are either igneous or metamorphic, whereas those above are sedimentary.

NORMAL FAULT Fault in which the hanging wall has moved down relative to the footwall.

OCEANIC RIDGE Continuous, elongate, seismic mountain range that rises from the ocean floor and is interconnected in a worldwide system. This is thought by many geologists to be a zone where oceanic crust is created.

OCEANOGRAPHY Discipline devoted to the study of oceans.

ORE Naturally occurring material from which one or more valuable minerals may be extracted for a profit.

OROGENIC BELT Elongate region on the earth that has been subjected to folding, thrusting, metamorphism, and igneous activity.

OROGENY Process of formation of fold mountains that involves compression to produce thrusts, folds, and features of a mountain system.

ORTHOQUARTZITE Clastic sedimentary rock that is composed almost exclusively of quartz grains.

OUTLIER Portion of a rock unit that lies detached or away from the main body. Connecting portions have been removed by denudation.

OUTWASH Material carried from a glacier by meltwater. It is laid down as stratified deposits.

OUTWASH PLAIN Gently sloping surface underlain by outwash.

OXBOW Abandoned meander caused by a neck cutoff. When filled with water, it is an oxbow lake.

OZONE LAYER Layer in the upper atmosphere composed of O_3, which blocks most of the sun's ultraviolet rays.

PAHOEHOE Type of lava having a ropy, undulating surface; characteristic of Hawaii.

PALEOGEOGRAPHY Study and description of the physical geography of the geologic past.

PALEOMAGNETISM Study of ancient magnetic fields of the earth, principally as the position of ancient magnetic poles.

PANGAEA Ancient supercontinent thought to be composed of all of today's continents.

PARABOLIC DUNE Dune with a shape that resembles a parabola with the concave side facing toward the wind.

PEAT Unconsolidated deposit of semicarbonized plant remains; considered an early stage of coal.

PEDIMENT Broad, gently sloping surface carved in bedrock and generally veneered with fluvial gravels; occurs between mountain fronts and basin bottoms in desert regions.

PEGMATITE Very coarse-grained igneous rock, usually with the composition of granite.

PELAGIC Refers to the environment of oceans; organisms whose environment is open ocean rather than bottom or shore areas.

PERIDOTITE Dark, phaneritic, ultramafic igneous rock composed principally of olivine and pyroxene.

PERMEABILITY Ability of a rock or any material to transmit fluids; equal to velocity of flow divided by hydraulic gradient.

PETROLOGY Study of origin, occurrence, structure, and history of rocks.

PHANERITIC Igneous texture in which grains are coarse enough to be identified without the aid of magnification.

PHYLLITE Fine-grained foliated regional metamorphic rock composed predominantly of fine-grained micas; intermediate between slate and schist.

PHYSICAL GEOLOGY Broad subdivision of geology that investigates materials of the earth's crust and the processes that affect them.

PIEDMONT GLACIER Glacier formed by coalescence of valley glaciers spreading over plains at the foot of mountains from which valley glaciers emerge.

PIRATE STREAM One of two streams in adjacent valleys that has been able to deepen its valley more rapidly than the other and has extended its valley headward, breaking through the divide and capturing the upper portion of the neighboring stream.

PLACER Mineral deposit formed by accumulation of weathering and erosion products at the surface.

PLAGIOCLASE Member of a group of light-colored minerals of the feldspar group that has the composition $(Na, Ca)(Si, Al)Si_2O_8$.

PLANETOLOGY The comparative study of the geology of planets in the solar system, their natural satellites, and other cosmic bodies of similar nature.

PLANKTONIC Describes a floating organism dwelling in the oceans.

PLATE TECTONICS Theory that divides the earth into a few large, rigid plates that float on a viscous substratum. Continents ride as passengers on plates.

PLAYA Flat-floored center of an undrained desert basin. The temporary lake formed in a playa is known as a playa lake.

PLUNGING FOLD Fold whose axis is not horizontal.

PLUTON Any intrusive igneous body.

POLAR WANDERING Concept that magnetic poles of the earth have migrated geographically during the earth's history.

POLLUTANT Substance that reduces the usability of water or air.

POLYMORPHISM Phenomenon that some materials of similar composition may possess more than one crystal structure, particularly at different temperatures and pressures. Diamond and graphite are polymorphs.

POROSITY Percentage of open space in a rock or other earth material.

PORPHYRITIC Texture of igneous rocks in which larger crystals are surrounded by crystals of smaller size.

PORPHYROBLAST Large crystal grown in a finer groundmass under conditions of metamorphism.

POTHOLE Hole ground in solid rock of a stream channel by sand and gravel caught in an eddy of turbulent flow and swirled for a long time over one spot.

PRIMARY STRUCTURE Structures formed when a sedimentary or igneous rock is formed. Cross-bedding, graded beds, and vesicles are primary structures.

PYROCLASTIC Pertains to rock texture of explosive volcanic origin.

PYROXENE Group of iron–magnesium silicate minerals with prismatic cleavage with faces intersecting at 87 and 93°.

QUARTZ Mineral with formula SiO_2; second most common mineral in the earth's crust. Quartz occurs in up to 12 different colors because of impurities and is seventh on the Mohs' hardness scale.

QUARTZITE Metamorphosed sandstone.

RADIOACTIVITY Spontaneous decay of certain isotopes into new isotopes; accompanied by emission of alpha particles, beta particles, or gamma rays and heat.

RADIOMETRIC AGE Age in years determined by a radiometric dating technique; determination of absolute age.

RADIOMETRIC METHOD Calculating an age in years for geologic materials by measuring the amount of a radioactive element and its decay product.

REACTION RIM Rim of one mineral around a core of another; formed by reaction of the outer surface of a solidified mineral with remaining magma.

RECESSIONAL MORAINE Ridge of till that marks a period of temporary stability during general wastage of a glacier and recession of its front.

RECTANGULAR DRAINAGE Drainage pattern in which streams have sharp right-angle bends, frequently because of joints in underlying bedrock.

REEF Ridge or moundlike sedimentary structure built almost exclusively of remains of colonial organisms.

REGIONAL METAMORPHISM Metamorphism occurring over many square miles (kilometers).

RELATIVE TIME Time that relates one event to another in a chronological order, but not in terms of the number of years since an event took place.

REMANENT MAGNETISM Orientation of magnetic minerals fixed at the time a rock was formed; may not parallel today's field because of movement of the continent that contains the mineral.

REPLACEMENT Substitution of one mineral for another.

RESERVES Amount of a material that is estimated or proved to be present in a deposit.

RESERVOIR ROCK Rock with sufficient porosity to contain hydrocarbons.

RESIDUAL DEPOSIT Economic deposit present on the surface, commonly resulting from *in situ* development by weathering.

RESIDUAL SOILS Soils that have formed from material on which they presently rest.

RHYOLITE Group of extrusive igneous rocks usually composed of orthoclase and quartz; often exhibit porphyritic texture. Rhyolite is the extrusive equivalent of granite.

RICHTER SCALE Logarithmic scale used to measure magnitude of an earthquake.

RÔCHE MOUTONNÉE Sheep-shaped rock that has been rounded by the action of glacier ice. Gentle slope faces the direction from which the ice came; steep slope points in the direction that the ice was moving.

ROCK Usually an aggregate of two or more minerals.

ROCK CYCLE Sequence of events leading to the formation, destruction, and reformation of rocks through such processes as erosion, deposition, and metamorphism.

ROCK GLACIER Mass of poorly sorted angular boulders that moves slowly downslope because of a permanent ice coating of boulders beneath the surface.

ROCKSLIDE Mass of bedrock that moves rapidly downslope over an inclined surface of movement. Bedrock usually breaks up immediately into many smaller fragments.

RUNOFF That part of precipitation which moves over the earth's surface in streams or sheets of water.

SALTATION Mechanism by which a particle moves by jumping from one point to another.

SANDSTONE Field term for a clastic sedimentary rock that contains particles between $\frac{1}{16}$ mm and 2 mm.

SAPROLITE Thick, thoroughly decomposed rock layer commonly formed by weathering of igneous or metamorphic rocks under humid and tropical or subtropical conditions. Most layering and structural features of the original rock are preserved.

SCHIST Strongly foliated, coarse-grained, regional metamorphic rock composed principally of micas, talc, or amphiboles.

SCIENTIFIC METHOD Method of inquiry in which a problem is identified, data are gathered, and a hypothesis is formulated and then tested.

SEA-FLOOR SPREADING Process whereby the ocean floor is thought to be moving apart in opposite directions from a central zone in the ocean floor (an oceanic ridge).

SEAMOUNT A flat-topped or peaked elevation above the sea floor, usually volcanic in origin.

SEDIMENT Loose particles, either organic or inorganic, that have settled out of suspension to form layers.

SEDIMENTARY ROCK Member of a major rock group that includes those formed from consolidation of loose sediment and those formed by precipitation, usually from sea water.

SEIF DUNE A very long longitudinal dune; may be as high as 300 ft (100 m) and as long as 60 mi (100 km).

SEISMIC WAVES Vibrations generated by an earthquake. Typical waves are *P* waves, *S* waves, and *L* waves.

SEISMOGRAM Record of an earthquake.

SEISMOGRAPH Instrument used to record earthquakes.

SEISMOLOGY Scientific study of earthquakes.

SHALE Clastic sedimentary rock that is composed of fine-grained particles; exhibits *fissility*.

SHEETING JOINTS Joints formed that are parallel to the surface by unloading of the surface by erosion or some other process.

SHIELD Ancient stable core of a craton.

SHIELD VOLCANO Broad, low volcano composed of basaltic rock.

SILL Concordant tabular pluton.

SILTSTONE Clastic sedimentary rock composed predominantly of silt-sized particles; lacks *fissility*.

SINKHOLE Depression in the surface of the ground caused by the collapse of the roof over a solution cavern.

SKEWNESS Measure of asymmetry of a frequency distribution.

SLATY CLEAVAGE Pervasive rock cleavage formed by parallel alignment of clays, micas, or other fine-grained platy minerals.

SLUMP Type of mass movement characterized by downward and outward (rotational) movement of a block of material.

SOIL All weathered material above solid bedrock capable of supporting rooted plants.

SOIL HORIZON Layer of soil that is physically distinct from those above and below.

SOIL PROFILE Vertical section of soil that displays different horizons.

SOLIFLUCTION Slow downslope movement of water-saturated soil, commonly moving over a frozen lower zone.

SORTING Measure of the range of particle size distribution in a sample.

SPECIFIC GRAVITY Weight of something divided by the weight of an equal volume of water.

SPHEROIDAL WEATHERING Concentric layers of weathered rock that develop under combined chemical–mechanical conditions of hydration and expansion.

STALACTITE Deposit of calcite that hangs from cave ceilings.

STALAGMITE Stump-shaped deposit (usually of calcite) that grows upward from the cave floor.

STOCK Igneous pluton whose area of exposure is less than 40 mi^2 (100 km^2).

STRATA Layers of tabular masses of sedimentary rock.

STRATIGRAPHY Branch of geology that studies form, arrangement, geographic distribution, chronologic succession, classification, and especially correlation and mutual relationships of rock strata.

STRATOSPHERE Layer of the atmosphere that lies above the troposphere and below the mesosphere.

STRATOVOLCANO Volcano that consists of alternating layers of lava and pyroclastics.

STREAM PIRACY Process by which one stream extends in a headward direction, cuts into the valley of a second stream, and then diverts the flow of the second stream.

STRIKE Trend (compass direction) of a feature. Strike of an inclined plane is the trend of its intersection with the horizontal.

STRIKE-SLIP FAULT Fault whose movement is predominantly in a horizontal direction, parallel to the strike of the fault.

SUBARKOSE Sandstone that is intermediate in composition between arkose and pure quartz sandstone.

SUBDUCTION Process in which one crustal block is thought to descend beneath another; associated with one crustal plate descending beneath another.

SUBGRAYWACKE Sandstone with less feldspar and more and better-rounded quartz grains than graywacke.

SUBLIMATION Process by which a solid material passes into the gaseous state without first becoming a liquid.

SUBMARINE CANYON· Steep valley coursing through the continental shelf or slope.

SUPERIMPOSED STREAM Stream whose present course was established on young rocks burying an old surface. With uplift this course was maintained, as the stream cut down through young rocks to and into the old rock surfaces.

SUPERPOSITION Basic rule in geology that a sedimentary rock layer is older than those that overlie it.

SYNCLINE Fold in which layers dip inward

toward an axis and in which youngest rocks are found in the center of the structure.

TABULAR PLUTON Slablike pluton whose dimension in one direction is considerably less than those in the other two directions.

TACONITE Initially low-grade (25%) iron ore composed of magnetite, hematite, and impurities.

TALUS Rock fragments that accumulate at the base of a cliff or other steep slope.

TARN Lake formed in the bottom of a cirque after glacier ice has melted.

TECTOGENE Downbuckled portion of an orogenic belt.

TECTONIC CYCLE Cycle that relates larger structural features of the earth's crust to gross crustal movements and to kinds of rocks that form in various stages of development of these features.

TECTONICS Study of the origin and development of larger structural features of the earth and their relationships to each other.

TECTONIC SEA LEVEL CHANGE Apparent raising or lowering of sea level on a portion of a continent, an island, or a portion of an ocean as a result of tectonic (usually compressional) forces.

TENSIONAL FORCES Forces tending to pull something apart.

TEMPORARY BASE LEVEL Base level that is not permanent, such as that of a lake or dam.

TERMINAL MORAINE Ridge of till that marks the farthest advance of a glacier. Also called *end moraine.*

TERRAE Lighter areas of the moon.

TEXTURE Size, shape, and arrangement of grains in a rock.

THRUST FAULT Fault in which the hanging wall has moved up relative to the footwall. Most thrust faults have a low to moderate angle of dip.

TIDES Regular rising and falling of the surface of oceans and other large bodies of water (and land) in response to unequal gravitational pull by the moon and sun on the earth.

TILL Unstratified and unsorted glacial drift deposited directly by a glacier.

TRANSFORM FAULT Fault that results from different spreading rates, commonly in the sea floor.

TRANSLATIONAL SLIDE Nonrotational rapid mass movement process in which material moves over a slide surface that is parallel to the ground surface.

TRANSPIRATION Process by which water absorbed by plants is evaporated into the atmosphere from plant surfaces.

TRANSPORTED SOILS Soils that have been derived from parent material and that have been moved and redeposited away from the site of the parent material.

TRANSVERSE DUNE Sand dune oriented with its long axis perpendicular to the prevailing wind.

TRENCH Narrow elongate depression of the sea floor commonly oriented parallel to the edge of a continent or an island arc.

TROPOSPHERE Layer of the atmosphere that is closest to the earth's surface.

TRUNCATED SPUR Beveled end of a divide between two tributary valleys where they join a main valley that has been glaciated. The glacier of the main valley has sheared off the end of the divide.

TSUNAMI Seismic sea wave produced by a shallow submarine earthquake. The waves produced have very long wavelength and small amplitude.

TUFF Compacted pyroclastic deposit of volcanic ash that may contain up to 50% sediment, such as sand or clay.

TURBIDITY CURRENT Rapidly moving density current (most commonly in water) caused by an excess of sediment in suspension near the bottom on a gentle to steep slope.

TURBULENT FLOW Fluid flow in which paths of particle motion are irregular with eddies and swirls.

ULTRAMAFIC ROCKS Igneous rocks that are composed mostly of ferromagnesian minerals, principally pyroxene and olivine.

UNCONFORMITY Boundary within a sequence of rocks that represents a break in deposition formed by uplift, erosion, and then continued deposition.

UNIFORMITARIANISM "The present is the key to the past"; assumes that physical laws now operating have always operated throughout the geologic past and that the origin of ancient rocks can be interpreted in the light of today's processes.

VALLEY GLACIERS Glaciers confined to valleys; sometimes called *mountain glaciers* or *alpine glaciers.*

VALLEY TRAINS Sand and gravel deposits laid down in valleys by streams at the end of glaciers.

VENT Opening in the earth's crust through which volcanic materials are extruded.

VENTIFACTS Smooth, polished stones produced by repeated abrasion of sand and silt blowing over rocks for a long time. Also called dreikanters.

VOLCANISM Process by which magma and gases are extruded onto the earth's surface and into the atmosphere.

VOLCANO Accumulation of igneous material around a vent.

WATER GAP Gap cut through a resistant ridge by a superimposed or antecedent stream.

WATER TABLE Upper surface of the zone of saturation for underground water. Also called *groundwater* table.

WAVE REFRACTION Bending of waves around an obstacle or against the shore.

WEATHERING Change that takes place in rocks in response to chemical and mechanical pressures that tend to break them down on the earth's surface.

WIND GAP General term for an abandoned water gap.

XENOLITH Inclusion, usually of country rock, within an igneous body.

YAZOO-TYPE RIVER Tributary that is unable to enter its main stream because of natural levees along its main stream.

ZONAL SOIL Soils whose origin have characteristics that can be related to climatic factors.

ZONE OF AERATION That point between soil and rock in which both air and percolating water are present.

ZONE OF SATURATION Underground region within which all openings are filled with water.

Index

SEDIMENTARY

Organic Origin	Chemical Origin	Fragmental Origin
Fragmental Texture	Crystalline Texture	Fragmental Texture

Layered Structure
[Stratified Structure]

Fragmental Origin

	Unconsolidated	Consolidated	
COARSE EQUIGRANULAR > 2 mm	TALUS	BRECCIA	angular fragments (stones to small stones)
	GRAVEL	CONGLOMERATE	rounded fragments of rocks (larger and smaller pebbles)
EQUIGRANULAR (2 – 0.02 mm)	SAND: grains of quartz, or grains of any sort of rock, or a mixture of two or more kinds	GRAYWACKE: grains of any sort cemented with clay	
		ARKOSE: grains of quartz, mica or feldspar cemented with kaolin	
		SANDSTONE: grains of quartz (with mica). According to the cement: argillaceous, calcareous, ferruginous, siliceous	
FINE EQUIGRANULAR < 0.02 mm	CLAY	MUDSTONE, SHALE: siltstone, claystone	clay minerals (or other particles) with limonite, coal matter, graphite
	LOESS		wind-blown particles, especially from clay

Chemical Origin

Rock	Composition	Texture
ROCK SALT	halite. Mixtures: halides, sulphates, clay, etc.	granular, dense, fibrous
POTASSIUM SALTS	sylvite, carnallite, kainite. Mixtures: same as above	— " —
ANHYDRITE and GYPSUM	anhydrite, gypsum. Mixtures: same as above	— " —
TRAVERTINE	Calcite; limonitic pigment	porous
GEYSERITE	opal + meta-coloids of SiO_2; limonitic pigment	porous
NOVACULITE	opal, chalcedony, quartz; iron pigment	dense
Fe ORES	• chamosite • siderite • hematite • limonite • magnetite	dense, seed-like, pea-like, bean-like, nodular
Mn ORES	carbonates, oxides and Mn hydroxides	— " —
BAUXITE	Al hydroxides + limonite or hematite	dense to pea-like

Organic Origin

Group	Rock	Composition	Texture
Siliceous Material	DIATOMACEOUS EARTH	microscopic diatom shells of opal. Mixtures: clay, peat	dense
	QUARTZ FLINT	opal — chalcedony — quartz, graphitic pigment	— " —
Calcareous	LIMESTONE	calcareous shells. Mixtures: SiO_2, clay, bitumen, limonite, hematite	dense, fragm., grained
	DOLOMITE	Ca, Mg carbonate, content of Mg primary, diagenetic or epigenetic	— " —
The Coal Series (carbonization →)	PEAT	carbonized vegetable matter	
	LIGNITE	— " —	
	BROWN COAL	— " —	
	BITUMINOUS COAL [BLACK]	— " —	
	ANTHRACITE	— " —	
The Petroleum Series	NATURAL GAS	mixture of hydrocarbons (gaseous state)	
	PETROLEUM CRUDE OIL	mixture of hydrocarbons (liquid state)	
	OZOCERITE	mixture of hydrocarbons (solid state)	
	ASPHALT	result of oxidation of petroleum	